Oil and Gas Traps

Oil and Gas Traps

Aspects of their Seismostratigraphy, Morphology, and Development

Malcolm K. Jenyon
Consultant Geologist

JOHN WILEY & SONS
Chichester · New York · Brisbane · Toronto · Singapore

Copyright ©1990 by John Wiley & Sons Ltd.
Baffins Lane, Chichester
West Sussex PO19 1UD, England

All rights reserved.

No part of this book may be reproduced by any means,
or transmitted, or translated into a machine language
without the written permission of the publisher.

Other Wiley Editorial Offices

John Wiley & Sons, Inc., 605 Third Avenue,
New York, NY 10158-0012, USA

Jacaranda Wiley Ltd, G.P.O. Box 859, Brisbane,
Queensland 4001, Australia

John Wiley & Sons (Canada) Ltd, 22 Worcester Road,
Rexdale, Ontario M9W 1L1, Canada

John Wiley & Sons (SEA) Pte Ltd, 37 Jalan Pemimpin 05-04,
Block B, Union Industrial Building, Singapore 2057

Library of Congress Cataloging-in-Publication Data:

Jenyon, Malcolm K.
    Oil and gas traps : aspects of their seismostratigraphy,
  morphology, and development / Malcolm K. Jenyon.
        p.   cm.
    Includes bibliographical references.
    ISBN 0 471 92549 7
    1. Traps (Petroleum geology) 2. Petroleum—Geology. 3. Gas,
  Natural—Geology. I. Title.
  TN870.5.J46   1989
  553.2'8—dc20   89-22529
                                                     CIP

British Library Cataloguing in Publication Data:

Jenyon, Malcolm K.
    Oil and gas traps : aspects of their
    seismostratigraphy morphology and development.
    1. Natural gas & petroleum deposits
    I. Title
    553.2'8

    ISBN 0 471 92549 7

Phototypeset by Dobbie Typesetting Limited,
Plymouth, Devon
Printed by Courier International Ltd, Tiptree, Essex

*To Staff and Fellow Undergraduates,
Honours School of Geology,
The Victoria University of Manchester,
1950–1953*

Contents

Preface   xiii

Acknowledgements   xv
List of Abbreviations   xvii

PART I—INTRODUCTORY REVIEW

CHAPTER 1   *Source Rocks and Petroleum Migration*   3

1.1   The Basic Requirements   3
1.2   Petroleum Source Rocks   3
  1.2.1   The Raw Materials   4
  1.2.2   Classification of Kerogen   6
  1.2.3   Recognition of a Sedimentary Source Rock   9
    1.2.3.1   Identification   9
    1.2.3.2   Thermal maturation   10
    1.2.3.3   Geochemistry   14
  1.2.4   Source Rocks and Lithology   16
    1.2.4.1   Shale-type source rocks   17
    1.2.4.2   Carbonate source rocks   20
  1.2.5   Source Rocks and Geological Time   22
  1.2.6   Dissenting Voices   22
1.3   Petroleum Migration—Introduction   24
1.4   Primary Migration of Petroleum   25
1.5   Secondary Migration of Petroleum   30
  1.5.1   Hydrodynamic Factors   32

CHAPTER 2   *Reservoir and Seal Formations*   35

2.1   Reservoir Rocks   35
  2.1.1   Introduction   35

2.2	Porosity	35
2.3	Permeability	36
2.4	Sandstone Reservoirs	37
	2.4.1 Factors Affecting Sandstone Porosity	40
	2.4.2 Diagenetic Processes in Sandstones	42
2.5	Carbonate Reservoirs	47
2.6	Some Comments on Fracture Porosity	51
2.7	Reservoir Drive Mechanisms	53
2.8	Cap-rock Seal Formations	55
	2.8.1 Lithological Aspects	55
	2.8.2 Capillary Characteristics of Seals	56
	2.8.3 Diagenetic Seals	57
2.9	Summary: a 'Working Model'	59

PART II—STRUCTURAL TRAPS

CHAPTER 3 *Fold-related Traps* — 63

3.1	Structural Traps—Introduction	64
3.2	Fold-related Traps in Hydrocarbon Exploration	64
3.3	Morphology and Terminology	64
	3.3.1 Symmetrical, Asymmetrical, and Overturned Folds	64
	3.3.2 Axial Plunge and Aspect Ratio	65
	3.3.3 Open and Closed Folds	66
	3.3.4 Concentric and Similar Folding	67
	3.3.5 Faulted Folds	69
3.4	Compressional Anticlines	71
	3.4.1 Simple Anticlinal Traps	71
	3.4.2 Thrust-related Anticlines	78
	3.4.3 Simple Anticline: Complex Overburden	81
3.5	Uplift Anticline	82
	3.5.1 Introduction	82
	3.5.2 Uplift Anticlines with Chalk Reservoirs	85
	3.5.3 A Middle East Giant	93
3.6	Drape Anticlines	93
	3.6.1 Introduction	93
	3.6.2 North Sea Examples, British sector	95
	3.6.3 Libya—a Draped Carbonate Reservoir	103
	3.6.4 Drape Traps in Ecuador	103
3.7	Some Other Noted Anticlinal Traps	105
	3.7.1 Hewett Gasfield	105
	3.7.2 Wilmington Field, Southern California	107
	3.7.3 The World's Largest Gas Trap	111

CONTENTS

CHAPTER 4 Fault-related Traps — 112

4.1 General Comments — 112
4.2 Introduction — 112
  4.2.1 Fault Types and Geometry — 112
  4.2.2 Normal Fault — 114
  4.2.3 Reverse Fault — 115
  4.2.4 Strike-slip Fault — 117
  4.2.5 Thrust Fault — 122
  4.2.6 Growth Fault — 123
    4.2.6.1 Morphology and terminology of growth faulting — 124
  4.2.7 The Fault 'Plane' as a Seismic Reflector — 128
    4.2.7.1 Dip of the fault plane — 129
    4.2.7.2 Thickness of the fault 'plane' — 133
  4.2.8 Diffractions and other Hyperbolas — 133
  4.2.9 Graben and Block Faulting — 142
4.3 Examples of Fault Traps — 144
  4.3.1 Beatrice—an Upthrow Fault Trap — 144
  4.3.2 Fahud—another Upthrow Trap — 146
    4.3.2.1 Basin margin faults — 147
    4.3.2.2 An example from offshore Norway — 148
  4.3.3 Planar Step Faults — 152
  4.3.4 A Note on Block Fault and Complex Fold–fault Traps — 154
4.4 Examples of Growth-fault Traps — 155
  4.4.1 The Hibernia Oilfield — 155
  4.4.2 Eugene Island Block 330 Field — 157
  4.4.3 Two Malaysian Fields — 159
  4.4.4 Reservoir Details in a Roll-over Structure — 162
  4.4.5 Generalized Model of Growth-fault-related Traps — 164

CHAPTER 5 Traps Related to Plastic Deformation of Salt and Shale — 166

5.1 Introduction — 166
5.2 Physical Properties of Salt Rock — 166
  5.2.1 Deformation of Halite — 169
  5.2.2 Salt Rock and Petroleum — 170
5.3 The Initiation of Salt Movement — 171
5.4 Salt Rock as a Seal — 173
  5.4.1 Salt-sealed North Sea Traps — 174
5.5 Effects of Syndepositional Mobility — 181
  5.5.1 Traps Related to Pre-piercement Salt Structures — 185
5.6 Diapirism and Post-piercement Salt Movement — 190
  5.6.1 Primary Rim Syncline Inversion, and 'Turtle Structures' — 195
  5.6.2 Trap Types Related to a Salt Diapir — 198
  5.6.3 Cap Rocks, Overhangs, and Diapiric Shale — 201
    5.6.3.1 Overhangs — 204
    5.6.3.2 Diapiric shale — 207

PART III—STRATIGRAPHIC TRAPS

CHAPTER 6 *Unconformities and Buried Topography* — 213

- 6.1 Introduction — 213
- 6.2 Unconformities — 213
- 6.3 Unconformity Surfaces and Buried Topography — 214
    - 6.3.1 The Unconformity as a Reflection Event — 216
    - 6.3.2 Planar and Non-planar Unconformities — 218
    - 6.3.3 Reflection Configuration above an Unconformity — 219
- 6.4 Buried Topography — 229
    - 6.4.1 Positive Features — 229
    - 6.4.2 Negative Features — 234

CHAPTER 7 *Porosity and Pinchout Traps* — 239

- 7.1 Introduction — 239
- 7.2 Porosity Traps — 239
- 7.3 Pinchout Traps — 240
    - 7.3.1 Bed Thinning in Seismic Data — 240
        - 7.3.1.1 Seismic resolution (horizontal and vertical) — 242
    - 7.3.2 Seismic Example — 244
    - 7.3.3 'Erosional Pinchouts' — 246
    - 7.3.4 Stratigraphic Traps in Aeolian Rocks — 247

CHAPTER 8 *Carbonates, Evaporites, and Reefs* — 248

- 8.1 Introduction — 248
- 8.2 Carbonates — 249
    - 8.2.1 Non-reefal Carbonates — 252
    - 8.2.2 Non-reefal Carbonates as Source Rocks — 253
    - 8.2.3 Special Aspects of Carbonate Reservoirs — 254
        - 8.2.3.1 Chalk — 254
        - 8.2.3.2 Stylolites — 256
        - 8.2.3.3 Limestones v. dolomites — 257
        - 8.2.3.4 The carbonate 'signature' in seismic data — 257
    - 8.2.4 North Sea Examples — 259
- 8.3 Evaporites — 264
    - 8.3.1 The Deformational Behaviour of other Evaporites — 265
    - 8.3.2 Dissolution and Collapse — 269
    - 8.3.3 Salt-edge and Areal Dissolution — 273
    - 8.3.4 Remnant Features and Subsidence Drape — 281
- 8.4 V_P and V_S: a Concluding Note on Carbonates — 284
- 8.5 Reefs — 285
    - 8.5.1 Identification of Carbonate Build-ups in Seismic Data — 285

CONTENTS   xi

    8.5.2   A Shelf-margin Bioherm   286
    8.5.3   Lithology, Porosity, and Seismic Velocity   288
    8.5.4   Larger Features   291
8.6 Concluding Comment   295

CHAPTER 9 *Deltas and Fans*   296

9.1 Introduction   296
9.2 The Development of Deltas   297
    9.2.1   Deltaic Reservoirs   298
    9.2.2   Seismic Identification of Deltaic Systems   300
9.3 Submarine Fans: Introduction   308
    9.3.1   Brae Oilfield—a Fan-complex Reservoir   309
    9.3.2   Identification in Seismic Data   310
        9.3.2.1   Effects of burial compaction   312
        9.3.2.2   Longitudinal seismic section of fan   316

PART IV—COMBINATION AND COMPLEX TRAPS

CHAPTER 10 *Combination and Complex Traps*   319

10.1 Introduction   319
10.2 Combination Traps   320
    10.2.1 Permeability Effects   320
    10.2.2 Faults and Block Faults   324
    10.2.3 Tilted and Eroded Fault Blocks   328
    10.2.4 Other Brent-type Examples   334
10.3 Complex Traps   337
    10.3.1 Introduction   337
    10.3.2 Thrust Belts and Hydrocarbon Traps   339

PART V—SOME SPECIAL SITUATIONS

CHAPTER 11 *Some Special Situations*   347

11.1 Introduction   347
11.2 Direct Hydrocarbon Indicator Events   347
    11.2.1 Shear-wave Studies of Direct Hydrocarbon Indicator Events   357
11.3 The Bottom Simulating Reflection Event   357
11.4 Astroblemes   363
11.5 Igneous Features in Seismic Data   368
    11.5.1 Introduction   368
    11.5.2 Hypabyssal Intrusions   369
    11.5.3 Volcanic Features   372

References and Bibliography   375

Index   390

Preface

As intimated by the subtitle, the object of the book is to study some aspects of the seismostratigraphy, morphology, and development of oil and gas traps.

A decade ago, the use of so many seismic illustrations in a book with this title would have required lengthy explanations and perhaps some apologies. It is believed that this is no longer the case, as working geologists are now in the habit of using seismic data on a day-to-day basis. In the study of hydrocarbon trapping, there is mutual benefit to be gained from the use of seismic illustrations. The seismic data throws light on the geology, and in turn, the geology illuminates the seismic data; there are lessons to be learned by the geophysicist (in both the acquisition and processing stages) from the geological features that figure on the seismic section, although this is outside the terms of reference of the present work.

The book is intended to be useful to all practising geologists and geophysicists in both the industrial and academic spheres. The emphasis has been on the morphological aspects of traps, an important element being the identification of characteristic seismic responses to the geology of the various hydrocarbon traps being studied. However, since many factors other than the morphology govern the type and quality of the trap situation, some discussion of petroleum geology is also desirable.

With this in mind, the first two chapters form a review of current concepts regarding basic aspects of oil and gas traps—source rocks and petroleum migration being covered in the first chapter, with a brief discussion of reservoir and seal formations in Chapter 2. As always, in compressing a major subject into so few pages, much important and significant material has been omitted. It is hoped that the comprehensive References and Bibliography at the end of the book will go some way towards redressing the deficiency.

As the author's main area of interest and source of seismic examples is the North Sea, it is to be expected that a preponderance of the illustrations and examples are from that area. For this, no apologies are made, since it is firmly believed that geological lessons learned in one area may be applicable anywhere (which is, after all, only a kind of lateral-thinking version of Lyell's Uniformitarianism). However, in order to demonstrate this, a considerable effort has been made to introduce a quite substantial number of examples from other areas.

M. K. Jenyon
Petts Wood, Kent, England

Acknowledgements

It is with pleasure that I express my gratitude to the following people who gave unstinted assistance during the writing of the book.

To former colleague Dr A. A. (Al) Fitch I owe much for his encouragement in this and previous endeavours. In the present case, I thank him once again for his painstaking readings of first draft material, and for the many valuable suggestions that have been incorporated in the final version.

To my fellow member of the JAPEC(UK) Organising Committee for almost a decade, Dr G. D. (Douglas) Hobson, a distinguished Petroleum Geologist, author and editor of many technical works who also kindly read first draft material and offered many useful comments and suggestions, as he has done for previous publications.

To the Directors of Seismograph Service (England) Ltd (SS(E)L) for their permission to reproduce herein many examples of seismic data proprietary to the Company, and to all the other authors, publishers and companies who gave permission for the reproduction of illustrations and material from the literature. Individual acknowledgements have been made in figure captions, and citations are included in the text and in the References and Bibliography at the end of the book for all such material used.

To Mrs Bernice A. Farrell for the preparation of many of the illustrations, and for her skill and conscientiousness over a number of years.

Finally, my thanks go once again to my wife Ann and daughters Amanda, Elizabeth, and Joanne for their interest and support during the writing of the book.

List of Abbreviations

API	American Petroleum Institute (crude gravity (°))
bbl	barrels
Bbbl	billion (US) barrels
BPD	barrels per day
BSR	bottom simulating reflection
CAI	conodont alteration index
CPI	carbon preference index
D	Darcy (permeability measure; see also mD)
DHI	direct hydrocarbon indicator
ESR	electron spin resonance
EUR	estimated ultimately recoverable reserves
ft/s	feet per second (seismic velocity)
GOC	gas–oil contact
GOR	gas/oil ratio
GPOC	gas-prone organic carbon
GWC	gas–water contact
H/C	elemental hydrogen/carbon ratio
LOM	level of organic metamorphism
Ma	million years (BP)
MCF	thousand cubic feet
MCFGPD	thousand cubic feet gas per day
mD	milliDarcy
MM	million
MMbbl	million barrels
ms	millisecond (one-thousandth of a second)
m/s	metres per second (seismic velocity)
m.y.	million years (also see Ma)
O/C	elemental oxygen/carbon ratio
OPOC	oil-prone organic carbon
OWC	oil–water contact
%Ro	percentage vitrinite reflectance

SOC	sapropelic organic carbon (= OPOC)
TCFG	trillion cubic feet of gas
TAI	thermal alteration index
TMI	thermal maturity index
TOC	total organic carbon
TTT	total (one-way) transit time on acoustic log
TWT	(seismic) two-way time
V_P	(seismic) compressional (P-wave) velocity
V_S	(seismic) shear (S-wave) velocity

NB There is a certain lack of standardization of abbreviations in the oil and gas industry and the literature. Some terms have two or more different forms (e.g. md and mD for milliDarcy), while others are used so rarely that an abbreviation seems superfluous. The above list is a small selection out of a large field.

PART I
INTRODUCTORY REVIEW

CHAPTER 1

Source Rocks and Petroleum Migration

1.1 The Basic Requirements

In order that a hydrocarbon accumulation may evolve in any specific geological setting, certain basic requirements must be fulfilled that call for the presence of the following:

(i) a hydrocarbon source rock;
(ii) a reservoir rock;
(iii) a migration route from source to reservoir;
(iv) a trap configuration, formed before migration is completed;
(v) a sealing formation (cap rock).

It will be clear that each of these factors may take a wide variety of forms. The permutations and combinations thus possible lead to a great diversity in types of trap. The existence of particular trap types will be influenced strongly by such factors as the depositional and tectonic environments in the areas and segments of the stratigraphic column in which they are found.

The principal subject to be addressed in this work is requirement (iv)—the configuration of the trap. A wide variety of trap types will be studied with the help of illustrations from seismic data, well information, and other geological and geophysical sources. However, all the other factors mentioned above will, of necessity, enter into the discussions very frequently with regard to specific trap types. Because of this, the remainder of the present chapter is devoted to a brief review of source rocks and primary and secondary hydrocarbon migration. In the next chapter, there will be a similarly brief review of current thinking on reservoir rocks and cap-rock seal formations.

1.2 Petroleum Source Rocks

It is a salutary thought that, in spite of the massive scientific and technological advance made over the past century, there is *still* no firm consensus as to the originating processes responsible for the development of those hydrocarbons referred to as 'petroleum', whether in solid, liquid, or gaseous state. Some would even say that we still do not know the basic origin of petroleum with any certainty.

During the past decade, there has been growing speculation and interest in the possibility, under special circumstances, of the accumulation in the crust of 'abiogenic' gas derived from the mantle. This interest has culminated recently in the drilling of a deep test borehole at the Siljan Ring Complex in Sweden, the results from which have still to be evaluated at the time of writing. It should not be thought, however, that the concept of abiogenic hydrocarbons is new. The 'cosmological' occurrence of hydrocarbons, as in meteorites and in the atmospheres of the gas planets, has long been known, as has their producibility from inorganic sources in the laboratory. However, the *types* of hydrocarbons encountered in such circumstances differ from those found in petroleum accumulations, and there are many other difficulties that are detailed in the literature.

Even the most partisan supporters of the abiogenic gas hypothesis would not attempt to refute the very strong physical, chemical, and geological evidence that leads most geologists to believe that the vast preponderance of petroleum discovered so far in the crust is of biogenic origin. This strong evidence points to petroleum in its various states having evolved (mainly) from the organic remains of terrestrial and aquatic plant life fossilized and altered under the elevated temperatures and pressures of burial beneath later sediments.

Although there is a general consensus as regards the biogenic origin of petroleum, there is no such agreement about the processes involved in the evolution of its various forms, their expulsion from the source beds (primary migration), and movement through permeable rocks to the reservoir trap (secondary migration). There is limited agreement as to some of the more likely processes that may be involved in the different stages, and these will be discussed.

1.2.1 The Raw Materials

The complex organic material *kerogen* is now generally believed to be the material from which petroleum is generated. Found very widely disseminated in fine-grained sediments—shales and some carbonates—it consists of the fossilized carbonaceous remnants of animals and plants (mainly the latter) of both aquatic and terrestrial origin, and is insoluble in organic solvents at room temperatures. It is by far the commonest type of organic carbon to be found in the crust, exceeding the estimated amount of coal by up to 1000 times, according to some investigators (see e.g. Weeks, 1958; Hunt, 1967). Its principal elementary components (C, H, O, and N) show some variation that is probably due to the diverse origins of the material.

This diversity of origins is indicated by its basic classification into *structured* and *unstructured* types. The structured type consists of compact material often with some remanent indications of erstwhile cellular structure, as well as spores and pollen grains. The unstructured type is of amorphous, sapropelic material probably mainly of algal origin.

When plants first emerged from the aquatic environment to colonize the land, they required structural support, which evolved as woody tissue (lignin), and also some means of avoiding dehydration, which evolved as waxy coatings (cutin), the latter also providing some defence against microbial attack. It is the woody tissue and waxy coatings that survive as structured kerogen. Aquatic plants need neither type of tissue, and generally (with some

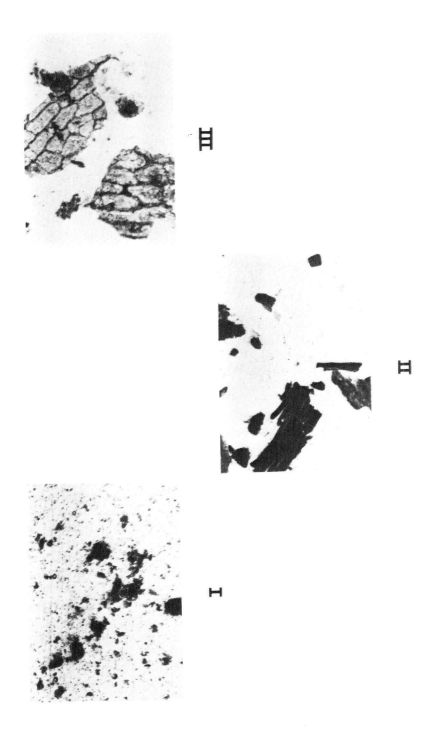

Fig. 1.1. Microscopic view of kerogen types I, II, and III. From Barker, 1981, reprinted by permission of American Association of Petroleum Geologists

animal remains) provide the material for the amorphous, unstructured type of kerogen (see Fig. 1.1).

Kerogen is believed to be formed from the remains of the biota by dynamothermal and chemical effects of a progressive nature attendant upon increasingly deep burial in fine-grained clastic sediments that typically eventually become dark shales, oil shales, and basinal carbonates.

The next stage in the process whereby petroleum is generated is fraught with controversy. In this book, the general 'working' view will be taken that this generation is brought about by the destructive distillation of kerogen under suitable conditions of temperature and pressure; this is probably the widest-held opinion amongst petroleum geologists. However, it must be recorded that there are difficulties in accepting this view, to the extent that some have come to believe that kerogen itself is *not* the originating material, but that it acts in some way as a 'carrier' of the substances that do, in fact, form the 'protopetroleum', which is the forerunner of the solid, liquid, and gaseous hydrocarbons that subsequently migrate into traps.

1.2.2 Classification of Kerogen

Kerogen has been classified into four basic types according to the material of which it consists. Type I is the amorphous, sapropelic matter of aquatic origin mentioned previously; type III is cellulose and lignin-based matter derived largely from the woody tissues of terrigenous plants; and type II may be either a mixture of types I and III altered by bacterial action, or waxy covering material (cutin) from land plants together with spores and pollen grains, or a combination of both. A type IV is also recognized that is equivalent to the *inertinite* in coaly material—woody tissue that has been oxidized beyond the possibility of petroleum generation—often graphitic in nature.

It is believed, with supporting evidence from field and laboratory, that type I kerogen, when sufficiently thermally matured with increasing burial depth, tends to generate oil (i.e. is oil-prone); type II tends to produce waxy oils, condensate, and 'wet' gas; whilst type III, the terrigenously derived woody material, tends to produce dry gas.

In a paper advocating greater application of geological information available on a worldwide basis to the problem of petroleum genesis, Hedberg (1967) pointed out that there appears to be a strong relationship between high-wax crude oils and fresh or brackish water origin of the source sediments. This was a remarkably prescient suggestion in the light of the later work relating gas generation with type II kerogen of terrigenous origin and a high content of waxy surface coverings. (Hedberg also suggested that much evidence points to the majority of oil and gas accumulations being associated with sediments of near-shore marine and paralic character.)

As might be expected, matters are not quite so simple and straightforward in practice. Many source sediments have mixtures of the different types of kerogen, including 'type IV, and therefore of the total organic carbon (TOC) content, only a part will be oil-prone or gas-prone. To be viable as a source, a sedimentary unit must have a minimum of 0.5–1.0% TOC, of which a significant proportion must be oil-prone organic carbon (OPOC) (sometimes referred to as sapropelic organic carbon (SOC)), or gas-prone organic carbon (GPOC). To give an idea of the proportions of organic matter required, a unit with <1.0% TOC would be considered a poor potential source rock, whilst a unit with >4.0% TOC would be classed as a good to very good potential source rock. Note the use of the qualifier 'potential'—factors other than the TOC bear on the generation of petroleum from a sedimentary rock. These include the degree of maturity of the organic matter in the sediment; this means whether it has been buried sufficiently deeply in the hydrocarbon 'kitchen' to have reached thermal maturity without having been 'overcooked' by too long an exposure to high temperatures; also the question of drainage in primary migration (see later) is of great importance—that is, how easily the 'protopetroleum' first formed can pass from the source beds into the permeable beds leading to the reservoir trap.

A further factor to be considered is related to the generation of gas. While the type of organic material may be of importance in some circumstances in the generation of 'dry' gas (i.e. the presence of the terrigenous 'woody' type of material), such gas may also be produced by advanced thermal maturation of the other types of kerogen. This 'thermal' gas is generated by the cracking of liquid/condensate petroleums owing to deep burial. There is a thermal 'oil window'—a range of burial depths and temperature—in any given set of geological circumstances, in which liquid petroleum will be generated. On deeper burial (or perhaps in some cases owing to extended periods at high temperatures, without increase in burial depth) the organic material/petroleum passes into the 'gas window' range of burial depths and temperatures, where only gaseous petroleum is generated. It should be noted that there are strong beliefs in the equivalence between effects on a source rock of an extended time period within a given temperature range, versus a shorter time period at higher temperatures. See, for example, Pusey (1973) and Connan (1974).

However, the views of Price (1983) on the influence of geological time on organic metamorphism differ from those just mentioned, and the difference is important. Price considers that: (i) the passage of geological time has no observable effect on organic metamorphism; (ii) vitrinite reflectance can be used as an absolute palaeogeothermometer; and (iii) the accepted models of organic metamorphism require radical rethinking. These points are based on mean vitrinite reflectance (Ro) data compiled from different basins with sediment burial times varying from 2 Ma to 240 Ma. The Ro data show a strong correlation of increase in Ro with increase in temperature, but *no* correlation of increasing Ro with increasing burial times for *any* temperature interval.

The implications of Price's study seem to be that petroleum generation is in many, perhaps most, cases more dependent on what might be thought of as 'pulses' in the palaeogeothermal gradient owing to major geological events with accompanying high heat flow—such as incipient or aborted rifting, extrusive/intrusive igneous activity, hydrothermal activity, and orogenesis. Some excellent plots of vitrinite reflectance values v. burial temperatures and times that support the arguments are included in the paper.

Price's ideas are borne out in some measure by studies of the thermal maturity of source-quality Carboniferous units in the Ouachita Mountains, USA. Houseknecht and Matthews (1985) found that in the western two-thirds of the Ouachita outcrop belt, thermal maturity contours run parallel to structural trends with vitrinite reflectances varying from less than 0.5% to over 3.0%. Such values are compatible with the geothermal range attained at various burial depths and indicate that strata are within the oil and wet-gas windows. Where older and younger strata are brought together by faulting, the older rocks are more thermally mature than the younger rocks *at the same depth*. The older rocks have been subjected to higher temperatures.

In the eastern one-third of the outcrop belt, however, Houseknecht and Matthews (1985) found that thermal maturities are significantly higher (Ro = 2.0–5.0%, within the dry-gas window), with thermal maturity contours cutting across the structural grain. Strata on either side of thrust faults display identical maturity levels. The authors ascribe this different situation to a thermal overprint in the eastern area related to Mesozoic rifting and intrusive activity in the Mississippi embayment.

Baird (1986) deals with some of the matters mentioned above in an interesting recent study of the Kimmeridge Clay source rock in the Norwegian North Sea. This study leads to conclusions regarding the distribution of the various kerogen types in this area. Maturation and source-rock evaluation by means of vitrinite reflectance (see later), original hydrocarbon indices, and TOC determinations indicate that type I kerogen was deposited in the deeper graben areas, type III on the shallow shelf areas, and type II (the dominant type present) in the remainder of the graben and basinal areas. This is shown in the sketch map of Fig. 1.2. The original paper should be referred to for an account of the detailed methods and calculations used.

The study also showed from vitrinite reflectance measurements that oil expulsion (generation) began, on average, at burial depths of about 11,000 feet (3340 m), corresponding to a value of vitrinite reflectance (%Ro) of 0.5 in this study. Maximum oil generating levels were reached at depths of about 13,500 feet (4100 m).

It is of interest to note that at an earlier time, the frequent occurrence of gas at greater depths than oil, and a mistaken belief that oil gravities become lighter with increasing depth,

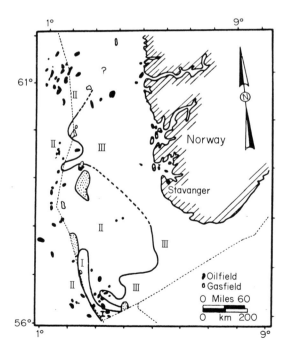

Fig. 1.2. Zonation of kerogen types in a shelf-to-basin setting, offshore Norway. Stippling indicates gas field. From Baird, 1986, reprinted by permission of American Association of Petroleum Geologists

influenced theories of petroleum generation by linking petroleum gravity inversely with depth of burial of the source sediment. Of interest is a paper by McIver (1967).

As actual lithostratigraphic examples of rocks, the organic content of which consists of different kerogen types, Tissot (1977), in an excellent résumé of geochemical methods, quotes the Green River Shales in USA as containing mainly type I kerogen; the Silurian of North Africa, Jurassic of Western Europe, and the Irati Shales of Brazil as containing mainly type II; and the Lower Mannville Shales of western Canada, and Upper Cretaceous of the Douala Basin, Cameroon, as containing mainly type III. It is noted that the hydrogen/oxygen ratio decreases progressively from type I through to type III kerogen, and this is conveniently expressed in a van Krevelen-type diagram, in which the atomic H/C and O/C ratios that result from elemental analysis are plotted.

1.2.3 Recognition of a Sedimentary Source Rock

The determination of the viability of a sedimentary unit or sequence as a source rock comprises: (i) its initial identification as a potential source; and (ii) the determination of the degree of thermal maturity of the organic content of the sediments.

1.2.3.1 *Identification*

Identification of a potential source must include: (i) the determination of the total organic content in the rock, (ii) identification of the type(s) of kerogen present; and (iii) assessment of the drainage capacity of the potential source in its setting with associated rock types. Discussing these in turn:

(i) *Total organic content* (TOC). This is usually determined by burning off the organic content and weighing the resulting carbon dioxide in some specific reaction compounds to determine the TOC.
(ii) *Types of kerogen present*. The particular types of organic matter present can be determined by different methods:
 (a) petrographic analysis of separated and prepared material;
 (b) destructive analysis of the elementary composition of the material, with special emphasis on the C and H content.
(iii) *Assessment of drainage capacity*. This is carried out by constructing a model of the source and associated sediments using stratigraphic and well information, taking into account such factors as fracture zones and faults, high porosity zones, and embedded stringers of more or less permeable material. For a concise treatment of this matter see, for example, Cornford (1986).

Another approach to the identification of source rocks uses suites of downhole logs (e.g. gamma-ray, sonic, density, neutron logs). The method is of interest since it lends itself to making quantitative (volumetric) as well as qualitative assessments of source units; such assessment is increasingly recognized to be of importance in the appraisal of a potential source rock, as it bears on the matter of drainage capability just mentioned. The basis for the method lies in the high level of natural radioactivity possessed by many shales with a large organic content. Figure 1.3 shows a gamma-ray

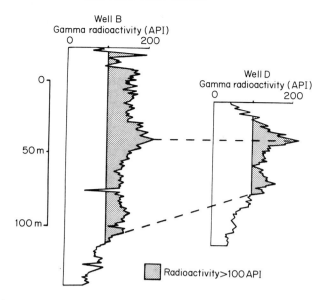

Fig. 1.3. Gamma-ray log correlation of the Kimmeridge Clay between two wells in the North Sea. Reproduced by permission of the Geological Society from 'Hydrocarbon generation and migration from Jurassic source rocks in the East Shetland Basin and Viking Graben of the northern North Sea', J. C. Goff, *J. Geol. Soc. London*, **140** (1983)

log correlation of the Kimmeridge Clay, a well-known source rock in the North Sea, between two adjacent areas.

1.2.3.2 Thermal maturation

The degree of thermal maturity (organic metamorphism) of the organic content of a potential source rock is determined by various methods. Figure 1.4 is a diagram showing the progressive stages of metamorphism of organic matter with increasing burial depth and temperature, and indicates the so-called 'oil and gas windows'—the depth–temperature ranges over which oil or gas tend to be generated (NB the depths and temperatures shown are approximate only).

Tissot (1977) notes that three successive stages of organic metamorphism are distinguishable with increasing burial depth and temperature, these being generally coincident with the three stages shown in Fig. 1.4.

(i) *Diagenesis stage:* marked by loss of oxygen. An immature stage, in which mainly carbon dioxide, water, and some methane and heavy hetero-compounds are generated. This stage corresponds to the 'immature' level of metamorphism shown in Fig. 1.4.
(ii) *Catagenesis stage:* marked by loss of hydrogen. The O/C ratio is unaffected or may even increase slightly. This corresponds to the main stage of petroleum generation— liquid oil, wet gas, and condensate—the 'mature' level of Fig. 1.4
(iii) *Metagenesis stage:* the deep to very deep burial stage. The H/C ratio is about 0.5 and decreases only slightly. Only methane is generated. The 'metamorphic' level of Fig. 1.4.

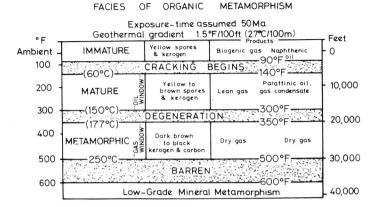

Fig. 1.4. Diagram showing progressive stages of organic metamorphism with increasing burial-depth-temperature, and indicating the oil and gas generation 'windows'. From Fuller, 1982, after Evans and Staplin, 1971. Reproduced by permission of the Joint Association for Petroleum Exploration Courses (UK)

See also Welte *et al.* (1975) and Wright (1980) for considerations of time and temperature in organic maturation.

Clearly, the determination of the maximum palaeotemperature to which the source rock has been exposed is of prime importance in deciding whether it is viable. It can also be of importance to relate the time period of maturation in any area with structural developments, such as growth faulting, change of rate of subsidence, episodes of rapid sedimentation (as in a delta), and development of overpressuring in some units, etc. To do this, diagnostic palaeothermometric characteristics are required, and the two most commonly used methods are:

(i) *Vitrinite reflectance.* Landes (1966) noted the investigation in the USSR about that time of what was termed 'coal reflectance' as an index of thermal alteration of sediments.

Some types of organic material—notably, but not solely, vitrinite in coal—develop increasing reflectance when heated, and this effect is irreversible. This increase can be measured and checked against a set of standard values, being expressed as %Ro (percentage vitrinite reflectance in oil) with values between 0.5 and 3.0 being typical. This is demonstrated in Fig. 1.5, which is a curve showing %Ro related to burial depth for Mesozoic coals and mudstones. This has been used to establish a present-day maturity–depth gradient in parts of the northern North Sea East Shetland Basin and Viking Graben (Goff, 1983). In order to utilize such a curve, it is necessary to make the assumption that some of the organic material disseminated through sediments behaves like the vitrinite in coal. This assumption appears to be valid, to an approximation, based on cross-checks by other methods.

The stages of thermal metamorphism already discussed were defined in terms of vitrinite reflectance by Tissot and Welte (1978) as follows:

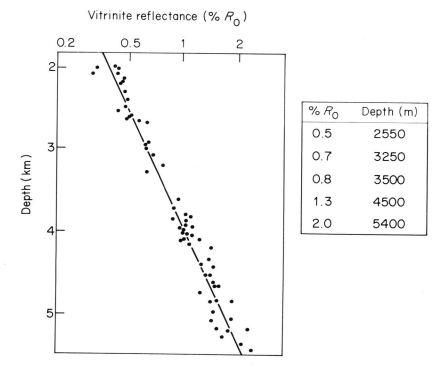

Fig. 1.5. The vitrinite reflectance value (%Ro) related to burial depths of some Mesozoic coals and mudstones. ●, Jurassic coal. Reproduced by permission of the Geological Society from 'Hydrocarbon generation and migration from Jurassic source rocks in the East Shetland Basin and Viking Graben of the northern North Sea', J. C. Goff, *J. Geol. Soc. London*, **140** (1983)

   (a) Diagenetic stage (immature source rock)   Ro < 0.5%
   (b) Catagenetic stage—(the main oil window)   Ro < 1.3%
                 (wet gas/condensate zone)   Ro ⩾ 1.3% ⩽ 2.0%
   (c) Metagenetic stage (thermal, or dry gas)   Ro > 2.0%.

(ii) *Spore/pollen coloration*. A similar kind of system exists for the colour of kerogen as related to burial depth and temperature increase—palynomorphs, such as spores and pollen grains, undergo a change of colour from pale yellow through yellow, orange, and brown to black, representing progressively higher maximum temperatures to which the material has been subjected (see indications in Fig. 1.4, column 2). Originally, colour changes in *any* kerogen samples were used, but as different components of the material have different colours, the method was imprecise. Now only colour changes in palynomorphs are observed, since these bodies all tend to be effectively colourless before being heated. The diagram in Fig. 1.6, from Cooper (1977), shows curves of spore colour changes with increasing temperature, utilizing the spore colour index of Barnard *et al.* (1978) and the thermal alteration index of Staplin (1969).

Cooper (1977) mentions other methods for determination of the stage of maturation of organic matter that can be used with varying degrees of success. One of these is the

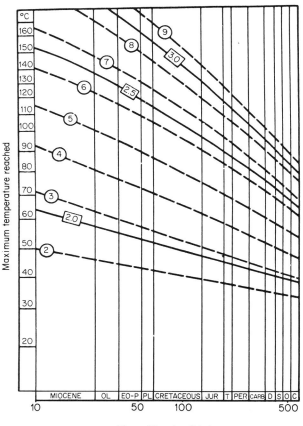

Fig. 1.6. Curves of spore colour changes (----: Barnard et al., 1978) and thermal alteration index (———: Staplin, 1989). Reproduced by permission of Elsevier Applied Science Publishers Ltd (from Cooper, 1977)

measurement of electron spin resonance in kerogen samples. This is based on the changes in chemical composition of kerogen that take place during thermal maturation. The kerogen sample is placed in the path of a microwave, under the influence of a magnetic field that can be varied. At specific magnetic field strengths, the free electrons will resonate and alter the microwave frequency. Measurements by an electrometer determine the number of free electrons per gram, the location of the resonance point, and the width of the signal.

Other methods include the determination of palaeotemperatures by fluid inclusion thermometry; the study of clay mineral diagenetic effects—some conversions, such as the montmorillonite/smectite to illite conversion take place over known temperature ranges; increase in carbon content in ratio with volatiles such as hydrogen; and ratio of TOC to residual carbon.

Studies of thermal maturity indicators in Carboniferous shales of the Ouachita Mountains of Oklahoma and Arkansas, USA (Guthrie et al., 1986) led to some interesting observations.

With regard to clay mineral diagenesis it was noted that bulk clay mineralogy can be useful as a general indicator of thermal maturity, but cannot be used for quantitative indications of specific maturity levels. However, specific diagenetic effects—namely the illite sharpness ratio and crystallinity index—can be used to provide quantitative indices. Weaver (1960) noted that the intensity and sharpness of the 10-Å illite peak on X-ray diffractograms increases with thermal metamorphism. The illite sharpness ratio is defined as diffractogram peak height at 10.0 Å divided by peak height at 10.5 Å, and increases with increasing illite crystallinity. Kubler (1968) defined the illite crystallinity index as the width of the 10.0-Å peak at one-half the maximum peak height; the index shows a *decrease* with increasing illite crystallinity. It was further shown that the illite sharpness ratio increases, and the crystallinity decreases, with increasing mean vitrinite reflectance. This suggests that all three of these parameters are under the same controls of temperature and time. The relevance of the clay mineral parameters to shale maturity will be referred to in the next section, on geochemistry.

New maturity indicators are always being actively sought. For some time now, attention has been paid to organic fossil remains, particularly those of conodonts and graptolites. In this connection, a quite recent development is the study of scolecodont concentrates—see, for example, Goodarzi and Higgins (1987). Scolecodonts are organic fossil remains of jaw parts of certain types of polychaete annelid worms, and are found in both carbonates and shales. From microscopic examination, it has been found that the reflectance, refractive, and absorptive indices of scolecodonts increase with the increase of the Conodont Alteration Index (see Epstein *et al.*, 1977), and can thus be used to determine sediment maturity. As Goodarzi and Higgins point out, dispersion of the optical properties mentioned shows similar trends to those in vitrinite, natural bitumen, and graptolites, indicating similar molecular structural changes with increasing maturity.

Articles by Lopatin (1971) and Assefa (1988) on thermal maturity determination are worthy of study.

1.2.3.3 Geochemistry

So far little has been said about the role of geochemistry in studies of the origins of petroleum.

Much work has been carried out on the identification of the origins of organic content in actual and potential source rocks. As already discussed, physical methods are often adequate to produce credible results in the case of 'structured' kerogens. However, where 'unstructured' kerogens of type I (and sometimes type II) are concerned, involving amorphous sapropelic material, otherwise powerful techniques, such as the measurement of vitrinite reflectance (or the determination of the spore colour index), are useless. In such cases, geochemical methods must be used, and great advances have been made in this field in the last few decades.

An interesting account of some of the techniques developed is given in Moldowan *et al.* (1985), describing a number of methods that may be applied to distinguish between those oils generated from non-marine sources, and those generated by either marine shales or carbonate rocks, and Table 1.1 lists a series of indicators that can be utilized. Several of these are non-diagnostic but may be used as supporting evidence. Others give more positive indications, as in the case of the distribution of monoaromatized steroids.

Table 1.1. A listing of some geochemical indicators. From Moldowan et al., 1985, reprinted by permission of American Association of Petroleum Geologists

	Non-marine v. marine	Non-marine v. marine shale	Non-marine v. marine carbonate	Marine shale v. carbonate
C_{30} steranes	+ + + +	+ + + +	+ + + +	−
Sulphur (%)	+ +	+ +	+ + +	+ +
MA steranes	+ +	+ +	+ +	+
High molecular weight paraffin	+ +	+ +	+ +	−
Carbon preference index	−	−	+ + +	+
Pristane/phytane	−	−	+ +	+
Steranes/hopanes	+	+	+	−
Carbon isotope	−	−	−	−
Gammacerane index	−	−	−	?

+ + + + Definitive. + + + Strong indicator (may be affected by secondary processes). + Weak indicator. − Non-indicator.

Table 1.2. Kerogen density related to vitrinite reflectance. Reproduced by permission of *Journal of Petroleum Geology* (from Kinghorn and Rahman, 1983)

Sample	Kerogen type	Specific gravity range	Vitrinite reflectance (%)
MOS	I	<1.28	0.30
76	I	1.28–1.34	0.72
470	I	1.28–1.34	0.75
246	I	1.28–1.34	0.83
319	I	1.34–1.45	1.75
80/F	II	1.34–1.45	0.35
80/9	II	1.34–1.45	0.38
79/2	II	1.34–1.53	0.38
79/1	II	1.34–1.45	0.43
79/8	II	1.34–1.45	0.47
81/J	II	1.34–1.45	0.48
80/C	II	1.45–1.53	0.55
612	II	1.53–1.65	1.97
80/I	III	1.53–1.65	0.43
80/J	III	1.53–1.77	0.45
80/A	III	1.53–1.65	0.60
199	III	1.65–1.77	0.70
242	III	1.65–1.77	0.99
250	III	1.65–1.77	1.15
350	III	1.65–1.81	1.96
357	III	1.77–1.81	2.10
265	III	1.77–2.2	2.57
266	III	1.77–2.2	2.57
105	HLA	1.53–1.65	0.74
107	HLA	1.65–1.77	1.04
119	HLA	1.65–1.77	1.15
132	HLA	1.77–1.81	1.20
137	HLA	1.77–1.81	1.43
150	HLA	1.81–2.2	2.35

One novel parameter—the presence of C steranes in crude oil—has been found to be firmly diagnostic of source contributions by marine-derived organic remains.

Kinghorn and Rahman (1983) have fractionated kerogen concentrates into their components by differential and sequential gravity separation methods. The density of any kerogen component has been found to be related to its chemical composition, and hence to the nature of the original organic matter, and the level of maturation (organic metamorphism). Table 1.2 shows some results of the technique used, distinguishing between kerogen types I, II, III, and HLA (hydrogen-lean amorphous material). Unproductive carbon (inertinite) is also identifiable by density separation, and different amorphous kerogens can be separated and identified. The work determined that oil-prone (OPOC) amorphous kerogens have lower specific gravities than gas-prone (GPOC) amorphous material—a result that was not, perhaps, self-evident.

In the paper by Guthrie *et al.* (1986) referred to in the previous section, bitumen ratio values (i.e. bitumen/TOC) were plotted against various thermal maturity indicators—mean vitrinite reflectance, illite sharpness ratio, and illite crystallinity index, for carbonaceous shales in the Ouachita Mountains, USA. The resultant plots showed a field of data points in each case contained by a curve, the convex bulge of which approximately defines the liquid hydrocarbon window in the area. The analyses showed that shales in the western and central Ouachitas are mature, whereas those in the eastern Ouachitas are overmature. Petrophysical/geochemical studies of this kind can act as confidence builders for the use of clay minerals as maturity indicators in areas where vitrinite reflectance measurements are not practicable.

A recently described method (Ganz *et al.*, 1987) uses infrared spectroscopy for the rapid quantitative determination of bulk mineralogical composition of sediments, and the determination of kerogen type, maturity, and hydrocarbon potential of source rocks. This method is claimed to be superior to elemental analysis, and to correlate extremely well with vitrinite reflectance.

The importance of multiparameter geochemical studies in the correlation of source rocks with oils is brought home by the work of Peters *et al.* (1989) on the sources of oil in the Beatrice Field, Inner Moray Firth, UK waters. Biomarker and stable carbon isotope analyses show conclusively that the Beatrice oil could not have been derived from the classic (Upper Jurassic) Kimmeridge Clay source, as had been thought, but is a mixture of products from lacustrine Devonian dolomitic siltstones and marine Middle Jurassic rocks.

1.2.4 Source Rocks and Lithology

Twenty years ago, H. D. Hedberg (1967) suggested that it is possible to recognize two broad classes of rock-type assemblages to which petroleum occurrences appear to be genetically related. These are: (i) the shale–sandstone assemblage; and (ii) the carbonate assemblage. In the shale–sandstone assemblage, he placed, as typical examples, the probable source strata of such major accumulations as Burgan (Kuwait), the Upper Tertiary of the Louisiana–Texas Gulf Coast, the Upper Tertiary of the Niger Delta, and the Cretaceous of both Wyoming and Alberta. He recognized several subassemblages, such as the 'pure shales' (e.g. the Upper Devonian black shales of North America); the 'shale–sandstone–coal' group (e.g. the Wilcox of the Louisiana Gulf Coast); and a 'greywacke–mudstone–volcanics' group (e.g. the Katalla oil of Alaska, and the oil/gas of the Niigata area of Japan.

In the carbonate assemblage, Hedberg placed the Madison oil of the Williston Basin, the Mississippean gas of the Four Corners area, and the Zohar gas of Israel, amongst others. The subassemblages suggested were the 'carbonate–shale' source sequence (e.g. the Devonian fields of Alberta, and the Garzan Field in Turkey); and the 'carbonate–evaporite' or 'carbonate–shale–evaporite' sequences (e.g. the Ghawar Field of Saudi Arabia, and the Qum oil and gas of Iran).

With the benefit of hindsight, some might wish to argue as regards the individual examples quoted. However, others, such as Weeks (1958), have made similar suggestions as to major source-rock lithologies, although generally the twofold basic classification has been simplified to shales and carbonates. The term 'shales' is understood as being used as a portmanteau word covering other argillaceous types—clays, mudstones, etc.—in addition to shales *sensu stricto*. As a rough guide to the relative importance of the two basic types as source units, figures of about 50–60% of supposed world petroleum source units being of the shale group and about 40% being of the carbonate group are probably accepted by most petroleum geologists. The figure of 40% for the carbonate group is noteworthy since they comprise only some 16% of the sediments of continental and shelf basins, from which the vast majority of petroleum is produced (Ibe *et al.*, 1983).

1.2.4.1 Shale-type source rocks

As previously remarked, some 50–60% by volume of known source rocks consist of fine-grained argillaceous clastics—shales, clays, mudstones, etc.—containing widely disseminated organic material. Figure 1.7 from Cornford (1986) summarizes schematically the depositional environments on a passive continental margin that favour the accumulation of organic-rich source rocks of the clastic type under discussion. In the North Sea context, gas-prone source coals of Westphalian and Middle Jurassic ages were formed in delta-plain environments (9 in Fig. 1.7), while the Kimmeridge Clay Formation (Upper Jurassic) of oil-prone black shales accumulated in a restricted shelf–basin environment (1 and 5 in Fig. 1.7). In recent years, several studies of shale-type source rocks have been carried out, and some of these will be examined briefly.

The Anadarko Basin, Oklahoma, USA, is considered to be part of an aulacogen (Hoffman *et al.*, 1974; Walper, 1977) and contains a thick sequence of Palaeozoic sediments from Cambrian through to Permian. Petroleum productive zones are present at many levels, and one of the most important source intervals has been identified as the Woodford Shale, an Upper Devonian–Lower Mississippian black shale (see e.g. Hass and Huddle, 1965). In a basin where many deep wells have been drilled, the deepest gas-producing well having reached a depth of more than 26,000 feet (7900 m; Kennedy *et al.*, 1971), the Woodford Shale is the oldest unit to contain vitrinite (Ham *et al.*, 1973).

The thermal maturation of the Woodford Shale has been studied by means of vitrinite reflectance by Cardott and Lambert (1985). They found that maturation of the shale with respect to the generation of liquid hydrocarbons varies from immature ($<0.5\%$ Ro) to depths of less than about 5000 feet (1500 m), through mature (0.5–2.0 %Ro) at depths to 18,000 feet (5500 m), to post-mature (2.0–5.0 %Ro) from depths of 18,000–25,700 feet (5500–7800 m) in different parts of the basin.

In the deeper basin, mean random vitrinite reflectances of 4.89 %Ro and 4.29 %Ro were measured, indicating temperatures greater than 400°F (200°C) were involved—perhaps as

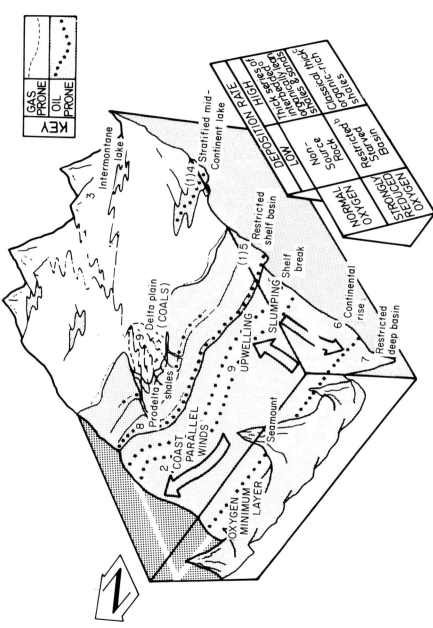

Fig. 1.7. Schematic showing depositional environments on a passive continental margin that favours accumulation of organic-rich clastic source rocks. Examples of the individual areas are as follows: (1) sometimes early rifting phase of major ocean—Cretaceous blackshales of South Atlantic, present-day East African Rift; (2) western side of continents; (3) Devonian Orcadian Basin, Scotland; (4) Permian Phosphoria Formation and Eocene Green River Shales (USA), and European Kupferschiefer; (5) Upper Jurassic North Sea; (6) north-west African Margin, Miocene; (7) Cariaco Trench; (8) Mississippi/Gulf Coast and Tertiary deltas of the Far East; Upper Carboniferous southern North Sea. Examples of deposition rate are: (a) deltaics; (b) some Palaeozoic blackshales; (c) Kimmeridge Clay Formation. Reproduced by permission of Blackwell Scientific Publications Ltd (from Cornford, 1986)

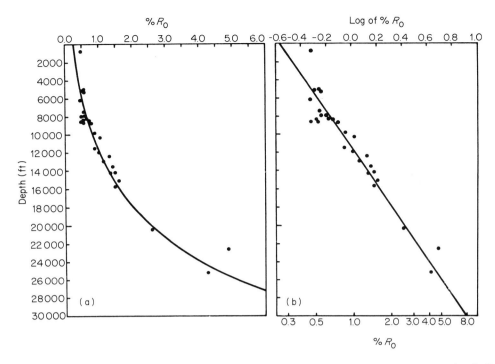

Fig. 1.8. Linear–linear (a) and log-linear (b) plots of vitrinite reflectance (%Ro) against depth (feet) for the Woodford Shale of the Anadarko Basin, Oklahoma, USA. From Cardott and Lambert, 1985, reprinted by permission of American Association of Petroleum Geologists

high as 600–800°F (300–400°C). The current normal geothermal gradient in the basin could have produced temperatures no higher than 450°F (250°C). The thermal anomaly thus revealed is, it appears likely, related to the Wichita orogeny during the (post-Mississippian) Pennsylvanian Period. Figure 1.8 shows linear–linear and log–linear plots of vitrinite reflectance (%Ro) against depth (feet) for the Woodford Shale, and includes two data points for the thermal anomaly (at 4.89 and 4.29 %Ro). Note that the higher value of %Ro is at the shallower depth of the two. The thermal anomaly maximum temperature 'pulse', occurring at some time after deposition of the Woodford Shale, would fit in well with the ideas of Price (1983) mentioned earlier, related to the maximum temperature dependence, and the burial time independence, of source maturation. See also the approach on time and temperature adopted by Lopatin and Bostick (1973).

Various studies have determined that the kerogen in the Woodford Shale is a mixture of types I, II, and III. No definitive quantitative analysis of kerogen type related to basinal situation has been carried out to date, as far as is known, although depositional zonation in the basin has been established from microfossil studies (Urban, 1960).

The roll-over anticlinal traps associated with deltaic growth faulting in the Niger Delta are well known to petroleum geologists. Preliminary descriptions of an onshore (Bomu) and an offshore oilfield (Okan) were given by Frankl and Cordry (1967). In both cases it was suggested that the hydrocarbon source lay in the shales of the alternating sand–shale sequence of the Agbada Formation of Tertiary age. Although some (e.g. Evamy *et al.*, 1978) believed that the Agbada Formation shales were immature, and that the source of most oils in the Niger Delta structures is to be found in the underlying (Tertiary) Akata Formation of marine shales, others (e.g. Lambert-Aikhionbare and Ibe, 1984) are of the opinion that the Agbada Shales are the major source.

It is clear that in the Agbada Formation, with its interbedded shales and sands, drainage paths from the shales into the sands are short, and from this viewpoint the formation must be regarded as an excellent potential source, provided the shales can be shown to have reached maturity.

Recent work by Nwachukwu and Chukwura (1986) included an organic-matter survey in three recent exploration wells. It was shown that kerogens in all three wells are mature, and pyrolysis indicated that the kerogen is largely (63%) of type III, the balance being of type I and II with some recycled coaly material. Mean TOC values for the wells are 2.6%, 2.6%, and 2.4%, respectively, with a range of 0.2–6.5% over all three wells. Vitrinite reflectance was used as a measure of kerogen maturity, together with thermal alteration index (TAI) as determined by Staplin (1969); of the two methods, the vitrinite reflectance measurements are the more precise.

The study concluded that about 3000 feet (900 m) of mature Agbada Shales are present in the western part of the Niger Delta, and the TOC, kerogen type, and maturity indicators suggest that these shales could well have been the major oil source in at least this western area of the delta.

1.2.4.2 Carbonate source rocks

There are two principal associations of petroleum with carbonate rocks that come to the attention of the petroleum geologist. One is the association with biohermal reef carbonates, and the other is the association with bedded carbonates (whether biostromal or otherwise). In the case of biohermal reef carbonates, normally the association is restricted to hydrocarbon trapping mechanisms—for example, porosity traps or drape structures—provided by these. In the case of non-reefal bedded carbonates, however, in addition to sometimes providing reservoir traps, these may also be petroleum source rocks.

In earlier times—and indeed to a limited extent at present—many petroleum geologists had some difficulty in accepting that carbonate rocks can provide viable sources of petroleum. The objections usually include the ideas that: (i) carbonates have insufficient total organic carbon (TOC) content; and (ii) they lack a credible expulsion mechanism to effect primary migration, as compared to that in shales. It is felt that there is a fundamental connection between petroleum generation/migration and clay mineralogy and diagenesis, and that apart from marginal conditions in some impure limestones and marls, this connection is lacking in carbonates.

In contrast to such ideas, it has been demonstrated that in certain areas—notably the Middle East, where some of the largest known petroleum accumulations occur—the source units *must be* chemical sediments. This includes locations where marine carbonates make up the vast proportion of the Phanerozoic stratigraphic section, with few, or no shales of source-rock quality being present (see e.g. Owen, 1964; Palacas, 1984).

Taylor (1986) mentions basinal Zechstein carbonates in the North Sea that are dark, compact argillaceous–shaly micrites, mostly deposited well below wave-base and commonly under anoxic conditions. The potential of these as source rocks is limited only by their thinness. Their organic content is probably largely of algal, sapropelic origin; this is particularly so in the later stages of the carbonate phases of evaporite cycles, when salinity becomes too great to be tolerated by biota other than some algae, phytoplankton, etc. Such organisms, living in the well-lit, warm, and aerated surface-water layers will, at death, sink into the highly saline, anoxic bottom waters and be preserved. Although the sedimentation rate in the basinal depths will usually be low, the reducing environment will ensure preservation for a sufficiently long period for burial to take place. Such an environment is likely to have been present during the deposition of great thicknesses of carbonates in some Middle East areas, and would result in a preponderance of liptinitic type I kerogen (oil-prone) in the sediments.

Some problems associated with the acceptance of carbonates as viable source rocks involve the observations that while the organic content of Recent carbonate sediments, as in the Bahamas and Florida Keys area, is high (generally between 0.98 wt.% and 5.23 wt.% TOC in samples, as quoted in Ibe *et al.*, 1983), the same is not true of ancient carbonate sediments. In the latter, the TOC values are consistently much lower than the TOCs for ancient shales. Ibe *et al.* (1983) are of the opinion that this is due to a much higher degree of convertibility of the TOC in carbonates into petroleum hydrocarbons than is the case in shales. These authors also suggest that the apparent lack of an expulsion mechanism in carbonates is not significant, since *in situ* generation of petroleum takes place in carbonates that have adequate primary or secondary porosity. This would avoid any inefficient primary and secondary migration processes, and could explain the occurrence of many of the unusually rich petroleum accumulations in carbonate in the Middle East, for instance. It would also explain occurrences of petroleum in carbonates that are completely encased in impermeable sealing units. On the other hand, it must be commented that *some* migration is needed to give the segregation required to form an accumulation, since the concentration in the pore spaces of the source zone will still be very low at the start.

If the carbonate unit has good primary or secondary poroperm characteristics, of course, then short-path migration into suitable units in the immediate vicinity may occur. In the El-Ayun oilfield in the Gulf of Suez, Egypt, for example, a Lower Eocene limestone shows oil staining, and there is production from the immediately overlying Lower Miocene Gharamul limestone formation (Elzarka and Younes, 1987). It is possible that the oil staining is due to upward migration through the Eocene carbonate from underlying Turonian and Palaeocene shales. However, as remarked by the authors, the Eocene limestone has sufficient content of (mainly type I) kerogen at the necessary LOM (level of organic metamorphism, as defined by Hood *et al.*, 1975) to be viable as a source rock in itself, and may well have been such.

Taking a pragmatic view, it seems likely that kerogen-type zonation will occur in a basin with carbonates as it seems to do with shales. Basinal carbonate source rocks can be expected to contain a preponderance of type I sapropelic kerogen (oil-prone), while continental- and shelf-type carbonates are likely to contain more type II and III kerogen, depending on the specific depositional environment. This seems to hold good in Middle East areas of thick marine basinal carbonates that are generally oil-prone, and is probably true in other areas.

1.2.5 Source Rocks and Geological Time

It is of interest to consider whether any relationships of significance can be ascertained from a study of the occurrence and relative abundance of hydrocarbon source rocks as related to the passage of geological time.

Using data drawn from 'giant' hydrocarbon fields world-wide (since the source rocks for these have generally been identified), Grunau (1981, 1983) comes to the conclusions that: (i) the Cretaceous is the most important period for the occurrence of oil source rocks; (ii) the Tertiary is the most important for gas source rocks; and (iii) the Tertiary also comes out as the most important period for source rocks of *all* hydrocarbons in place. Grunau points out that the latter result is strongly influenced by the generation of large volumes of gas hydrates by Tertiary source beds.

Grunau also makes the point that whilst the generation of oil and gas is closely related in the catagenetic 'window', gas is also generated in the diagenetic and metagenetic 'windows' in which effectively no oil generation occurs. It is therefore important to include purely gas source rocks in the occurrence–time relationship. A further complication is the large amount of biogenic ('bacterial') methane gas generated under low-temperature conditions by anaerobic bacteria. Rice and Claypool (1981) estimate that gas of this shallow origin (typically generated in swampy and deltaic environments) may account for over 20%—and perhaps as high as 30%—of world natural gas resources.

There may well be links between the time-related occurrences of source rocks, and variations in the climatic and depositional conditions in various geological periods—such variations from the Jurassic through the Cretaceous to the Tertiary being well established. However, much more data needs to be gathered if adequate studies of these matters are to be carried out.

1.2.6 Dissenting Voices

Before concluding this cursory review of hydrocarbon source rocks, some mention must be made of those views that are not in accord with some or all of the thermochemically oriented theories underlying matters discussed so far.

As regards explanations for the variation in produced petroleum types, according to received ideas the factors of importance are: (i) the variation in original kerogen types involved; (ii) the basinal location (linked with factor (i)); (iii) the burial history of the source sediments; and (iv) the possibility of fractionation during primary or secondary migration.

Others, however, consider that minerals associated with the organic material in source sediments may have a greater influence on petroleum type than hitherto imagined. Tannenbaum *et al.* (1986), for instance, suggest that the differing adsorption capacities of such minerals as illite, montmorillonite, and calcite, have a great influence on migration of polar and high molecular weight compounds generated by kerogen breakdown. This is believed to result in mainly heavy oils with some gas being generated by carbonate source rocks (due to negligible adsorption by calcite of the high molecular weight compounds), and light oils and gases being produced by source rocks containing expandable clays with catalytic and adsorptive properties.

As mentioned earlier, there has always been a school of thought in favour of an abiogenic source for hydrocarbons, and taking cosmological evidence alone, there would appear to be a strong *prima facie* case for this. However, crude oil shows overwhelming evidence of biogenic origins: optical activity; presence of porphyrins; predominant presence of compounds with an odd number of carbons (although it should be mentioned that the aliphatic fraction of many crudes has a CPI of about 1.0, that is, little or no odd carbon preference); frequent trapping in sedimentary lenses completely surrounded by impermeable seals, etc. On the other hand, the case for natural gas being of biogenic origin is less secure, and so it is hardly surprising that those who advocate an abiogenic origin have directed their attentions mainly towards gas to provide supporting evidence. A concise summary of the main arguments on both sides is given by MacDonald (1983), for example.

Although about 30% of the natural gas existing in proved commercial accumulations shares the traps with oil, and is therefore presumed by association to be of biogenic origin, it is possible that much of the remaining gas could be abiogenic.

Gold and Soter (1980, 1982) have put forward a persuasive case for outgassing—that is, large volumes of deep Earth gas (originating in the mantle, or at the mantle–crust boundary (the Moho)) having migrated upwards into the crust at certain preferred locations. These locations may include sites of deep fracturing such as astroblemes (impact craters—see discussion in Chapter 11). The outgassing of methane, which could be of abiogenic origin, has been observed at submarine vents on the East Pacific Rise (Welham and Craig, 1979), and from the Nyiragongo lava lake near Lake Kivu, East African Rift (Gerlach, 1980) (see also Coveney *et al*. (1987) for H in Kansas).

Apart from those mentioned above, others have also advanced arguments for abiogenic hydrocarbons, including Porfir'ev (1974) in the USSR, and Giardini and Melton (1983) in the USA. The latter authors believe that general recognition of a juvenile origin for hydrocarbons has been prevented by the presence of 'traces' of biologically derived compounds and fossils that are 'intrinsic' to the sedimentary nature of "source rock" and reservoirs'. The implication that much of the evidence for biogenic origins is purely superficial would, it is believed, not be acceptable to most petroleum geologists, but these authors' arguments should be studied.

Aside from abiogenic theories of hydrocarbon origins there are also suggestions that the thermochemical degradation of kerogen may not be the principal route by which petroleum is produced, but that this is provided by bacteria in the diagenetic 'window' previously discussed. For instance Zhang Yi-gang (1981) proposes a cool, shallow origin owing to microbial genesis, especially for the non-marine oils of China (for instance in the Songliao Basin, as described by Yang Wanli *et al*. (1985)). It is envisaged that the sedimentary organic matter is first transformed into microbial lipids, which after migration and accumulation are finally converted into petroleum at low temperatures (within the range 50–80°C. It is also proposed by Zhang Yi-gang that the petroleum potential of a basin may be enhanced by the ascent of volcanic hydrogen gas into the basinal sediments, which would favour an anaerobic environment.

It is stressed that an open mind must be maintained on the subject of abiogenic gas. All the evidence is not yet in, although it is noted that as far as is known, no major commercial gas accumulation has yet been claimed to be of abiogenic origin, where such a claim can be supported by hard evidence.

As regards the cool, shallow origin of petroleum by bacterial action, there seems no reason to doubt the operation of this mechanism—indeed it is known that much shallow-sourced gas in the diagenetic window must originate from a process of this kind. However, it is another matter to propose this seriously as a mechanism by which the bulk of all petroleum has been produced; it is not thought that such a view would gain general support, since it appears to cast doubt on very well-founded determinations of thermal maturity carried out on source sediments throughout the world by several independent methods, as described earlier.

In a discussion on the origins of petroleum, Chapman (1976) considers the following matters.

Kidwell and Hunt (1958) described studies into what is referred to as petroleum 'of a sort' that is being generated in Recent sediments at very shallow depths (e.g. 35 m) in the Orinoco Delta.

Data on clay compaction suggests that *if* early primary migration of petroleum occurs along with the water expelled by compaction, this must take place at shallow depths, since the *rate* of liquid expulsion decreases with increasing depth—i.e. more liquid is expelled during burial to 1000 m than during burial from 1000 m to 2000 m, etc. (Note: In argillaceous source rocks particularly, the fact that the major part of the compaction and water expulsion that takes place on burial occurs at depths much shallower than those within the 'window' in which oil generation takes place, is a strong argument against the compaction–water-expulsion relationship as an important factor in the mechanism of primary migration, as will be noted later.)

Many petroleum accumulations can only have begun to be of importance after sediment was buried to depths of at least 1000–2000 m *after* the formation of a trap. Research into petroleum content of clays in California suggests a late-generation conclusion—the composition of hydrocarbons in the fine-grained source rocks does not approach that of the accumulated petroleum until depths of about 4000 m are reached (Philippi, 1965).

With these and other matters in mind, Chapman (1976) formulates the following questions: (i) Does the absence of petroleum at shallow depth in an area indicate that it was only generated deeper, or that there are no shallow source rocks? (ii) Should a source rock contain petroleum of the same composition as that expelled? . . . or do some components migrate more readily than others? . . . or does petroleum in the reservoir alter during burial? (iii) Does the presence of hydrocarbons at shallow depth in the Orinoco Delta mean that it will become a significant accumulation, given the right conditions, in a few million years' time? . . . or is it part of the immense quantities of petroleum considered to be lost through lack of a trap?

No definitive answers to these, and many other questions, can be given at the present time. They will remain largely unanswered until an unequivocal understanding of the mode (or modes) of origin of petroleum is reached. It should thus be borne in mind that the current thermochemical concept of petroleum genesis is only a 'working hypothesis', which may be abandoned or substantially modified when a model is presented that better fits all the known facts.

1.3 Petroleum Migration—Introduction

The process of migration, or movement of oil and gas from the locations at which they are generated in the source rocks to those locations where they are eventually trapped (or escape to the surface) is seen as being separated into two distinct stages.

The first of these, in which the gaseous or liquid petroleum moves from the site of generation in the source rock into what might be called the 'carrier' or 'conduit' formation(s), is referred to as *primary migration*.

These 'carrier formations' may form part of the eventual reservoir rocks themselves, or they may be interposed between the latter and the source rock. In either case, movement of the petroleum out of the source rock and through the carrier formations is referred to as *secondary migration*.

The processes and controls governing secondary migration are reasonably well understood. However, the same cannot be said of primary migration. The manner in which newly formed petroleum moves out of the source rock involves problems that still puzzle petroleum geologists, for reasons that will become apparent as the various hypotheses current are discussed in the following section.

1.4 Primary Migration of Petroleum

Possible causes of the initial movement of petroleum out of the source rock into the conduit or reservoir units are noted by Magara (1977) as: (i) forces resulting from sediment compaction and the concomitant expulsion of compaction water; (ii) the effect of water expansion due to temperature increase with increasing burial depth; (iii) osmotic effects related to developing salinity gradients; and (iv) the water release attendant on the progressive dehydration of clay minerals with increasing burial depth and temperature (the montmorillonite–illite conversion).

One of the first problems to be encountered in any attempt to relate primary migration to water movement—whether by compaction, clay mineral dehydration, or other means—is that of solubility. It is a fact that within the temperature window known to be normal for oil generation (i.e. about 60–150°C), oil is practically insoluble in water, although gas is very soluble and could be carried in solution during primary migration. If oil is *not* in molecular solution in water, then the questions arise—in what form *does* the oil migrate, and how, if at all, does water influence the movement if oil is not in solution? Another form of solution that has been suggested is *micellar solution* (see Baker, 1962); this is believed by Magara (1977) to be implausible for several reasons. Magara mentions other possible causes of primary migration, for example, capillary pressure, buoyancy, diffusion, and generation of hydrocarbons—especially gas, etc.—but feels that water movement of some kind must be involved. It is believed that oil migrates in a continuous oil phase along with the formation water in a compacting shale; Dickey (1975) states that 'oil will move along with water . . . if it occupies about 20% or more of the pore volume'.

Magara's (1977) model for primary migration in the oil phase is shown in Fig. 1.9. The upper schematic shows relative permeability v. degree of compaction in a shale; as the latter compacts, the relative permeability to water decreases, and that to oil shows an increase. Although there is an increase in the *relative* permeability to oil, the *absolute* permeability of the shale will decrease progressively as water loss continues and the shale compacts (middle sketch). Oil migration in the oil phase will reach a maximum at an intermediate stage of compaction, then decline as the absolute permeability of the shale decreases (lower sketch).

Magara, who has made many important and thought-provoking contributions to the study of hydrocarbon migration, believes that fluid flow owing to sediment loading may

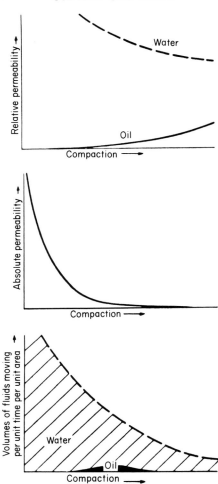

Fig. 1.9. Sketches illustrating a model of shale compaction and oil migration. Reproduced by permission of Elsevier Applied Science Publishers Ltd (from Magara, 1977)

play an important part in primary migration (Magara, 1987). He demonstrates that fluid pressure in excess of hydrostatic pressure can be generated during sediment loading, resulting in horizontal motion of the compaction fluid, which may be essential for driving hydrocarbons towards a trapping position during the primary stages of migration. It is pointed out that during secondary migration in a reservoir, globules of oil or gas are able to move from a structurally lower point to a higher point if the buoyancy is high enough to overcome the capillary restrictions of the reservoir rock. During the primary migration stage, however, the buoyancy is usually considered to be too low to overcome the relatively high capillary pressure of most fine-grained source rocks. 'Under such conditions, the concentration of oil or gas may be possible if there are strong horizontal hydraulic and aquathermal forces related to sediment loading.'

Interesting discussions of oil-phase migration can be found in Hobson (1954) and Dickey (1975), and the former author has also commented (Hobson, 1980) on another hypothetical mechanism of primary migration, namely transfer by wick-type action, presumably involving adsorption. We are also reminded in the latter paper that when the relatively high pressures and temperatures that occur during the formation of oil are involved, substantial volumes of gas could be dissolved in the oil phase, and would move with it as well as in aqueous solution.

Another difficulty, apart from solubility, that must be faced in any attempt to explain the primary migration mechanism is where compaction-related processes are invoked. The main problem here lies in the fact that in source rocks of the clay/shale type, the larger part of any compaction and loss of porosity on burial takes place at depths shallower than that required for the top of the 'oil window' to occur when an average geothermal gradient is present. Through the 'oil window' depth interval, only relatively minor compaction effects are present.

In spite of a lack of general agreement as to the mechanisms of primary migration of oil, a consensus seems to have emerged on certain points. These are that: (i) gas probably migrates primarily in aqueous solution (and possibly also in solution in emigrating oil); and (ii) oil migrates in a continuous phase, perhaps assisted in some cases by movement of water (particularly 'structural' water expelled from clay mineral crystal lattices, as in the montmorillonite–illite conversion).

The somewhat different approach of Momper (1980) must be mentioned here. Momper puts forward a strong argument that neither compaction nor free water is required to cause expulsion of oil from the source rock, but that the mechanism of oil generation is itself adequate to effect this. Momper likens the conditions within the source rock at the period of oil generation to the interior of a pressure cooker. The source rock, closed off by impermeable seals and raised to a high temperature, becomes overpressured when oil droplets begin to be exuded by the organic material (see Fig. 1.10). Two characteristics of the source rock are recognized by Momper as being of great importance: (i) heterogeneity; and (ii) anisotropy. The heterogeneity relates to the organic matter, which is not disseminated uniformly through the source rock, but is concentrated in bedding surfaces, lamination planes, etc., in the more or less fissile argillaceous source rock. Anisotropy refers to permeability in the source rock—relatively high along bedding planes and laminae, but poor to very poor perpendicular to the bedding, except where vertical fractures occur to form migration paths across the bedding planes. During overpressuring, the bedding and laminar planes tend to open up intermittently with pressure build-ups, and oil forming from the organic matter along the bedding planes is expelled (probably in a series of pulses).

There may be sufficient organic matter within a potential source rock for oil to be generated, but insufficient for an adequate level of overpressure to build up for oil expulsion. Examples are known of source beds like this, which have not produced any commercial oil. Before oil expulsion can begin from a source rock, the rock itself must become saturated by hydrocarbons. Momper, and others, state than 15 bbl/ac-ft. of oil must be generated in the source rock before primary migration will take place.

Momper's work results in a number of important propositions related to generation and primary migration. The principal points are:

(i) Initially, the generation of hydrocarbons from kerogen causes a volume reduction of organic matter within the source rock, and hence an increase in porosity. This storage volume must be taken up by the early generated petroleum (cf. the 15 bbl/ac-ft. mentioned above) before pressurization leading to expulsion can take place. With continuing petroleum generation, the residual kerogen, and the newly generated petroleum, represent a greater volume than the original kerogen—hence overpressuring develops leading to expulsion.

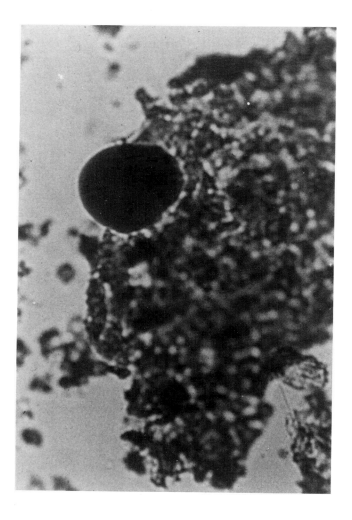

Fig. 1.10. An oil droplet being exuded by kerogen in this microscopic view. An anecdote has it that this effect was first noticed accidentally—and with amazement—when the heat of a microscope's illumination system caused evolution of crude oil from a kerogen specimen. The observation was kept strictly secret for some time afterwards by the oil company concerned. From Momper, 1980, reprinted by permission of American Association of Petroleum Geologists

(ii) Shale-type oil source rock includes over 30% of clay-sized minerals, including quartz. The particle size and mineral content result in the 'impermeable seal' effect, confining the overpressuring within the source-rock system.
(iii) On initiation of primary migration, the petroleum fluid moves along oil-wet laminations and bedding planes, laterally, and along fractures and faults vertically or at an angle to the bedding planes.
(iv) Water is not an important factor in the internal movement of oil within a source rock, being adsorbed or isolated in pores; carbon dioxide—a product of kerogen breakdown—may, however, be an assisting factor in this movement.
(v) Internal migration will stop (temporarily) when the internal pressure is no longer capable of opening the lateral pathways in the rock. Sealing and reopening may occur repeatedly. The process operates like a 'pump', with a succession of expulsion 'pulses' moving relatively small volumes of petroleum each time they occur.

Fuller (1982), who gives a more detailed discussion of Momper's work, notes that hydrocarbon expulsion from a source rock takes place in two main phases: (i) oil expulsion, seen at 83% carbonization of kerogen in the source material (oil-generating kerogen contains >7% hydrogen, with an atomic H/C ratio greater than 1.0); and (ii) gas expulsion, at carbonization levels of 88–91% in the source material (gas-generating kerogen usually has 6.8% hydrogen or less, and an H/C ratio of about 0.8 or less).

Some recent work also suggests that self-expulsion of petroleum occurs. Mackenzie *et al.* (1987) suggest that in Kimmeridge Clay source rocks of the South Viking Graben, North Sea, primary petroleum expulsion took place due to excess pressures (once the petroleum-rich phase filled most of the pores of the source rock), enhanced by capillary forces. Capillary pressure differences caused oil spontaneously to drain from mudstones into adjacent sandstones.

The concept put forward by Momper and others that oil generation itself may cause the propagation of fracture systems in a source rock, thus enabling expulsion to occur, is analysed by Ozkaya (1988). He reaches the conclusion that a continuous fracture network is necessary throughout the source rock for primary migration to take place, and that oil-generation–induced-fracture propagation is possible if certain conditions are satisfied. These conditions relate to the concentration and particle size and shape of organic matter in the source rock (oil-induced fracture propagation is not possible if kerogen is finely disseminated through the source rock); the impermeability of kerogen until a small fraction has been converted to oil; and the TOC of the source, which must be greater than 1% or 2%. A close relationship is thus suggested between organic matter content and expulsion efficiency of source rocks.

Before leaving the subject of primary migration, the work of Price (1981a,b) should be noted. Earlier it was stated that the movement of oil in (molecular) aqueous solution was effectively ruled out owing to the low aqueous solubility of crude oil within the temperature range covering the oil window. Price (1981a,b) reports the measurement of the aqueous solubility of oil over the temperature range 100–400°C in the presence of gas, and goes on to put several propositions, the principal ones being:

(i) Primary petroleum migration in aqueous solution *can* take place within the temperature range 275–375°C with gas in the sediment pore waters.
(ii) Observation of fine-grained rocks from deep wells shows that C15+ hydrocarbons

can survive over long geological time periods at temperatures above 200°C. Laboratory work suggests that in closed, pressurized, water-wet systems, extensive carbon–carbon bond breakages of hydrocarbons are detectable only above 375°C.

(iii) Extrapolation of existing data in metamorphic petrology suggests that water losses of only 5–10 vol.% of the sediments would occur over the burial temperature range 275–375°C.

With these, and other points, Price refutes the most serious arguments put forward against such high-temperature generation and primary migration of petroleum. If various assumptions and extrapolations included in his argument are justified, then there may well be a case for at least some generation of petroleum in this high temperature range.

As remarked by Selley (1983), although an absorbing subject to speculate on, the exact mechanisms of primary petroleum migration are mainly of academic interest, with weighty arguments for and against the various processes mentioned here.

1.5 Secondary Migration of Petroleum

As noted previously, the mechanisms involved in secondary migration are reasonably well understood; the various pore fluids involved are subject to controls of buoyancy/gravity, capillarity, and pressure gradients on local and regional scales, as well as some of the effects discussed in the previous section. (Capillarity, or capillary pressure, is that force in water-wet rock that resists the passage of the oil phase migration, and is proportional to interfacial tension; that is $P_c = 2\gamma\cos\theta/r$, where P_c is the capillary pressure, γ is the interfacial tension, θ is the contact angle between oil and water, and r is the radius of the pore throat. Goff, 1983). The part of the expression $\gamma\cos\theta$ is related to the property of a surface that has been termed 'wettability'.

Some workers believe that the operation of these processes may extend beyond the 'secondary migration' stage of movement. Indeed, there is a categorical statement in a recent paper on the subsurface movement of petroleum fluids that 'The separation of . . . movement by previous authors into different processes, called primary migration, secondary migration, and accumulation, is artificial. Nearly all petroleum flow in the subsurface occurs by the same mechanism—bulk transport driven by gradients in petroleum flow potential. Capillary forces dominate over viscous forces, and the flow is non-turbulent'. As yet it is not clear whether this view meets with general acceptance amongst petroleum geologists, but see a further comment on this later.

Magara (1977) points out that when hydrocarbons have (by whatever means) been expelled from a source rock into a reservoir formation, they meet with very different physical conditions to those extant in the source rock. These conditions include larger pore spaces, lower capillary restriction, less 'semi-solid' water, and lower fluid pressures. These factors are conducive to the confluence, interconnection, and enlargement of the hydrocarbon globules, with a significant increase in the buoyancy force. This creates a tendency for the hydrocarbon to move to higher structural levels in the reservoir.

In passing, it is commented that the considerable physical differences between source and reservoir rocks, as noted by Magara, are probable causes for unease in the minds of some petroleum geologists at the categorical statement quoted above that the differentiation of hydrocarbon movement into primary and secondary migration is 'artificial'. Also, clearly, such a view is not in accord with the ideas of Momper (1980), mentioned earlier in regard to primary migration. The conditions proposed by Momper as being present in the source rock 'pressure cooker' are quite different to those in reservoir or carrier-bed conditions outside the impermeable seal surrounding the source, and must lead to different modes of movement of the hydrocarbons. Whether Momper's whole argument is accepted or not, it seems unlikely that the buoyancy force, which appears to be a major factor in movement through the carrier and reservoir beds, can have a similar importance in anisotropic, laminated argillaceous source formations. In these, apart from the locations of vertical fractures, any expulsion mechanism must be adequate to cause lateral movement of hydrocarbon globules. If this is the case, it would seem possible that differentiation of movement into primary and secondary migration may be fundamental, rather than artificial.

Pratsch's (1983) paper on secondary migration of gas in north-west Germany affirms the belief that the most important factor in secondary migration is the buoyancy effect operating on a less dense hydrocarbon molecule in an environment of more dense water molecules in water-filled porosity. He suggests that for different basinal configurations (circular or elongate; symmetrical or asymmetrical; straight or curved axis; single or multiple depocentres) there are preferred secondary migration pathways for hydrocarbons, and within some basins, directions of 'focusing' of migration. He notes that 'Hydrocarbons migrate updip unless extreme pressure differentials prevent this', and quotes as examples of causes of such abnormal pressure differentials the influx of fresh water from basin flanks or local vertical pressure gradients from high-pressure undercompacted clays to underlying normal-pressured porous layers. He also remarks that hydrocarbons will migrate vertically or laterally, depending on geological conditions, and that permeability-enhancing fracturing that aids migration is common on both local and regional scales.

Price (1980) also points out the prime importance of major faults of large throw to the process of secondary migration from deep sources into shallow reservoirs and traps.

A recent study underlines the probable importance of vertical pathways in secondary migration. The generation, migration, and accumulation of oil in three Lower Miocene reservoirs in the El-Ayun Field, Gulf of Suez, Egypt has been studied by Elzarka and Younes (1987). Trace element contents in the crude oils and sediments were used to examine the process of oil migration from probable source to reservoir rocks. Turonian shales in the trough of the depositional basin are identified as the probable source rocks. Primary migration is believed by the authors to have been initiated by the effects of overburden loading during the deposition of Senonian and Palaeocene sediments, although the exact mechanism is not detailed. Both lateral migration of oil and connate water, and vertical migration mainly through fault pathways, is postulated, leading to generally updip migration from the source rocks in the basin trough to the reservoir sandstones, during the secondary migration period.

In addition to the buoyancy force, other effects (some of which have already been mentioned) may be active—compaction of the sediments, the aquathermal effect, clay dehydration, etc., may all affect secondary migration—either by assisting the generally upward flow of hydrocarbons, or by acting against it. In the latter case, structural and hydrodynamic complexities may arise that cause hydrocarbons to move out of structural highs in a downdip direction, perhaps somewhere to be held in hydrodynamic equilibrium in a non-structural trap location, or to be lost altogether.

1.5.1 Hydrodynamic Factors

Consideration of the secondary migration of petroleum through carrier/reservoir beds into the trap location within the latter must lead to discussion of the possibility of hydrodynamic traps being formed. Such traps, believed by some to result from the effect on petroleum accumulations of moving formation water, have always been a subject of some controversy, being held on the one hand as a class of trap distinct from structural and stratigraphic trap categories, or on the other hand as figments of the imagination, or misinterpretations of other physical effects.

The subject of the movement of formation fluids has been studied in considerable depth by many workers, some of whom will be mentioned here. Hobson (1954) and Amyx *et al.* (1960) discuss the influence of capillary pressure on multiphase flow and trapping; Hubbert (1953, 1967) and Dahlberg (1982) study potentiometric surface gradients related to the configuration of the oil–water contact surface in traps; and Berg (1975), developing these ideas, makes a predictive calculation of oil-column height in a stratigraphic trap. A useful summary of the subject is given by Davis (1987).

The question of whether any important present-day hydrocarbon traps can be ascribed solely to hydrodynamic effects is a vexed one. Magara (1981) summarizes some very telling arguments against the idea. He reminds us that there are two different types of water movement in a sedimentary basin—sediment source water movement and meteoric water movement. Sediment source water moves principally: (i) from a shale to a sandstone or other porous bed; and (ii) from a basin centre to its margins, or from deeper to shallower parts. The movement of such water may be of some significance in the primary migration of hydrocarbons, being involved in compaction expulsion of fluids from the shales.

On the other hand, the movement of meteoric water takes place primarily under gravity within sandstones and other permeable formations, and from basin margins towards basin centre, although the water must eventually find its way out to the surface by some means. This movement is taking place at the present, and may or may not have developed in the geological past. It is probably unimportant in primary hydrocarbon migration, but may affect the movement and trapping conditions of hydrocarbons in a reservoir formation: this is what is normally meant by 'hydrodynamic trapping'.

Magara's most telling arguments for discounting hydrodynamic trapping as an important mechanism include:

(i) In most interior sedimentary basins, the present hydrodynamic patterns are considered to have been developed only a short while ago, geologically speaking, and most petroleum accumulations were formed a long time before the present hydrodynamic flow patterns.
(ii) In oil or gas zone in reservoirs, the relative permeability for water is virtually nil. It is therefore almost impossible to move water through these zones even though there is some absolute permeability.
(iii) A tilted oil–water (or gas–water, presumably) contact is sometimes used as evidence of hydrodynamic conditions. However, such a tilt could also result from other causes, such as pore geometry and size, caused by changing lithology, stress, and diagenesis. Also, sealing faults that compartmentalize the oil–water contact could also give the impression of a tilted contact (see Fig. 1.11).

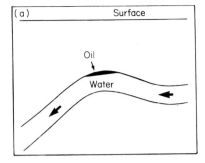

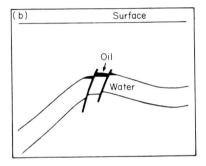

Fig. 1.11. An apparently tilted petroleum–water contact (a) that could be explained by the contact being compartmentalized into short, horizontal segments by faulting (b). Reproduced by permission of *Journal of Petroleum Geology* (from Magara, 1981)

These, and other points in Magara's paper, are cogent arguments, and it seems likely that in line with his conclusions, we may suppose that many apparent hydrodynamic traps are really misinterpretations of the physical realities in various situations. However, is it possible that hydrodynamic effects could still be significant in the case of some traps, and can counter arguments be marshalled?

As regards the first point made above, petroleum currently trapped in one location may originally have been trapped elsewhere, and have shifted recently owing to neotectonism (tilting, subsidence effects, etc.). As regards relative water permeability of oil-saturated reservoirs, these may be only part-filled, or occur only in the upper part of an already water-saturated formation, in which case water may move beneath, but in contact with, the petroleum-filled porosity. With reference to tilted liquid–liquid (or gas–liquid) contacts, some interesting examples of these have now been seen in seismic data, and are worth discussing.

In the Fig. 1.11 example, the argument that seems to be implicit is that drilling tests would just sample the different levels of the oil–water contact in the trap, and the assumption could incorrectly be made that the contact was tilted. Alternatively, if seismic data are involved, it seems to be implied that the resolution of the data would be insufficient to differentiate between the situation in Fig. 1.11(b) and a real tilted contact.

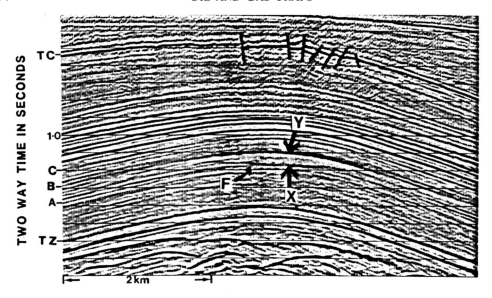

Fig. 1.12. An actual gas–water contact event (X) in seismic data. The top of the gas-filled porosity interval (proved by drilling) is defined by the 'bright spot' DHI event (Y). The basal flat spot X is interrupted by a fault (F) of small throw. Rest of key: TZ = Top Zechstein (top of salt interval); A, B, and C = three levels in the Triassic; TC = Top Chalk. Reproduced by permission of Seismograph Service (England) Ltd

In this connection, the seismic example in Fig. 1.12 is of interest. This shows an actual gas–water contact event, X, in a southern North Sea location where the gas accumulation has been proved by drilling. The event marked Y is the 'bright spot' defining the top of the gas-filled porosity interval, with X being the 'flat spot' gas–water contact event at the base of the same interval. A, B, and C are different levels in the Triassic, while TZ is the top of the Zechstein mobile salt interval producing the anticlinal structure, and TC the Top Chalk/Base Tertiary reflection group. In particular, the small fault F that is affecting the gas–water contact event should be noted. It is submitted that in consideration of the clear appearance of this small fault, and also the geometry of the situation (particularly noting the seismic stratal events within the A–B interval related to the shape of the 'flat spot' event X), it would seem unlikely that the obvious tilt on the contact event is due to faulting, or to other factors such as facies change or a series of 'microreservoirs', as has also been suggested.

From this, and other seismic evidence in the North Sea area, the writer believes the existence of hydrodynamic trapping mechanisms to be probable, although acknowledging the necessity for more information and verifiable observations before any firm conclusions can be drawn. Hubbert's (1967) paper on this subject is recommended as being of particular interest and relevance.

CHAPTER 2

Reservoir and Seal Formations

2.1 Reservoir Rocks

2.1.1 Introduction

Rocks within which petroleum is able to move, accumulate, and become trapped are reservoir rocks.

First and foremost, such rocks must be *permeable* to petroleum. This implies that they must be porous (although the reverse is not true; porous rocks need not be permeable—think of pumice, for example), and if they are not, they are of no use as reservoir rocks. Permeability implies not just that petroleum can migrate into a trap in the reservoir formation, but also that it can subsequently be *produced* commercially.

There are different types of porosity and permeability. Rocks may be porous because they have intergranule (interparticle) pore spaces, or fractures, or vugs, and in each of these cases the permeability will take different forms.

All sedimentary rocks have a certain porosity on deposition (primary porosity), and this decreases with burial depth, owing to compaction.

Apart from burial compaction, other factors can also reduce porosity (or change it) with time. Typical of such factors are various processes involved in diagenesis, such as cementation and solution, or clay mineral transformations.

Fracture porosity may occur in many different rock types. In recent years, considerable interest has grown in geophysical methods offering means of determining the aspect ratio and the orientation of cracks and fractures in rocks. Determination of fracture orientation can be important: (i) in planning the drilling and production of a new oil- or gasfield; and (ii) in secondary recovery operations.

2.2 Porosity

In any rock, some of the pore space is interconnected, and thus of importance with regard to its permeability (the ability of the rock to allow fluids to flow through it). Other pore spaces exist that are *not* interconnected, and therefore do not contribute to permeability.

The ratio of the *total* pore space (both interconnected and non-interconnected) to the total rock volume is known as the *absolute porosity*, which is expressed as a percentage:

$$\%\text{porosity} = (\text{pore volume}/\text{bulk rock volume}) \times 100$$

Clearly this is mainly of academic interest. What the petroleum geologist is interested in is the *interconnected* pore space; the ratio of this to the total rock volume is called the *effective porosity*.

Rock porosities vary appreciably according to a number of factors—lithology, depth, diagenetic effects, presence or absence of fractures, etc. Completely uniform porosity is probably never present in any rock; rather, the rock is made up of varying proportions with different porosities. Effective porosity of less than about 5% indicates a non-commercial situation normally, whilst over 25% is considered as excellent. The average tends to fall somewhere between 10% and 20%. Porosities are usually measured either on samples in the laboratory, or by downhole log determinations. The details of these determinations appear in any standard petroleum geology or reservoir engineering text.

2.3 Permeability

This allows the flow of fluids through the effective porosity in a rock without changing or damaging the rock fabric (i.e. flow without the application of undue pressure). The value measured is the *absolute permeability*. When, as in reality, a fluid does not completely saturate a rock, the permeability of the rock to that particular fluid in the presence of others (e.g. for gas in the presence of oil and water) is known as the *effective permeability* of the rock to the given fluid. The ratio of effective permeability to absolute permeability is known as the *relative permeability*, and the latter is a measure of how much the permeability of one specific phase has been reduced by the presence of another phase or phases.

The unit of measurement of permeability is the *Darcy*, named after Henri Darcy who carried out original work in this field in the mid-nineteenth century. Darcy's law can be expressed in an equation, as given by Russell (1960):

$$P = FVL/CD$$

where P = permeability in Darcys, F = volume of flow in cm^3/s, V = viscosity of fluid in centipoises, L = length of sample in direction of flow in cm, C = cross-sectional area of sample in cm^2, and D = pressure difference between faces of sample, in atmospheres.

The fluid flow has to be non-turbulent, and only a single fluid, which does not interact with the rock, must be passed through the sample at one time. The Darcy is too large for most practical purposes, and the unit commonly used is the milliDarcy (1 mD = 0.001 Darcy).

Levorsen (1967) gives as a rough guide to reservoir permeability: poor, <1.0 mD; fair, 1.0–10 mD; good, 10–100 mD; very good, 100–1000 mD—noting that permeability, as well as porosity, varies greatly both laterally and vertically in the average reservoir rock.

Permeability is usually measured in the laboratory by testing cores in an apparatus known as a *permeameter*, which records the rate of flow and the pressure drop of fluid through a sample of known cross-sectional area. The fluid normally used is (non-reactive) air or another dry gas, passed at very low pressure gradient so as to avoid the occurrence of any turbulent flow, which can result in serious errors.

Note that Darcy's law assumes that only one fluid is present in the porosity. This is usually not the case in a natural reservoir, where the porosity may contain oil, gas, *and* water in varying proportions. As noted, the ratio between the effective permeability to a given fluid at partial saturation and the permeability at 100% saturation (the absolute permeability) is known as the *relative permeability* (e.g. Levorsen, 1967).

2.4 Sandstone Reservoirs

Of the two principal lithological classes of reservoir rock, namely sandstones and carbonates, sandstones contain the greater proportion of the world's oil accumulated in fields that can be categorized as 'giant'.

It is remarked by Taylor (1977) that on the macroscopic scale, the typical sandstone reservoir (owing to depositional environment) is characterized by lateral rather than vertical development and dimensions. On the microscopic scale, the grain size, shape (roundness), and sorting set limits to porosity, permeability, and hydrocarbon saturation. Much published work suggests that sand porosities at or shortly after deposition fall in the range 30–50%, clustering markedly about a value of 40%. Most reservoir sandstones, on the other hand, have appreciably lower porosities than this—in commercial North Sea fields, for instance, commonly between 20% and 30%. The difference seems to be more the result of compaction rather than cementation (see section 2.4.1).

As regards grain size and sorting, Beard and Weyl (1973), working with mixtures of river sands, find that porosity is effectively independent of grain size for wet-packed sand of the same sorting, but varies from about 42% for extremely well-sorted sand to 28% for very poorly sorted sand.

The effect of grain size and sorting on permeability has been investigated by Krumbein and Monk (1942). They find that:

$$K \propto d^2 \exp(1 - 1.35\sigma)$$

where K is the permeability, d is the geometric mean grain diameter, and σ the standard deviation of the grain size about the mean.

In unconsolidated sands used for experimental purposes in these determinations, the maximum average measured value was 475 Darcys for extremely well-sorted coarse sand, reducing to 14 Darcys for a very poorly sorted coarse sand. Permeabilities in 'real Earth' sandstone reservoirs are much lower, being considered good if over 1 Darcy, and exceptional over 4 Darcys. The large difference is due partly to cementation. The presence of clay minerals particularly—which is common in real reservoirs—has a very significant effect on permeability (both because of the inherent blockage of pores and pore throats, and because of the hydrophilic property of clays that increases the volume of 'fixed' water in the porosity); with deeply buried reservoirs the compressibility of the pore spaces owing to burial-depth compaction, as mentioned previously, comes into play.

Depositional factors controlling reservoir properties and geometry are of great importance, and the paper by Taylor (1977) is strongly recommended—discussing in detail a number of examples of the influence of environment of deposition on reservoir character in sandstones. Figure 2.1 is reproduced from Taylor (1977) and gives an excellent résumé of sandstone reservoir characteristics for the principal depositional environments.

38

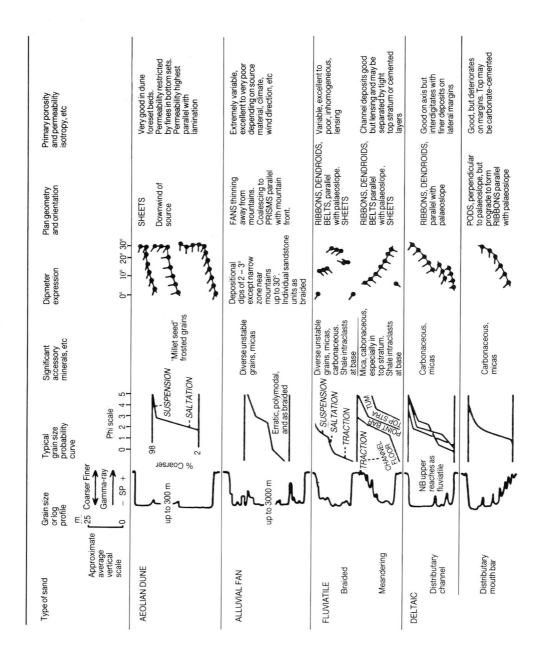

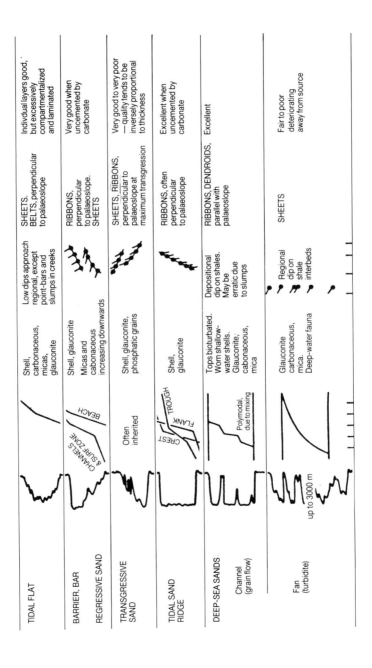

Fig. 2.1. Sandstone reservoir characteristics for the principal depositional environments. Reproduced by permission of Elsevier Applied Science Publishers (from Taylor, 1977)

2.4.1 Factors Affecting Sandstone Porosity

There are various factors that may conceivably influence the final form and amount of porosity in a sandstone reservoir. Some of these are more significant than others. Scherer (1987) lists the following parameters: age, mineralogy, maximum depth, sorting, grain size, rounding, sphericity, grain orientation, temperature, abnormal pore pressure, hydrocarbon saturation, and chemistry of formation water. He gives a useful selection of references for each of these. It is also noted that the creation of secondary porosity by leaching can be an important factor. Leaching creates porosity (Schmidt et al., 1977; Burley and Kantorowicz, 1986) by dissolution of: (i) chemically unstable framework components; and (ii) soluble cements.

Scherer concludes that the most important parameters influencing porosity are age, framework mineralogy (and particularly the content of detrital quartz), sorting, and maximum burial depth. Also, he suggests that because of the relationship between compaction and cementation, cement does not reduce porosity by an amount equivalent to the cement volume, and that porosity reduction by cement is normally only a small fraction of the total reduction.

This perhaps rather surprising conclusion (since many believe that major losses of primary porosity occur due to secondary cementation) seems to be borne out by the work of others. The relative importance of compaction and cementation processes in the modification of sandstone porosity has also recently been investigated by Houseknecht (1987). He points out that compactional processes embrace both purely mechanical compaction, and the effects of intergranular pressure solution (what might be termed 'chemical compaction'). Both these processes, together with cementation, framework grain dissolution, and cement dissolution, are noted as playing significant roles in modifying porosity in various sandstones. A technique whereby the relative importance of the compactional processes and cementation can be quantified is described, and applied to data for two reservoir sandstones—the Nugget Sandstone, Utah, and the Simpson–Bromide Sandstone of Oklahoma.

The conclusions reached by Houseknecht suggest that starting with an assumed original intergranular porosity volume of 40% for well-sorted sands, the porosity can be reduced to about 30% by mechanical compaction only, and from 30% to virtually 0% by chemical compaction (including intergranular pressure solution). Application of the method to the two real reservoir sandstones indicated that in 97% of the samples analysed, mechanical compaction and intergranular pressure solution together have destroyed more porosity than has cementation. This result is in general agreement with that of Scherer (1987) quoted earlier.

However, agreement may not be universal. Magara (1980) discusses the difference between the porosity–depth relationship in shales and that in sandstones. Figure 2.2 is a summary of shale porosity–depth relationships for 10 different areas in the world. All curves indicate that shale porosity decreases with increase of depth of burial, with the rate of decrease being rapid at shallow depths and slower at greater depths. The curves showing relatively high porosities at depth have probably been affected by abnormally high fluid pressures (overpressuring) in the shales.

Sandstones, on the other hand, apparently indicate a different situation. In Fig. 2.3, after Galloway (1974), two different sandstone porosity–depth relationships are shown,

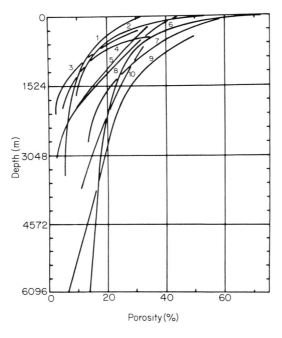

Fig. 2.2. Summary of shale porosity–depth relationships for 10 different areas of the world. Reproduced by permission of *Journal of Petroleum Geology* (from Magara, 1980)

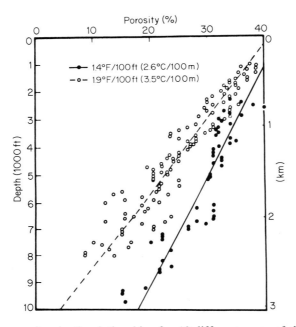

Fig. 2.3. Sandstone porosity–depth relationships for 10 different areas of the world. Reproduced by permission of *Journal of Petroleum Geology* (from Magara, 1980)

for areas with two different geothermal gradients, in the north-east Pacific arc. The relationships tend to plot as straight lines on normal graph paper, suggesting that the rate of sandstone porosity reduction is about the same whether at deep or shallow depths. This leads Magara to suggest that the porosity reduction is primarily controlled by chemical and mineralogical agents, rather than by physical agents. He says '. . . the rate of porosity reduction seems to be controlled neither by the pressure or stress applied per given contact area, nor by the mobility of pore-water if the sandstone is saturated by water'. The conclusion seems not to be in accord with the results of the work of Scherer (1987) and Houseknecht (1987) quoted earlier.

A further factor that can significantly affect the poroperm characteristics of a sandstone (or other) reservoir rock is the finite strain related to tectonic deformation mechanisms. The progressive evolution of a sandstone can be traced, as noted earlier, from its condition of primary porosity of about 40% through porosity reduction related to compaction, ductile grain deformation, pressure solution, and diagenetic effects such as clay mineral transformations and partial infilling of porosity by cements. Subsequently, some secondary porosity may develop owing to selective leaching of framework minerals and/or cements. Finally, deformation mechanisms, such as cataclasis and tectonically induced pressure solution, reduce porosity, while extension, fracturing, and brecciation enhance it.

Mitra (1988), who has made a study of this subject in the Central Appalachian Overthrust Belt, concludes that tectonic deformation mechanisms play an important role in reducing or enhancing the eventual porosity of potential reservoir rocks in structurally complex areas, such as overthrust belts (and perhaps to a lesser extent in less complex tectonically stressed areas?). Study of finite strain is recommended in analysing regional and local variations in reservoir potential. In particular, the relative timing of pressure solution and fracturing with respect to folding history determines the possibility of retaining open fracture porosity in a reservoir.

2.4.2 Diagenetic Processes in Sandstones

It is stated in Glennie *et al.* (1978) that '. . . minerals are only completely stable in the environment in which they are formed . . .', and this sums up the matter of diagenetic alteration in minerals and rock fabrics.

Numerous studies of the effects of various diagenetic processes on sandstone porosity and permeability have been carried out, some of the most noteworthy during the last two or three decades. As noted by Taylor (1978a) in his introduction to a collection of papers on this theme, good progress has been made in studies of the geometry and facies distribution of depositional environments likely to lead to desirable reservoir qualities *at the time of deposition*. However, no equally simple and generally applicable models of diagenesis had so far been proposed up to that time (and to a large extent, this still holds good), in spite of the important influence of diagenesis on reservoir properties.

Taylor resolves the problems into two aspects: what patterns of cementation, if any, can be attributed to the accumulation of sand in particular depositional environments, and what further changes can be expected to occur as a result of subsequent burial?

The effects of deep burial have been investigated, and a porosity–depth gradient method developed for the Gulf Coast area, USA, by Atwater and Miller (1965) and Maxwell (1964). It was found that in the relatively uniform conditions of sandstone deposition and burial

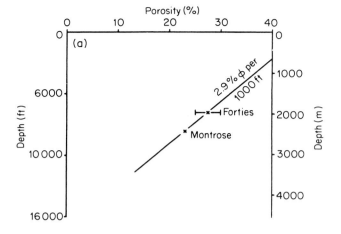

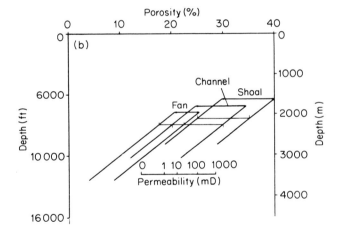

Fig. 2.4. In (a) a porosity gradient for Palaeocene sandstones is based on published data, while (b) shows porosity windows for various Palaeocene sand facies, constructed by using the gradient in (a) with porosity spectra for known depths. Reproduced by permission of the Geological Society from 'Porosity gradients in North Sea oil-bearing sandstones', R. C. Selley, in *J. Geol. Soc. London*, **135** (1978)

in that area, the average porosity decreased linearly with depth. Selley (1975a, 1978) in North Sea studies has evolved a family of porosity gradient curves for sandstone reservoirs from the Palaeocene to the Permian. Figure 2.4 shows (a) a porosity gradient for Palaeocene sandstones based on published data, with a point plotted to indicate the average porosity in the Montrose reservoir, and a bar indicating the range of porosities in the Forties sandstone reservoir, superimposed on the gradient. In (b), porosity windows for various Palaeocene sand facies have been constructed using the porosity gradient in (a) and porosity spectra for known depths. The thin horizontal lines indicate ranges of porosities of facies at the given depths.

The mineralogical changes responsible for the decrease in porosity with depth, concomitant with increasing temperature and pressure, include the precipitation of carbonates, secondary quartz growth (Waugh, 1970), sometimes feldspar developments, pressure solution of quartz (resulting in a reduction of pore volume), and the growth of authigenic clays. (The collection of papers on sandstone diagenesis in the *Journal of the Geological Society of London*, **135**(1), January 1978, is still relevant and interesting on many aspects.)

Taylor (1978b) suggests specific ways in which depositional environment may influence later diagenetic processes. An example is quoted of channel sands, where the sorting processes during deposition result in freedom from pore-restricting detrital clay, while precipitation of authigenic clay on the clean grain surfaces inhibits later secondary quartz overgrowths and, as burial becomes deeper, enables pressure solution to be effective. In channel floor deposits, moderate primary porosity and permeability resulting from depositional processes may be reduced mainly by widespread dolomitic cementation, which, where present, prevents significant later changes (an effect common in channel floor sands).

The relationship between diagenetic processes and eventual porosity values is well demonstrated by an actual study of reservoir quality in the (Lower Eocene) First Wilcox Sandstone of Livingstone Field, Louisiana, by Johnston and Johnson (1987). The oil in this field is produced from the named sandstone, which is a 40–50-feet-thick barrier island deposit (see Fig. 2.5). Diagenesis in the different depositional facies largely controls the occurrence and quality of the reservoirs. There was a significant reduction (65–75% of the

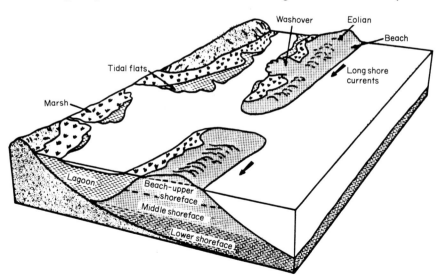

Fig. 2.5. Suggested depositional environment (barrier island deposit) of the First Wilcox Sandstone of Livingstone Field, Louisiana. From Johnston and Johnson, 1987, reprinted by permission of American Association of Petroleum Geologists

primary porosity) of porosity by clay and quartz overgrowths and by carbonate cement. Dissolution of the carbonate cement, and leaching of feldspar and other unstable grains, restored the porosity to 65–75% of the original values.

In this area, the highest values of the porosity so restored were located in the facies that had the highest primary porosity—the aeolian, beach, and upper shoreface sands. The initially less porous middle and lower shoreface sands developed little or no secondary porosity. Some later loss of permeability has been occasioned by reduction or blocking of pore throats by late-stage growth of diagenetic clays, such as kaolinite, illite, chlorite, and smectite.

The primary authigenic cements in the rock are calcite and kaolinite, with minor or trace amounts of quartz, pyrite, smectite, illite, chlorite, ferroan dolomite, and siderite. These authigenic cements rarely make up more than 10% of the bulk volume of the sandstone, and have no serious effect on reservoir quality. There are, however, some patches within the upper and middle shoreface sands where authigenic clays, particularly kaolinite, have a serious effect on reservoir quality, blocking pore throats and increasing irreducible water saturation.

An interesting study of porosity reduction in quartzose sandstone by quartz overgrowths has been carried out by Leder and Park (1986). A mathematical model is developed based on fluid flow through initially porous, unlithified sands in which the fluid phase is saturated with $Si(OH)_4$ and is always in equilibrium with quartz. As the circulating fluid migrates updip towards the basin margin, it cools and precipitates quartz in the pore spaces, causing a loss of porosity. Secondary dissolution porosity and other diagenetic processes (compaction, carbonate and clay mineral precipitation) have been neglected for the purposes of the model. Porosities calculated for the model correspond to measured porosities of simple quartz-cemented sandstones in 27 wells in North and South America.

In these circumstances, the rate of porosity decline was found to depend on the following variables in decreasing order of significance: burial rate, age, initial porosity, basin size (dip angle), fluid dynamics, initial permeability, and geothermal gradient.

The authors note that regardless of burial rates, thermally driven fluid velocities appear to be of the order of 10 cm/year at depths where hydrocarbons may be generated; generated oils would travel at these or higher velocities updip along the porous conduits. This implies that secondary migration of oil is efficient.

On the other hand, in arkosic sandstones, diagenetic alteration of feldspar minerals becomes significant as an influence on reservoir quality. Feldspar dissolution and decomposition can yield secondary porosity within detrital grains, and also result in enlarged intergranular pores. However, clay authigenesis and associated secondary quartz precipitation create adverse effects in terms of overall porosity/permeability of a sandstone (Tieh *et al.*, 1986). The latter authors conclude that although it was not possible in the sandstones studied (Upper Miocene arkoses in Kern County, California) to accurately account for the gains and losses resulting from these processes, in general, porosity destroyed in these sandstones by clay and quartz authigenesis outweighs that created by feldspar dissolution and decomposition. This appears to be a contradiction to predictions from simple volume-change calculations. The reasons for this are discussed in detail by Tieh *et al.*; one can be mentioned here, which is that although feldspar dissolution creates secondary pores, these commonly contribute only to microporosity holding irreducible water saturation. It is also noted that K-feldspar-dominated arkoses form better reservoir rocks than plagioclase-rich arkoses; the former contain coarse-grained kaolinite, whereas the latter tend to contain abundant authigenic montmorillonite, which has an adverse effect on permeability.

In contrast to the findings of Leder and Park (1986), relatively rapid deterioration of sandstone reservoir quality with increasing depth is categorized by Trevena and Clark (1986) in their study of the Pattani Basin, Gulf of Thailand, as being principally due to high geothermal gradients. They compare the situation with that in the Gulf Coast area of the USA.

In the Pattani Basin, geothermal gradients fall within the range 4.0–5.8°C/100 m, as opposed to 3.2–4.1°C/100 m in the Gulf Coast area. The relevant porosity decline gradients are quoted as 3.0–3.5%/300 m in the Pattani Basin, and 1.3–2.3%/300 m in the Gulf Coast.

Figure 2.6 shows the depth–mean-core-porosity gradient for the Pattani Basin sandstones, with porosity ranges shown as horizontal lines.

Reasons for the differences between the two areas are related to accelerated reaction rates resulting from higher temperatures at any given depth; these promote rapid dissolution of feldspars, and accelerated precipitation of quartz overgrowths and authigenic clays, for example. In addition, high temperatures can accelerate compaction processes by increasing ductility and pressure solution.

Dixon and Kirkland (1985) in studies of reservoir quality of Cenozoic sands in lower California, also found a strong positive correlation between geothermal gradient and porosity decline gradient.

Although the greater part of the foregoing discussion applies to sandstones formed in any depositional environment, there are some special characteristics of wind-laid sandstones, as noted by Fryberger (1986), that deserve mention.

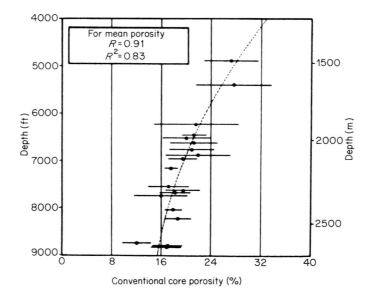

Fig. 2.6. The depth–mean-core-porosity gradient for the Pattani Basin sandstones, with porosity ranges shown as horizontal lines. From Trevena and Clark, 1986, reprinted by permission of American Association of Petroleum Geologists

Apart from conventional traps in aeolian sands, there are certain trap types that occur because of the special depositional environment. These are: (i) geomorphic traps in preserved and sealed topographic relief on aeolian sands—usually the tops of a dune field; (ii) diagenetic traps—aeolian sand deposits tend to undergo early cementation, providing internal seals for hydrocarbon traps; and (iii) system boundary traps, in which oil or gas is trapped at the updip depositional edge of the aeolian system deposits where they interdigitate with sealing, impermeable sediments of a different depositional environment.

The special nature of some diagenetic processes in aeolian dune sands has been well described for the (Lower Permian) Rotliegendes Sandstone of the southern North Sea by Glennie *et al.* (1978).

2.5 Carbonate Reservoirs

Although quantitatively somewhat less important than sandstones as reservoir rocks, carbonates can nevertheless be important—and even dominant—as reservoirs in some parts of the world, such as the Middle East.

Unlike sandstone reservoirs, in which primary intercrystalline porosity is often preserved to some extent and plays a significant role in the final reservoir characteristics, carbonate reservoirs are usually formed by secondary porosity in one of the following ways: (i) by secondary porosity resulting from leaching of either framework minerals or cement; (ii) intercrystalline porosity resulting from diagenetic effects, such as dolomitization. Fracture and vuggy porosity may also be important in these reservoirs.

Perhaps the best known type of carbonate reservoir and trap is the facies-bounded situation found in biohermal reef carbonates. Although primary porosity in reef carbonates can be very good—at least the equivalent of the mean 40% quoted for initial primary porosity in sandstones—burial is normally highly destructive of this, and the many producing fields in bioherms invariably owe their existence to the later development of secondary porosity in a part or parts of the reef structure.

To the exploration geoscientist working with seismic data, reef carbonates of the type just mentioned are often not difficult to observe, since the morphological aspects of biohermal reef structures often lend themselves quite readily to identification on the seismic section. This cannot be said, however, for non-reefal carbonates (either limestones, lime muds, or dolomite). The reasons for this will be discussed in some detail in Chapter 8, but the point can be made easily by referring to Figs 2.7 and 2.8. In Fig. 2.7, the seismic expression of a biohermal reef is seen, making a clear morphological feature in the data; a detailed explanation for this will be considered in chapter 8. In Fig. 2.8, the top of a non-reefal carbonate formation (in this case the top of the Carboniferous Limestone—Top Dinantian) is seen as a strong seismic reflection group of dominant low frequency; in this case, there is no assistance from morphology in identification—the strong acoustic impedance contrasts producing the group of events is not diagnostic, as it could be the result of many other juxtapositions of lithological units. Even if tending to be high, seismic velocities in carbonates cover a wide range (see Fig. 2.9), and overlap those of other lithologies.

Although difficult to identify by seismic means alone, bedded non-reefal limestones or dolomites can sometimes be recognized tentatively in seismic data by certain common facies associations, and this will be discussed again later.

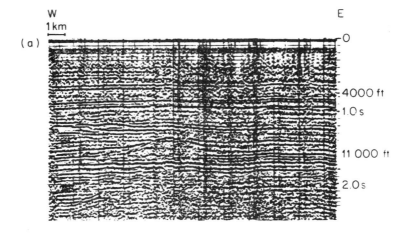

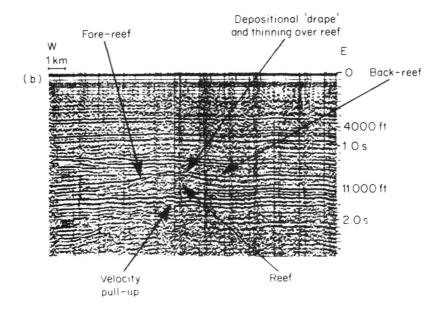

Fig. 2.7. The seismic expression of a biohermal reef, which is clearly identifiable in the section (from Jenyon and Fitch, 1985, Gebrueder Borntraeger Verlag., Stuttgart)

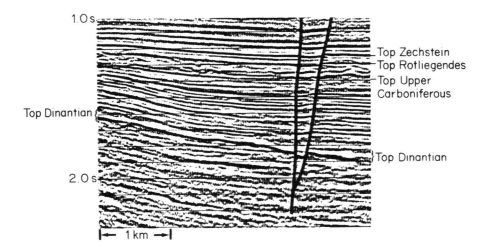

Fig. 2.8. The top of a non-reefal carbonate formation as seen in seismic data (the formation in this case being the Carboniferous Limestone (Dinantian)). Unequivocal identification of the lithofacies cannot be made in this case. Reproduced by permission of Seismograph Service (England) Ltd

In spite of what was noted earlier regarding carbonates and primary porosity (which is especially the case for reefal carbonates), some bedded carbonates *do* tend to retain their primary intergranular porosity to a great extent, with negligible effects from cementation. Illing *et al.* (1967) note that this seems to be particularly true of undeformed areas of continuous miogeosynclinal carbonate sedimentation in some parts of the Middle East and elsewhere, quoting as an example the (Lower Cretaceous) Minagish Oolite of Southern Kuwait. On the other hand, in folded areas and epicontinental regions, lime sediments tend to have lost their primary porosity and to have become tightly cemented by secondary calcite. In such areas, diagenetic secondary porosity becomes important, and dolomite becomes the chief carbonate reservoir rock.

Microdolomites, it is noted, are poorly permeable, but often highly porous; intercrystalline and vuggy porosity can be related to the process of dolomitization. Where they are fractured, and thus gain permeability, they may become adequate reservoir rocks. In many cases, they are interbedded with anhydrite that can act as an impermeable cap rock. Macrodolomites are usually less related to anhydrite, and the coarser porosity can create excellent permeable reservoirs.

Chalk, particularly if fractured or jointed, may act as a reservoir (as in the Dan Field in Danish waters of the North Sea); if not fractured, the cryptocrystalline porous (but relatively impermeable) chalk may act as an efficient sealing cap rock.

In the USA, limestone reservoirs outnumber dolomite reservoirs by a factor of 3:1, simply because limestones are more abundant than dolomites. In the eight States accounting for over 90% of carbonate reservoirs in the USA, dolomite reservoirs somewhat outnumber those in limestones. (NB There may be several reservoirs present in a single formation.) There is evidence that dolomitization of a limestone tends to improve the reservoir properties

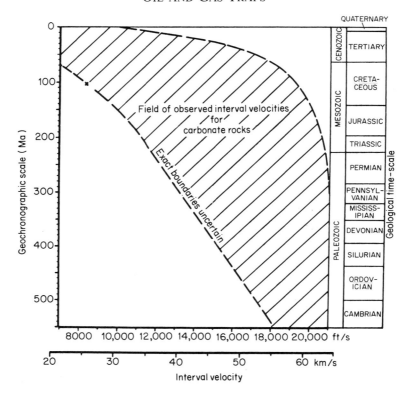

Fig. 2.9. Seismic interval velocities for carbonate rocks derived from well data, and plotted against geological age. The wide velocity range normally prevents the velocity parameter from being diagnostic of carbonates. From Bubb and Hatlelid, 1977, reprinted by permission of American Association of Petroleum Geologists

of the formation, and that effective fracture systems at reservoir depths are more likely to occur in dolomites than in limestones (Schmoker *et al.*, 1985).

The subject of dolomitization is dealt with in great depth in the literature, and so will not be pursued in any detail here. There is a general acceptance of the proposition that the bedded dolomites found in many evaporite basins are of secondary origin, having been limestones that were later dolomitized by late-stage Mg-rich brines. In some areas, the percolation of these concentrated brines through underlying rocks was irregular. An example is in the western Williston Basin in the USA, where the process led to the development of highly irregular and localized dolomitized reservoirs in the Red River 'C' zone carbonates, characterized by interdigitating limestones and dolomites (Longman *et al.*, 1983). The reservoir porosity developed mainly in porous dolomite and partly dolomitized limestones beneath a capping anhydrite at the top of the Red River 'C' formation. The localized nature of the process is believed to be due, at least partially, to subtidal downward migration of the Mg-rich brines through 'holes' in the capping anhydrite, the holes being the results of fracturing, minor faulting, or expulsion of water during compaction of underlying sediments.

In evaporite basins, there is a close association between carbonate rocks and evaporites—an association that springs from the very nature of the restricted basinal situation. An evaporite basin is really a basin of carbonate deposition where water circulation has become restricted for some reason, tectonic or otherwise, leading to evaporative drawdown. The carbonate rocks in such a basin are not evaporitic in origin for the most part, but form the 'pre-evaporite' series in the early period of each evaporite cycle (see e.g. Taylor, 1984; Jenyon, 1986d), which from the base upwards ideally consists of a clastic member marking the original marine transgression, followed by carbonates (limestones, dolomites), sulphates (gypsum, anhydrite), chloride (halite), and (sometimes) Mg-rich potash (bittern) salts. In fact, in older buried evaporite sequences, the limestones have been partly or wholly dolomitized by late-stage Mg-rich brines, and the gypsum has been converted to anhydrite. The dolomitization process, as mentioned previously, increases the formation porosity and often produces a viable reservoir.

The Zechstein of the southern North Sea includes in its lowest part, basinal carbonates (now dolomites) of probable reservoir quality that are, however, too thin to produce commercial hydrocarbons (the deposition rate of basinal carbonates being low, and the period of deposition during the Zechstein being relatively short). Elsewhere, however, as in parts of the Middle East, similar situations lasting much longer have produced carbonates of great thicknesses and of reservoir quality.

In the USA, the Permian Basin provides another example, in which a cyclic sequence of shallow-water carbonates (dolomitized) and evaporites prograde across the North-west Shelf towards the Midland and Delaware Basins. The lower San Andres Formation in Cochran and Hockley Counties, Texas, is one of the most prolific hydrocarbon-bearing formations of the Permian Basin. Most San Andres fields are stratigraphic traps controlled by a combination of dolomitized porosity and anhydrite plugging (Cowan and Harris, 1986). The generalized stratigraphic column from the Glorieta (San Angelo) upwards comprises subtidal marine limestones (carbonates), shoaling and supratidal dolomite–anhydrite cycles (carbonates, sulphates), and halite (chloride), which (together with the dolomitization, which implies the presence of late, concentrated Mg-rich brines) ties in reasonably well with the 'idealized' pre-evaporite–evaporite cycle mentioned previously. In this case the dolomitized reservoir intervals have developed sufficient thicknesses (either individually or cumulatively in the cyclic sequences) to be highly productive on a commercial scale.

In terminating this brief discussion of carbonate reservoirs, it is noted that there are many interesting problems regarding the development of reservoir and related characteristics in particular carbonate lithologies. Space precludes any detailed account of these matters here, but the paper by Dunnington (1967) on aspects of diagenesis and shape change (including folding by solution mechanisms) together with observations on flow barriers in stylolitic limestone reservoirs, and by Lucia and Murray (1967) on origin and distribution of porosity in crinoidal limestones, are recommended as examples of such topics.

2.6 Some Comments on Fracture Porosity

Fracture porosity in rocks normally has less to do with depositional environment than it has with local or regional stresses acting subsequently to deposition and compaction.

All rocks are subject to fracturing, faulting, the development of joint systems, and microfractures. However, perhaps because of their frequently more compact and brittle

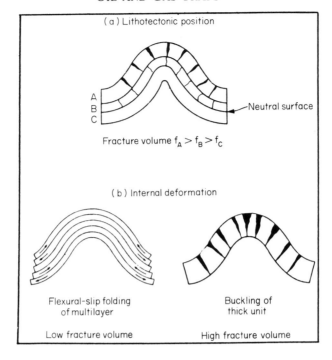

Fig. 2.10. Schematics showing how fracture volume may be dependent on lithotectonic position, and how the nature of internal deformation influences fracture potential in a fold. From Mitra, 1988, reprinted by permission of American Association of Petroleum Geologists

nature and relative lack of intergranular porosity, carbonate rocks and often found to have more important reservoir-quality fracture porosity and permeability.

Fracturing may develop within a tectonic stress field on local and/or regional scales. As an example of the former, it is instructive to consider the effects of lithotectonic position and the nature of internal deformation on fracture potential in a fold system. Figure 2.10, from Mitra (1988), shows how (a) fracture volume is dependent on the position of a stratum with respect to a neutral surface within an anticlinal fold; below the neutral surface shown, the stratum C is under compressional rather than tensional stress, which is a common situation in a compressional fold (although not in an uplift fold, where there may be no compressional stress zone). The sketch (b) shows how fracture volume is affected by whether flexural slip occurs along a large number of bedding planes within an interval, or whether a (much larger) fracture volume is present occurring in an interval of constant thickness that flexes as a single unit. It is interesting to note that if the right-hand sketch in (b) is regarded as a compressional (flexural) fold, then the fractures would be inhibited towards the concave side of the fold by a compressional stress zone; if it is regarded as an uplift fold, then, as noted above, the whole stratal unit could be in extension, and there would be no such inhibition of the inward development of the fractures towards the lower (concave) side of the stratum.

On the regional scale, it is relevant to note the emergence over the last two decades

(Nur and Simmons, 1969; Nur, 1971) of the extensive-dilatancy anisotropy (EDA) hypothesis, which proposes that there is a distribution of stress-aligned, fluid-filled microcracks pervading most crustal rocks (Crampin, 1985, 1987). Anisotropy here refers to the property of a solid material, such as rock, whereby the elastic behaviour varies with direction. This may be due to alignment of mineral grains, clasts, or cracks, for instance. Dilatancy refers to the volume increase resulting from the opening of microcracks in a rock specimen before the failure of the specimen under stress, as measured in the laboratory.

Anisotropy in rocks has in the past been measured by determining azimuthal differences in the value of compressional (P-wave) seismic velocities. In rocks known to have a system of cracks with a relatively high aspect ratio (i.e. referred to three mutually perpendicular axes, the crack is substantially elongate parallel to one axis as compared with its dimensions parallel to the other two axes), it can be demonstrated that the P-wave velocity is normal for the rock in the direction parallel to the long-axis of the aligned cracks, but lower than normal in a direction perpendicular to this. It is also known that aligned microcracks, even though microscopic in size and not affecting the bulk density of a rock, can still produce the effect of anisotropy on seismic velocity in the rock.

It has been known for some time that in the propagation of seismic shear waves (velocities of which are about half of the P-wave velocities in the same material) in rock, birefringence (double refraction, or splitting of the shear wave into two mutually perpendicular polarized components) can be observed to take place. This has been interpreted as the effect on the shear wave of the EDA in the rock resulting from aligned fluid-filled microcracks. The polarization directions of the shear waves are believed to change with slight changes in crack orientation, which would occur with changes in the direction of stress. There could also be changes of crack density, aspect ratio, pore-fluid content, and pressure.

Some possible implications of the ability to observe and monitor shear-wave splitting and the polarization directions include such matters as new approaches to earthquake prediction research (possibilities of observing significant stress changes in pre-earthquake and pre-volcanic eruption periods), and estimating the internal structure of hydrocarbon reservoirs. The developing techniques promise, in the latter case, to allow the detection of fracture and permeability trends using only seismic methods (Lynn, 1986). Such evaluations could lead to optimizing production strategies for secondary and tertiary recovery, for instance. Anisotropy and crack orientation observations bear directly on preferred flow directions and other properties of the reservoir. Detailed modelling of shear waveforms in three-component shear-wave VSPs (Crampin et al., 1986) yields accurate estimates of the internal structure throughout the subsurface surrounding the well, and not just in the immediate vicinity of the latter. Clearly this approach is potentially a powerful tool for use by the petroleum geologist in the future.

2.7 Reservoir Drive Mechanisms

Although it is not intended to stray deeply into the large field of reservoir engineering, a brief note on certain basic matters seems appropriate (for the benefit primarily of any readers unfamiliar with the subject), before leaving the matter of reservoir rocks for the time being.

When an oil or gas accumulation is discovered, two important calculations must be made by the reservoir engineer at an early stage. In simplified terms, these are:

(i) An estimate of the *oil or gas in place* in the reservoir. This is arrived at by:
   (a) estimating the total volume of rock containing the oil/gas;
   (b) multiplying (a) by the best estimate of mean effective porosity;
   (c) multiplying this product by the best estimate of oil/gas saturation (i.e. the proportion of oil or gas in the pore spaces relative to other fluids).
(ii) An estimate of the volume of recoverable oil or gas. This is obtained by multiplying the figure for the oil or gas in place (c) by an estimated *recovery factor* (which is often no more than 20% for oil). Gas volumes are always given at s.t.p. (15°C and 760 mm of mercury).

The recovery factor is very variable; one of the most important influences on it is the type of reservoir *drive* mechanism available to produce the petroleum. The principal types of drives encountered (Cole, 1969) are:

(i) *Depletion drive* (also referred to as solution gas drive). There is no free gas cap in the reservoir. Production is due to gas liberation from solution in the oil in the reservoir, causing bulk fluid expansion and expulsion of the oil. Regarded as the least efficient drive mechanism, it often results in a poor recovery factor (see Fig. 2.11).
(ii) *External gas drive*. The energy for production is the result of an expanding free gas cap displacing the oil ahead of itself as it expands because of pressure reduction. The recovery factor depends on the size of the gas cap, and the displacing efficiency of the gas (see Fig. 2.11).
(iii) *Water drive* (the most efficient driving force). Water displaces the oil, invading the oil zone in the reservoir as a result of pressure reduction allowing expansion of the water in the adjacent aquifer. Pressure-production history for a typical water drive is shown schematically in Fig. 2.11.

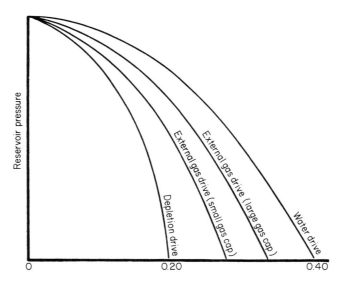

Fig. 2.11. Schematic production histories of reservoirs with the principal drive mechanisms. The ultimate recovery fraction of the original oil-in-place is plotted against the reservoir pressure. Not to any particular scales (from Cole, 1969)

It should be noted that in reality, production from most reservoirs is a combination of two or more of the above mechanisms together with the action of gravity segregation and/or capillary forces.

Reservoir performance in any field is closely and continuously monitored, using as a basis for control some form of the material balance equation, expanded to include a very large number of different factors (and estimates of unknowns) involving reservoir characteristics and production methods, including positioning and densities of multiple wells. In this way, early estimates of reservoir performance are continuously updated. An explanation of the calculations involved in the material balance equation can be found in any reservoir engineering handbook.

2.8 Cap-rock Seal Formations

2.8.1 Lithological Aspects

The ideal cap rock, or hydrocarbon trap sealing formation would have two principal characteristics:

(i) complete impermeability to gaseous or liquid hydrocarbons;
(ii) ductility—the property of deforming plastically under stress (see e.g. Jenyon, 1986a)

In the 'real Earth', a wide range of lithological types form cap rocks. In areas that have undergone little or no tectonic deformation, any rock with sufficiently high capillary entry pressure (see later), whether porous or not, would be suitable. This includes some carbonates, evaporites, and fine-grained argillaceous clastics. They are adequate in this environment because they have not been subjected to fracturing.

Where deformational stress has occurred, however, the field of possible lithological types able to act as efficient seals is considerably narrowed. Certain evaporites and shale/clay rocks that are able to deform plastically under stress, perhaps sealing fractures and faults, may be adequate. Undoubtedly the best of these is salt rock, consisting mainly of the mineral halite. A more unusual, but in certain circumstances equally effective type of seal can be provided by gas hydrates in the section. Diagenetic seals within sandstones and carbonate reservoirs may be present in some areas, and various types of fault seal may be critical in some trap situations.

Reliable statistics are not easily come by, but Grunau (1981) states that for 176 major gas accumulations world-wide, about 62% of the cap-rock seals are shales, and about 38% evaporites (considering the much greater relative abundance of shales over evaporites in the crust, this says something about the quality of evaporites as seals). Thicknesses vary from 20 m to several hundred metres. Thickness can be an important parameter in seal formations; although a relatively thin bed may have the required low permeability, it is unlikely that it will maintain the seal over an extended area in spite of minor faults, fractures, etc. This is more likely with a thicker unit. As suggested earlier, Grunau also notes that optimal conditions for seal preservation occur in areas that have had a comparatively simple geological history. It is mentioned that of the world's 25 largest gas fields, 21 are in simple, cratonic settings and four in fold belts. Those in the fold belts all have evaporite seals (and probably would not have survived otherwise).

Downey (1984) points out the necessity to look at a formation being assessed for its seal properties on two different scales—the 'micro' and the 'macro'. It is highly unlikely that a sample of the 'micro' properties of a seal (for example, in a core sample) are invariant over the whole area of the sealing surface, perhaps of many hundreds of square kilometres. It is the weakest point of the sealing surface that is of particular concern, and as such, often difficult or impossible to isolate.

In practice, the great majority of efficient seal formations are evaporites, fine-grained clastics, and organic-rich rocks; they have high capillary entry pressures, are laterally continuous, maintain stability of lithology over large areas, are (or can become) relatively ductile, and constitute a significant proportion of the fill of sedimentary basins. Ductile lithologies tend to flow or 'heal' plastically when deformed, and clearly this can be an excellent property for a seal. Downey lists the following seal lithologies in order of decreasing ductility:

>   Salt rock
>   Anhydrite
>   Kerogen-rich shales
>   Clay shales
>   Silty shales
>   Carbonate mudstones
>   Cherts

The property of ductility requires to be contemplated with caution; some rock types—evaporites particularly—may be ductile at moderate to large burial depths, but become brittle at shallow depths (salt rock being the most noteworthy).

2.8.2 Capillary Characteristics of Seals

The sealing capability of any cap-rock seal formation is measured by the capillary pressure (entry pressure, or displacement pressure) of the rock.

According to Purcell (1949), this can be expressed as:

$$Pd = 2\gamma \cos\theta / r$$

where Pd = the capillary pressure, γ = the hydrocarbon–water interfacial tension, θ = contact angle (conventionally measured through the denser fluid—water), and r = mean radius of largest pore throats. This is illustrated in the sketch of Fig. 2.12.

The capillary pressure increases with *decrease* of the throat radius of the largest connected pores, with *decrease* of the contact angle, and with *increase* of the hydrocarbon–water interfacial tension (Downey, 1984). The capillary forces of the seal formation confine any hydrocarbons within the reservoir as long as the buoyancy forces of the static hydrocarbon column (the product of the column height and the density difference between hydrocarbons and reservoir pore water) do not exceed the capillary entry pressure of the seal.

In the view of Watts (1987), cap-rock seal formations can be divided into two main classes: (i) the 'membrane' seals; and (ii) the 'hydraulic' seals.

With membrane seals, the dominant trapping mechanism is the capillary displacement pressure of the seal. These cap rocks act as a kind of 'membrane' in which the weakest point is the largest interconnected pore throat, as discussed by Berg (1975). Berg suggests

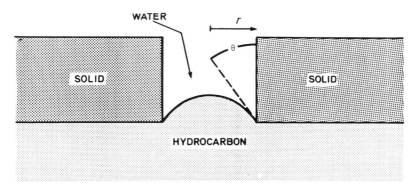

Fig. 2.12. Sketch illustrating the physical meaning of the parameters that, together with the hydrocarbon–water interfacial tension, are used to calculate capillary pressure in a cap-rock seal formation. After Downey, 1984, reprinted by permission of American Association of Petroleum Geologists

that most stratigraphic traps may be sealed in this way, where updip facies change causes a reduction in maximum pore-throat size until trapping occurs. See also Schowalter (1979) for cap-rock seals, and Smith (1966, 1980) for fault seals.

An alternative, less-considered cap-rock seal type, the hydraulic seal of Watts, occurs where the capillary displacement pressure is so high (effectively infinite for real hydrocarbon columns) that seal failure can occur only by fracturing of the cap rock, or wedging-open of, and migration along, faults or fractures. This kind of mechanism has also been considered recently by, for example, Goff (1983) and Ozkaya (1988), the latter by computer simulations.

Watts also mentions fault-related seals of two types: (i) sealing faults, where the fault plane or zone itself (being of finite—and sometimes substantial—thickness) acts as the seal; and (ii) juxtaposition fault seals, where a sealing unit is juxtaposed with a trapped hydrocarbon accumulation across the fault plane. These two types are illustrated in the sketches in Fig. 2.13. In sketch (a) the sealing fault situation is seen, with filling material in the fault zone of sufficiently high entry pressure acting as the seal, while in sketch (b) the juxtaposition fault seal is represented, where a seal formation with sufficiently high entry pressure inhibits the escape of hydrocarbons from the reservoir across the fault plane.

2.8.3 Diagenetic Seals

In many sandstone and carbonate units, diagenetic sealing can inhibit movement and escape of reservoir hydrocarbons. According to Schmidt and Almon (1983), processes that create these trapping seals include: (i) chemical compaction through pressure solution of silicate and carbonate minerals; (ii) concentration of insoluble clay minerals and organic matter during chemical compaction; (iii) cementation by authigenic minerals; (iv) volume increase in rock constituents resulting from hydration or replacement; (v) coalescive recrystallization; (vi) mechanical deformation of ductile constituents; (vii) emplacement of immobile organic residue derived from crude oil and natural gas. Sealing cements include silica minerals, clays, zeolites, carbonates, sulphates, chlorides, and other minor mineral groups.

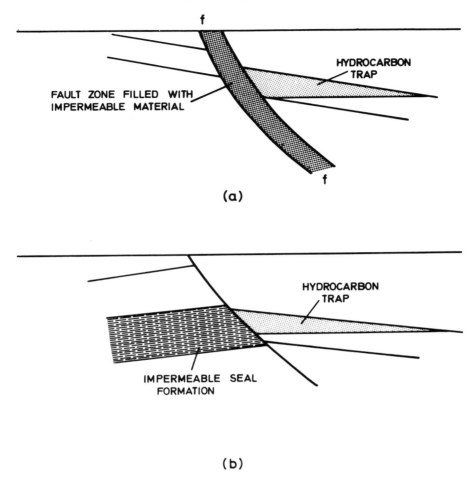

Fig. 2.13. Sketches of two types of fault seal, as discussed by Watts (1987). In (a) the sealing fault is shown, while in (b) the 'juxtaposition' fault seal is shown

Schmidt and Almon give an interesting example of this type of sealing in the Recinus Cardium 'A' pool of Alberta. 'In this field, lateral, top, and bottom seals have formed as a result of mesogenetic (concurrent with burial) cementation at the margin of the sand body. The limiting pressure in these seals is approximately 550 psi (3790 kPa) (mercury against air), which translates to a pore throat radius of 0.195 μm. This seal could effectively withstand a hydrocarbon–water injection pressure of 51 psi (352 kPa), which represents a trapped hydrocarbon column of between 200 feet and 500 feet (61 and 153 m). This clearly indicates that diagenetic seals can be tremendously effective.'

It is important to regard a cap-rock seal not as the static, dead-end of a trap, but as involving a fluid dynamic situation—the 'membrane' mentioned earlier in connection with the work of Watts and others. Berg (1975) points out that capillary pressure differences between oil and water in small pores can result in oil being trapped in a seal situation that allows water to pass through the seal 'membrane'. In such a case, the seal is not a completely impermeable barrier to flow. Diagenetic alterations in sandstones and carbonates can form such barriers, as in the trap mentioned in the previous paragraph.

Cant (1986) discusses the processes mentioned previously as being related to the situation in the sandstones of the Spirit River Formation of Alberta, Canada. Sandstones with a maximum burial depth of 3000–4000 m have measured permeabilities in the range of 0.0001–0.5 mD, overlapping with shale permeabilities. These sandstones have undergone extensive quartz-overgrowth cementation with minor carbonate and clay cementation. Experimental work by Walls (1982) shows that when core samples of these rocks are subjected to compressive stresses approaching those of the *in situ* sandstones, the narrow sheet-like (and crack-like?) pores between quartz overgrowths closed, and permeabilities declined sharply. Increased water saturations also markedly reduced the permeabilities to gas. These results are taken to suggest that partially water-saturated, tight sandstones in the subsurface may have effective or actual permeabilities less than one-tenth those indicated by conventional core analysis (Walls *et al.*, 1982). On such evidence, together with the Cardium Formation example of Schmidt and Almon quoted earlier, it can reasonably be concluded that tightly cemented sandstones can act as seals to hydrocarbon traps, and that similar considerations probably hold good for some carbonate formations.

Another interesting matter raised by Cant (1986) is that saturation of a rock by hydrocarbons is known to cause a cessation of the normal range of diagenetic effects. However, where the hydrocarbons are sufficiently degraded, they form immobile bitumen (which itself can form a seal against further hydrocarbon leaks, as has happened with the Athabasca Sands). It is possible—even probable—that this process may explain some of the 'phantom' apparent gas–water or oil–water contact events seen occasionally in seismic data. When tested by drilling, these features have proved to be void of 'live' hydrocarbons, although producing identical effects to the 'flat spot' event frequently seen at the base of real gas-filled porosity (examples of which will be shown later).

2.9 Summary: a 'Working Model'

This book is not intended as a textbook for beginners either in petroleum geology or seismic interpretation. A basic knowledge of these multidisciplinary subjects is assumed. The primary purpose of Chapters 1 and 2 has been to present a brief overview of current concepts and hypotheses in petroleum geology, within which there are many conflicting opinions and unproved controversial suggestions. This presentation is intended to give the various concepts a 'local habitation and a name' so that they can be referred to easily, if and when necessary, in the body of the work during discussions of individual types or classes of trap.

Readers, however, are entitled to demand more of the author of a book of this type than that he sets out, without bias, comment or commitment, numerous conflicting ideas, leaving them with the task of deciding which of these they can accept. It would be side-stepping a legitimate expectation if he did not declare his own opinions and choices amongst the ideas put forward to explain the generation, migration, and trapping of petroleum.

To this end, the following 'working model' is proffered as one possible route from source to trap. This represents the writer's current opinion—but it is not guaranteed to be the same in 12 months' time.

While keeping an open mind on the possibility of abiogenic petroleum generation, the author believes that the main *corpus* of evidence is in favour of biogenic origin for the vast majority of the oil and gas so far discovered. Seemingly, oil and gas are generated by thermochemical means from bacterially altered organic material derived from plant (mainly) and animal remains of marine (mainly) and terrigenous origin. Amorphous sapropelic marine organic material is oil-prone, while terrigenous, waxy plant material is gas-prone, although the final product is strongly influenced by the depth of burial (and hence the maximum temperature) undergone by the organic material.

As regards the mode of generation of petroleum, this is believed to take place by the destructive distillation of the basic organic material known as kerogen, largely in fine-grained, organic-rich argillaceous (and some carbonates) source sediments. Expulsion of the crude oil (with some dissolved gas) from the source rock into the carrier bed takes place owing to overpressuring within the (impermeably sealed) source rock by volume increase involving the evolution from kerogen to relict kerogen plus generated oil. Pressure build-up occurs during this process, all driven by the temperature increase concomitant with increasing burial depth of the source rock. Along the lines of the model suggested by Momper (1980), porosity heterogeneity and permeability anisotropy in the source rock, together with the overpressure, lead to expulsion of crude oil droplets along bedding planes, laminae, and some vertical fractures and joints, into the carrier bed, probably in a series of pulses.

It is also believed, with Price (1983), that geological time is relatively unimportant in organic metamorphism, but that the most important factor in the maturity level attained by the organic carbon content of the source rock is the maximum temperature reached (depending on the geothermal gradient and the maximum depth of burial).

In the matter of porosity reduction with increasing burial, it is believed that mechanical and chemical compaction have more important effects than does secondary cementation; following this it seems logical that the steepness of the geothermal gradient in an area is perhaps the most important factor in the rate of decline of porosity with depth, both from the point of view of chemical compaction processes (particularly the influence of pressure solution on pore-volume reduction) and also on reaction rates in diagenetic transformations, which may reduce, or enhance, poroperm characteristics.

On matters of secondary migration, and of cap-rock seal formations, there is much less controversy (albeit still much to be learned), and the various factors outlined in the present chapter as being of importance are all believed to be valid in a variety of circumstances, with no mutually exclusive ideas of any importance having been presented.

This, then, is a highly simplified model of the progress of petroleum from its raw material, through source and carrier formations, to the trap. Several of the aspects mentioned briefly are fraught with controversy still, and many even quite basic concepts require unequivocal verification. It is offered in order to provide a coherent guideline to which later discussions may be referred. The future will reveal how close the model is to reality.

PART II
STRUCTURAL TRAPS

CHAPTER 3

Fold-related Traps

3.1 Structural Traps—Introduction

In this and the two following chapters, there will be a discussion of hydrocarbon traps that have been formed by tectonic stress of a localized nature. Chapter 3 deals with anticlinal traps produced by compressional folding, by uplift, and by drape over older tectonically created features. In Chapter 4, traps produced by faulting are considered, while Chapter 5 deals with some of the traps arising from deformation of the overburden by mobile materials, such as salt or shale.

This division into three apparently separate classes is artificial, and adopted only for convenience. Some traps are created by two, or even three of the folding, faulting, and mobility processes. In certain cases, one particular process is the sole cause of the trap, with the others playing no part, or at most a subordinate one, in which case it is easy to classify the structure. An example of this would be the Dan Field structure in Danish waters of the North Sea, which is a faulted dome. With or without the faults, the dome would still constitute a trap, and so it is included in the fold-related traps in this chapter. Frequently, however, two or more of the named processes are involved with equal importance in the creation of a trap. In such cases, an 'editorial' assignment to a class is made—usually because the feature illustrates some characteristic of the class.

In these three chapters—and indeed in the book overall—no general classification of structures or traps is used. Such systems as exist may be studied in standard works on structural and petroleum geology. Rather, features have been gathered into convenient groups (see the chapter headings) in which the individual examples can be illustrated by seismic data in most cases, or if not by illustrations derived from seismic data and/or drilling information. Thus the underlying groupings might be considered to be based on seismic environments rather than conventional geological classifications. Subgroupings have been used, as in this chapter, to separate structures genetically.

It is believed that this scheme is rational given that one of the primary objectives of the book is to act as an aid and reference to those using seismic data as part of an interpretational project in hydrocarbon exploration. Another important aim of the work is to show petroleum geologists, production geologists, reservoir engineers, and others the amount and type of information that they should expect to be able to extract from good seismic data. This, in turn, enables the user to exercise his own quality control, which is not necessarily the same as that exercised by the professional geophysicist.

3.2 Fold-related Traps in Hydrocarbon Exploration

In this context, the mention of fold-related traps normally implies that the fold involved is an anticline.

The anticlinal trap is the classical trap type in petroleum geology; since the earliest days in petroleum exploration, the 'anticlinal theory'—the idea that natural gas and oil will tend to migrate to, and accumulate in, the highest part of a reservoir bed—had proved to be a successful finder of oil, and in those early days had begun to persuade those who said that 'geology never filled an oil tank' that they might be wrong.

Levorsen (1967) notes that this concept became the most important in oil exploration for a long period, and it was not until 1934 that McCoy and Keyte put into words the recognition that a broader structural theory was closer to the truth—'. . . commercial deposits of oil and gas are associated with structural irregularities in porous sedimentary rocks, the most important irregularity being the anticline or dome'. Notwithstanding this broadening of the basic ideas, the anticline in its many different forms remains the most important type of structural trap.

It should be understood that although singled out for a separate chapter because of its basic importance as a trap configuration, the anticlinal fold is peculiar to no specific depositional or tectonic environment. It may occur as at least an element in traps covered by many of the situations that appear in the chapters that follow, involving a variety of depositional, tectonic, and seismic environments.

3.3 Morphology and Terminology

In order to investigate the morphological variation in anticlines and to become familiar with some of the basic terminology associated with them, several groups of features with related characteristics will be discussed.

A consideration of the dynamics of development of anticlinal structures leads to their division into three basic classes—those due to: (i) compressional folding; (ii) uplift; and (iii) drape over older features. Examples of these three distinct types of fold will be considered later, but first, morphological variations that may be seen in all three groups will be examined.

3.3.1 Symmetrical, Asymmetrical, and Overturned Folds

In Fig. 3.1, schematics of a progressively deformed anticlinal fold are indicated. Sketch (a) shows a *symmetrical*, upright structure. The dashed line (AP) represents the trace in cross-section of the *axial plane* of the fold—this plane being a surface equidistant from the two fold limbs. The trace of the AP in this case is vertical.

Sketch (b) shows an *asymmetrical* anticline, with one limb dipping more steeply than the other, and the AP inclined to the vertical. In sketch (c), an *overturned* anticlinal fold

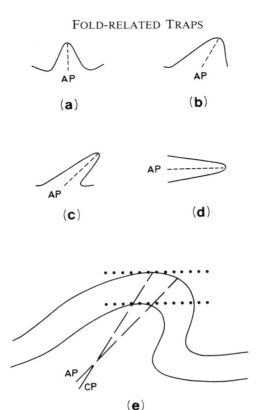

Fig. 3.1. Schematic sketches of progressive deformation of an anticlinal fold (a–d), and an indication of the difference between the axial plane (AP) and the crestal plane (CP) as discussed in the text

is seen; the left-hand limb is *normal*, but the right-hand limb is overturned or inverted, having been rotated through the vertical. The fold in sketch (d) is a *recumbent* anticline, with an AP that has been rotated to (or very close to) the horizontal.

In sketch (e) an overturned fold similar to that in sketch (c) is shown, with the AP marked, and the trace in cross-section of the *crestal plane* (CP) also shown. The crestal plane is a surface containing all crestal lines through the fold, the crestal lines being the loci of the highest points on the fold surface at any formation level. The dip of the beds is in opposite directions on each side of the crestal plane, which coincides with the axial plane in the upright, symmetrical fold of type (a), but does not in types (b), (c), and (d).

In areas of complex geology, where fold structures may have been inverted, it can be convenient in the description of features to refer to the *facing direction* of a fold. This is the direction along the axial surface in which the stratal units become *younger*. Thus, a normal, upright anticline is said to be facing upwards, while an inverted syncline is facing downwards.

3.3.2 Axial Plunge and Aspect Ratio

Some anticlines continue for long distances with the fold axis essentially horizontally, as in Fig. 3.2(a), forming a long ridge-like feature. In others, however, the fold axis *plunges*

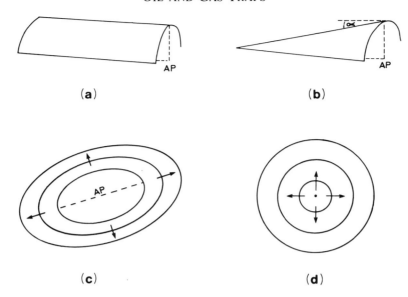

Fig. 3.2. Axial plunge (pitch) sketched in (b), double-plunge (c), and an anticlinal pericline, or dome (d)

or *pitches*, dipping at an angle (α in Fig. 3.2(b)) to the horizontal. Elongate anticlinal features in which the fold axis plunges in both directions are known as *double plunging anticlines* (Fig. 3.2(c)). The limiting case is that of the *dome*, sketch-contoured in Fig. 3.2(d), where the contour lines are circular, and dips are everywhere away from the central point. Such a structure is also known as an anticlinal *pericline*. A circular basin with dips everywhere *towards* a central point is a synclinal pericline. In effect, the anticlines in Figs 3.2(a), (c), and (d) can be regarded as a series of decreasing aspect ratio.

3.3.3 Open and Closed Folds

Figure 3.3(a) shows an *open fold*, where dips are generally quite small, and where all deformation movement within the fold is accommodated by *bedding-place slip*, with no plastic deformation being involved. A *closed* or *tight* fold, on the other hand (Fig. 3.3(b)) shows steeper dips, with an element of plastic deformation involved (see also *similar* folds, below). At the extreme, closed-fold deformation produces *isoclinal* structures, in which the limbs and axial planes of the folds are approximately parallel.

The degree of 'openness' of a fold can be quantified in some cases by measuring the interlimb angle—the smaller angle between the two limbs of the fold—if necessary by constructing tangents to the inflection points on the limbs in a profile of the fold. Fleuty (1964) classifies folds in this way as follows:

Interlimb angle	*Fold description*
180–120°	Gentle
120–70°	Open
70–30°	Close
30–0°	Tight
0°	Isoclinal

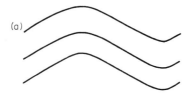

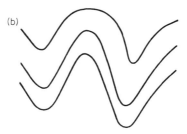

Fig. 3.3. An open fold (a), and a closed, or tight, fold (b)

3.3.4 Concentric and Similar Folding

Folding in which the traces of the bedding surfaces in cross-section form arcs of circles with a common centre is known as *concentric* (or sometimes *parallel*) folding. Ideally it develops during deformation of competent strata that accommodate to the folding by bedding-plane slip. Thicknesses of stratal units remain constant normal to the bedding planes, and no plastic deformation occurs within the fold. Because of its geometry, a fold of this type dies out both upwards and downwards in the section (see Fig. 3.4(c)). Upwards, the anticline develops progressively less curvature. Downwards, it develops progressively *more* curvature and passes into a cusp with infinite curvature. Further downwards the cusp flanks show less dip, and tend to die out. The appearance of this downward cusp development is well illustrated in Fig. 3.5. The shapes described here appear in the reverse sense for synclines.

A special case of concentric folding is a *box fold*, which is due to layer-parallel compression, and in which the fold angles approximate 90°, giving the structure a rectangular profile. A box fold can also be an example of, or the result of, *conjugate folds*, in which a pair of asymmetrical folds having an opposed sense of symmetry also have their axial surfaces dipping towards, and intersecting, each other.

In *similar* folding (Fig. 3.4(b)), deformation takes place plastically, with movement along S-surfaces, so that stratal units thicken in the hinges of the fold, and thin in the limbs. Stratal thickness is (ideally) constant in directions parallel to the plane of the fold. The folding is sinusoidal, keeping a constant form both upwards and downwards in the section, indefinitely.

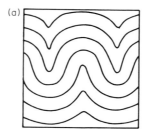

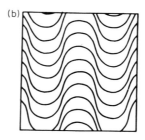

Fig. 3.4. Concentric (or parallel) folding (a), and similar folding (b)

Usually, 'real Earth' folds are made up of both competent and incompetent units interbedded, and deformation results in a combination of concentric and similar folding.

The 'reality' of an approximately concentric fold is shown in Fig. 3.5, which depicts a sequence of cyclic sediments in North Africa that have been subjected to some compressional deformation. In the lower centre of the photograph, a small concentric fold is seen, dying out rapidly downwards (the upper part of the progression cannot be seen due to an erosional hiatus represented by the small unconformity). The roughly concentric form of the strata involved, and their individually constant thicknesses are clear.

The seismic section in Fig. 3.6 shows an example of what is effectively similar folding. In the anticlinal fold at the centre of the example, a marked thickening in the hinge of the fold, and a thinning of the limbs, can be seen particularly well within the interval marked A. This interval consists largely of incompetent argillaceous rocks (Triassic mudstones) that have deformed plastically during folding. The structure is in fact an uplift fold produced by underlying mobile Zechstein salt, rather than a purely compressional fold, but illustrates the shapes involved.

One variant of similar folding is the angular or straight-limbed fold (also variously termed the chevron or zig-zag fold), as opposed to folds showing curvature in profile. They are normally held to be similar folds since they tend to maintain the same form both upwards and downwards in the section, with thickening in the hinges. It is noted that individual stratal units in these folds often retain a constant thickness (except in the immediate location of the hinge).

Chevron folds are common in some areas, including the Appalachian coalfields in the USA, and coalfields in the Ruhr (West Germany) and Belgium in the European area (Sherbon Hills, 1972).

Fig. 3.5. A 'real' example of a small concentric fold in cyclic sediments. Reproduced by permission of Seismograph Service Corporation

Supratenuous folds, as the name implies, are features in which the stratal units thin across the fold axis. This usually implies some syndepositional growth of the structure. Such folds can be seen in the overburden above a developing salt structure. They can also be due to differential compaction following deposition over buried hills, as in the Plains-type folds of mid-Continental USA.

Another type of feature encountered is the rheomorphic fold. This is a flow-structure developed in a rock that is undergoing plastic deformation. Often flow-folds have thickened hinges and thinned limbs, sometimes exhibiting shearing parallel to the maximum principal stress axis. Flow-folds are often picked out by competent bands within a salt body that has undergone plastic deformation, and also in metamorphic zones.

3.3.5 Faulted Folds

In some circumstances, continuing or renewed stress acting on an anticlinal fold leads to fracture and shearing of one of the limbs—brittle fracture if the structure consists largely of competent lithological units, or ductile shear and thinning of the limb if mainly incompetent units are involved.

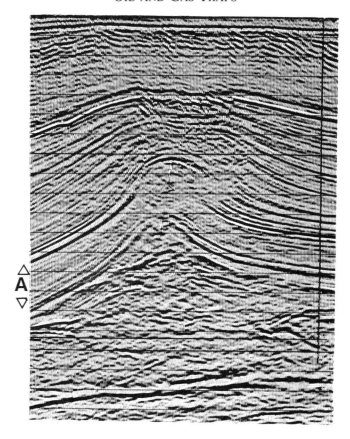

Fig. 3.6. The appearance of similar-type folding on the migrated seismic section—in the interval A, the fold is thickened in the hinge and thinned in the limb (particularly on the left). Reproduced by permission of Seismograph Service (England) Ltd

It is often the case that such stress acts on a symmetrical anticline, causing it to become asymmetrical, or overfolded, and in the limit, recumbent, with one limb shearing. The remaining part of the sheared limb, together with the hinge and the unsheared normal limb, may tend to slide along the plane of thrusting or reverse faulting—indeed, a fault of this type, closely associated with folding, is sometimes known as a *slide*, particularly in cases related to the ductile shear zones found in metamorphic areas (Park, 1983), such as the Sgurr Beag Slide in the Scottish Highlands.

A fold–fault situation of this type can lead to trap configurations dependent on the fold, the fault, or a combination of both. Where the fold is the dominant element, the fault may act as a conduit for hydrocarbons migrating into the structure; where the fault is sealed, the geometry may be such as to increase the total volume of the trap.

An example of such a structure, in which the fold is the dominant trapping element (although not the only one) is shown in Fig. 3.7, which is a schematic profile of the Yushashan oilfield in the Chaidamu Basin, Peoples' Republic of China. Multireservoir

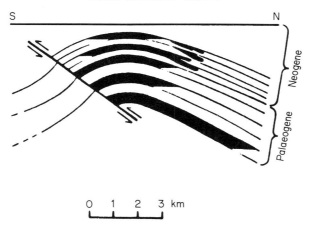

Fig. 3.7. A schematic profile of Yushashan oilfield in the Chaidamu Basin, People's Republic of China, showing the multireservoir trap formed by a reverse-faulted anticlinal fold. Note that the trap is partly structural and partly stratigraphic. From Quanmao and Dickinson, 1986, reprinted by permission of American Association of Petroleum Geologists

traps, as seen in this example, are common in this area. The folds are related genetically to basin-bounding thrust systems, and the frequently present reverse faults locally either produce closure, or as here, define downdip reservoir limits (Quanmao and Dickinson, 1986). Apart from the structural trap situation, some lensoid sandstone reservoirs are indicated at or near the crest of the anticlinal fold, serving as structural–stratigraphic traps.

3.4 Compressional Anticlines

To the best of the writer's knowledge and belief, the structures discussed in this part of the chapter have all been produced by some form of compressional stress. However, in the case of certain anticlines, it is not possible with the present state of knowledge to say this with certitude. In the Middle East, for instance, there are many very large, closed anticlinal and domal features, which, it is inferred, owe their origin to deep-seated salt movement—but to date this cannot be verified beyond doubt. As it happens, this lack of evidence for the genetic origin of structures in many cases is purely academic from the viewpoint of hydrocarbon exploration; such features may or may not be traps, and their mode of origin is not relevant. This is not always so. Sometimes it may be vital to determine the mode of origin of a structure in order to plan a development drilling programme, or to extend exploration either laterally or vertically in an area.

Because of the low resolution of seismic data relative to exact borehole information, it will be instructive to see, in some instances, how the apparently simple seismic expression of a trap situation develops startling complexity when 'fleshed-out' by borehole information.

3.4.1 Simple Anticlinal Traps

Perhaps it is an error ever to refer to any geological feature as 'simple', but in the first few examples shown here, it is at least true of the seismic expression of morphology.

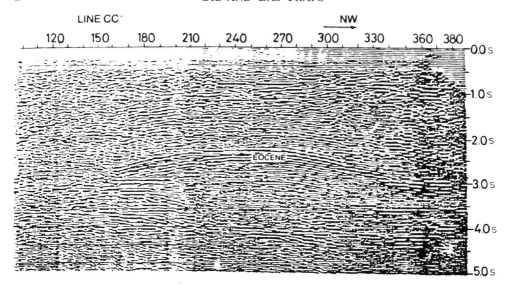

Fig. 3.8. A simple anticline produced by compressional tectonics, Potwar Plateau, northern Pakistan. The reservoirs include Eocene carbonates, and these dense, compact rocks are probably producing the good 'stand-out' of the anticlinal band of events, the crest of which is just below 2.0 s in the centre of the section. This is a very noisy, low-quality seismic section by today's standards. Owing to detector-ground and source-ground coupling problems, land seismic survey results are commonly of lower quality than marine seismic data: in the latter, both detectors and source are immersed in the same medium—water. From Khan et al., 1986, reprinted by permission of American Association of Petroleum Geologists

A very simple Tertiary anticline in the Potwar Plateau area of northern Pakistan is seen in the seismic section of Fig. 3.8. According to Khan et al. (1986) the structures in the region are the direct or indirect result of compressive stresses produced by post-marine-sedimentation tectonics. The objective reservoirs are carbonates of Eocene to Permian age, and clastics of Jurassic to Cambrian age, under a cover of Miocene–Pleistocene fluvial sediments. The quality of this seismic example is not good, and suggests that it was produced some years ago. In spite of this, the broad, gentle anticlinal fold can be seen standing out well from the strong background noise. This good 'stand-out' is often typical of the seismic response to carbonates—these compact, dense rocks can have high seismic interval velocities, and produce excellent acoustic impedance contrasts with enclosing sequences of other lithologies.

Another example of a simple anticlinal fold trap is the Eakring oilfield structure located in the English Midlands. The structure is a N–S-trending fold in Carboniferous rocks unconformably overstepped by Permian strata (see Fig. 3.9), with Carboniferous Limestone forming a west-dipping flank into the Nottinghamshire–Derbyshire coal basin.

This is one of several small oilfields discovered during the late 1930s that have declined since the Second World War (Tiratsoo, 1984) and are no longer productive. The reservoirs in the Eakring structure are several coarse-sandstone members of the Millstone Grit (Namurian), together with a small accumulation in the crest of the Carboniferous Limestone fold, which is probably a zone of fracture porosity resulting from extensional stress at the hinge of the slightly asymmetrical structure. The oil-bearing intervals are quite shallow (about 2000 feet below sea level).

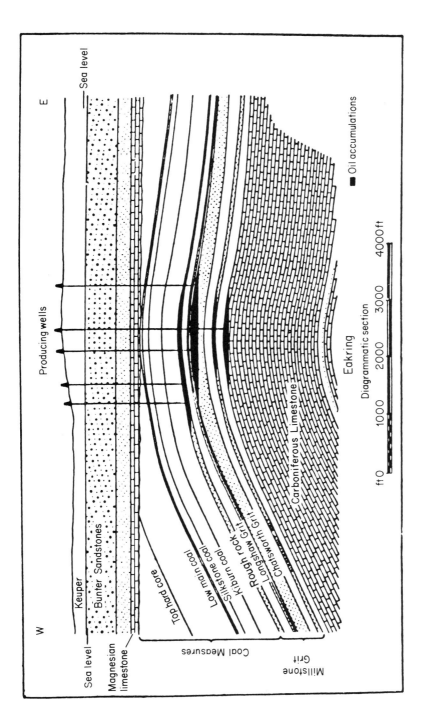

Fig. 3.9. The anticlinal trap of the Eakring oilfield in the English Midlands. One of the reservoir zones is in the Carboniferous Limestone (Dinantian). This rock is normally very hard, compact, and of low poroperm characteristics. Here, it has probably developed fracture porosity in the crestal region of the fold owing to extensional stress-zone fracturing (which has also probably enhanced porosity and permeability in the Millstone Grit reservoirs above). Reproduced by permission of Oxford University Press (after P. E. Kent, in V. C. Illing (Ed.), *The Science of Petroleum*, Vol. vi, *The World's Oilfields*, Part I, *The Eastern Hemisphere*, 1953)

74

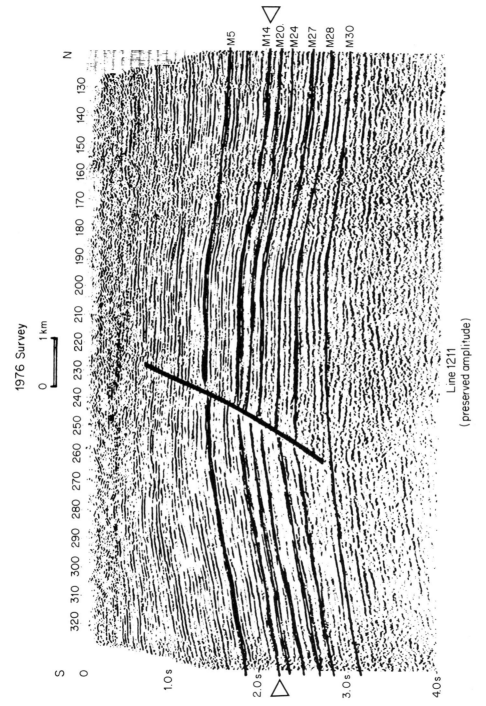

Fig. 3.10. A N–S section following the long axis of the slightly elongate domal anticline of the Handil Field, Indonesia. From Verdier *et al.*, 1980, reprinted by permission of American Association of Petroleum Geologists

Production at Eakring (together with that from a smaller accumulation in a separate culmination on the same trend at Duke's Wood) peaked in 1943 with an annual total of 602,855 US bbl, but by 1962 (the latest date for which figures were available at time of writing) had declined to 123,788 US bbl annually, in spite of the application of secondary recovery methods (Kent, 1953, and information from Dr G. D. Hobson).

It is of interest to note that oilfields of this type and size were, until more recently, the most important onshore fields in the UK. However, from the late 1950s, a series of more important finds (e.g. Kimmeridge, Wareham, Wytch Farm, Humbly Grove) came to light, which culminated, after proper appraisal, in the Wytch Farm Field with estimated recoverable reserves of between 100 and 200 MMbbl of oil in Lower Jurassic and Lower Triassic reservoirs. Output in 1981 alone was about 1.1 MMbbl from sandstone reservoirs in the structure (which is a northward-titled, E–W-trending fault block (Tiratsoo, 1984). Although very small by world standards for onshore fields, it is significant in UK terms because production costs are but a small fraction of those for the large offshore North Sea fields in the UK sector.

Another good example of a (seismically) simple anticlinal trap is provided by the Handil Field of East Kalimantan, Indonesia. A north–south seismic section across the field (approximately along the axis of this elongate anticline) is seen in Fig. 3.10, and a depth-structure contour map at one level (arrowed in Fig. 3.10) is shown in Fig. 3.11.

The field is divided into northern and southern halves by a large E–W fault throwing down to the south, which decreases in throw with increasing depth.

The structure has been studied by Verdier *et al.* (1980), who note that Handil is located in the Kutei Basin, and is one of several oilfields discovered that are related to middle and late Miocene depositional patterns. In Handil these indicate the presence of a Miocene delta, the position of which coincides with the modern delta of the Mahakam River. The structure is one of a succession of long, narrow anticlines (in this case, 10.5 km long by 4.5 km wide) believed to result from the compression caused by eastward gravity sliding of the sedimentary cover towards the depocentre. This sliding has in places been accompanied by shale diapirism (there are known to be overpressured zones below the objective section in the Handil Field). The sediments above basement consist mainly of shales below the clastic-rich shallower sediments that form the reservoir beds here. The sliding mass abutted against a N–S-trending basement high, and the resultant crumpling was transmitted upwards into the shallow sediments, on trends parallel with the high.

About 150 sandstone reservoirs have been found at Handil between depths of 450 m and 2900 m. They are tidal to fluvial deltaic plain deposits of mid- to late Miocene age, and most contain oil with a gas cap. The depositional environment is identified as one of channel fills, tidal bars, etc., and the field can be divided vertically into six superimposed zones corresponding to changes in depositional environment and/or physico-chemical oil characteristics. Depth-structure contour (isobath) maps show a displacement of the anticlinal crest some 3 km to the south-west from deeper zones to the surface.

Shown in Fig.3.12 is a schematic N–S depth-section of the field, indicating the multiple reservoirs and the distribution of hydrocarbons. As a scale comparison, the interval between the levels marked R6 and R27 on this schematic represent the time interval on the seismic section in Fig.3.10 between the shallowest marked horizon, and the fifth horizon downwards. The level of the R14 depth-structure contour map in Fig. 3.11 is indicated by the open arrows, as it is in Fig. 3.10.

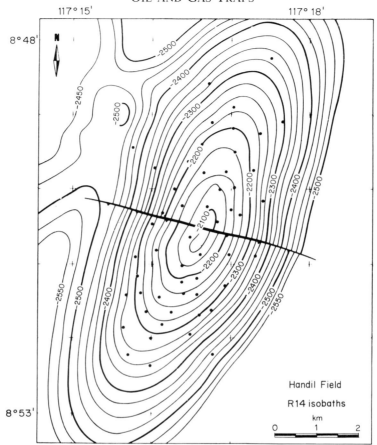

Fig. 3.11. A depth contour map of the Handil Field structure (at the level arrowed on Fig. 3.10. From Verdier et al., 1980, reprinted by permission of American Association of Petroleum Geologists

The sandstones are interbedded with shales and coaly beds (lignites), and the continuity of some of the latter across the field has assisted in correlation of the sandstone units. The TOC in the shales varies between 1.5% and 4%, averaging 2.2%, whilst in the lignite beds, it ranges from 40% to 70%. The organic-rich beds comprise about 10% of the sedimentary column in Handil. However, it is not likely that those interbedded with the sandstones are the source beds for the hydrocarbons present. Maturation determinations by vitrinite reflectance indicate that the main pay zones are located at or just above the top of the oil window in this area, suggesting that the trapped hydrocarbons are above their zone of generation, and must have migrated. It seems feasible that they were generated in a similar type of sedimentary section, but at a deeper level, in this vicinity.

Recent seismic data defines the position of the main fault, and also some subsidiary shallow faults in the southern half of the field, but not sufficient to define the sand bodies (at least by standard seismic processing routes).

Interpretation of the seismic and drilling data has revealed that vertical closure in the structure increases with increasing depth, from 100 m in the shallow part of the section

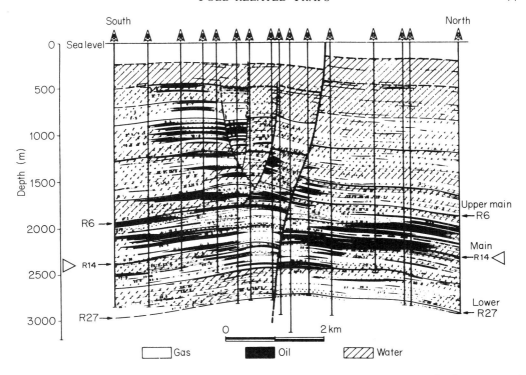

Fig. 3.12. A schematic depth-section constructed from well data, with the same horizon arrowed as in Fig. 3.10. Note the complex detail of the distribution of reservoirs and hydrocarbons, and compare with the simple picture given by the seismic data in Fig. 3.10. From Verdier *et al.*, 1980, reprinted by permission of American Association of Petroleum Geologists

to over 300 m in the deeper part below 2000 a depth. This is a relatively common occurrence. In this, and other instances, it is linked to the observation that stratal thinning takes place across the crest of the anticline (supratenuous folding), indicating early growth of the Handil anticline contemporaneously with reservoir deposition.

It is interesting and instructive at this point to compare the somewhat simple appearance of the seismic two-way time section in Fig. 3.10 with the complexities of the 'real Earth' structure as revealed by drilling results, and indicated in Fig. 3.12.

The shallow faulting south of the main fault depicted in the Fig. 3.12 schematic is less apparent, although just detectable in Fig. 3.10. However, the fine detail of reservoirs and hydrocarbon accumulations seen in Fig. 3.12 is clearly far beyond the resolving capabilities of the seismic section.

It is salutary for the seismic interpreter to recognize that even with the present—and what is regarded as advanced—state of seismic technology, careful correlation from drilling results—cores, samples, biostratigraphy, and other geological methods—will always yield a level of detail to which the seismic method cannot (yet) aspire. On the other hand, the less-dense sampling methods of the seismic survey can rapidly and relatively cheaply yield large-scale structural and stratigraphic evaluations that would be quite impracticable or impossible to attempt by purely geological means. This complementary nature of the various geoscientific disciplines is becoming more widely appreciated today, and serious attempts are now being made to attack exploration problems in ways that involve genuinely integrated studies.

The Handil structure also encourages the consideration of other points that can be related to matters discussed in the previous chapter. Apart from the intrinsic porosity and permeability of the reservoir sands, it seems likely that reservoir quality may well have been improved by the stress pattern imposed on the geological section over the zone of maximum curvature across the crest of the anticline. Extensional stress in this zone, perhaps acting individually on the many reservoirs above some neutral surface in the fold (refer to Fig. 2.10) may have enhanced the overall poroperm characteristics. This structure brings home strongly the attractions to the explorationist of a lithostratigraphic column with interbedded sands (of good porosity and permeability) and shales/organic-rich rocks with the ability to deform plastically to some extent and provide seals in a situation where deformation has occurred.

It is also of interest to consider the nature of the faulting present, from the viewpoint of sealing properties. Since there are no reports of serious leakage from the structure, it may be assumed that the major (and also minor) faults are sealed, at least above the reservoir. Referring particularly to the major E–W fault present, it seems a reasonable speculation that this is of the 'sealing fault' type discussed in section 2.8. It appears unlikely with the multiplicity of reservoirs and seal formations present of differing (and generally not large) thicknesses, that a 'juxtaposition' fault seal would be effective in this situation.

3.4.2 Thrust-related Anticlines

Some of the oldest exploited oil accumulations in the world (worked for several centuries by the local populace) occur in Assam and Burma, and many of these are of compressional anticline type, often associated with reverse faults and thrusts. Most of these fields can be illustrated only by line-drawn sections based on drilling information, as no seismic data are available.

The Digboi Field in Assam is of particular interest since it is one of many small oilfields discovered at an early date (1890, although probably known for many years before this because of surface seeps) and exploited in its early days by what would today be referred to as 'alternative technology'. A local method of drilling shallow wells for oil had been developed that resembled the use of a primitive cable tool rig. A large diameter hollow bamboo tube is suspended by a lever-arm arrangement from a bamboo platform on which stands the 'tool-pusher' and his assistant. The tube is dropped suddenly into the bottom of the hole that is being made, the assistant places a hand over the open top end of the tube to act as a 'valve', and the tube is raised far enough to be swung to one side out of the hole, when the assistant removes his hand, and the mud and water inside the bottom of the tube are released. The technique is most effective for the digging of shallow holes (in many places, oil is found in the area only a few feet below the surface). It may still be in use today—it certainly was within the last 30 years, having been witnessed by a colleague of the writer.

A section across the Digboi Field is shown in Fig. 3.13, and indicates the presence of a tight asymmetrical anticline with a thrust fault (the Naga Thrust) truncating the steeper northern limb. Oil sands occur in the thin development of Surmas, and in the Tipam

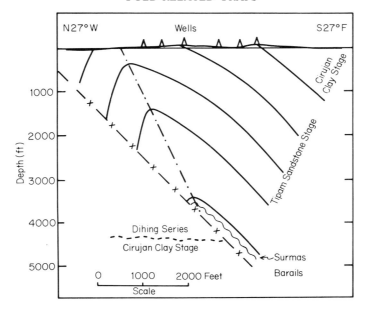

Fig. 3.13. Section across the Digboi oilfield, Assam Basin, showing the thrust-related asymmetrical anticline forming the trap. Reproduced by permission of Oxford University Press (from G. F. Wilson and W. B. Metre, in V. C. Illing (Ed.), *The Science of Petroleum*, Vol. vi, *The World's Oilfields*, Part I, *The Eastern Hemisphere*, 1953)

Sandstone stage. The source rock may be at deeper levels in the Barails Formation. The oil-bearing horizons range in depth from the surface to 5000 feet, but the sands are discontinuous and show marked lateral variation in reservoir properties. An unusual point is that 'several sands that outcrop at the surface have given excellent production in wells only a few hundred feet away . . .' (Wilson and Metre, 1953). In the case of one sand, oil retention in it was determined to be due to paraffination of the outcrop, but elsewhere, *no such paraffination is detectable*, and the sand is water-bearing for some distance from the outcrop. Thus, although it has not previously been mentioned in this context, there seems to be a real possibility that where paraffination has not occurred, cases of hydrodynamic trapping may be present in the Digboi area.

Within the regional setting, the Digboi structure is a local culmination at the north-east end of the Jaipur Uplift. In 1922, deeper drilling on the structure revealed further reservoirs in Miocene sandstones, and production from the field was increased substantially, reaching a maximum during the Second World War.

Structures of this type can be difficult to identify in seismic data. An example of a stacked seismic section (upper) and the equivalent FK-migrated section across a reverse-faulted anticline is shown in Fig. 3.14. Purely from the unmigrated (stacked) section, a normal fault might be inferred, with the right-hand half of the fold being taken for diffractions. However, the migrated section shows a low-angle reverse fault, with a backthrust and 'pop-up' structure developing, although the exact location of the reverse fault plane is difficult to place. Note the large horizontal displacements caused by the migration, particularly the shift in position of the backthrust. It is also worthwhile pointing out that there is normally

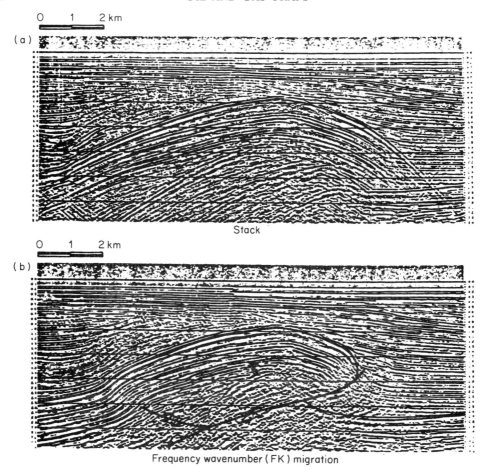

Fig. 3.14. A stacked (a) and FK-migrated (b) seismic section across a reverse-faulted anticlinal fold similar to those shown in Figs 3.13 and 3.16, to show the seismic response to this type of structure. Reproduced by permission of the European Association of Exploration Geophysicists (from Goudswaard and Jenyon (Eds) 1988)

considerable vertical scale exaggeration in a seismic section—typically 2:1 or 2.5:1, so that the real fault-plane dip of the main fault in this example is substantially less than appears in the section. Hence the categorization of the feature as a low-angle reverse fault may be somewhat inaccurate, and the movement here could reasonably be regarded as thrusting. This example, by courtesy of AGIP, is from the *Seismic Atlas of Structural and Stratigraphic Features* published recently (1988) by the European Association of Exploration Geophysicists; this publication contains many interesting and instructive seismic examples.

In Burma, similar structures occur, and at Yenangyaung, 'oil has been obtained from hand-dug wells . . . for several centuries.' The Yenangyaung structure is shown in plan in Fig. 3.15, and is an elongate, slightly asymmetrical anticline with numerous faults traversing the axis of the structure in the dip direction, but no significant faults along the

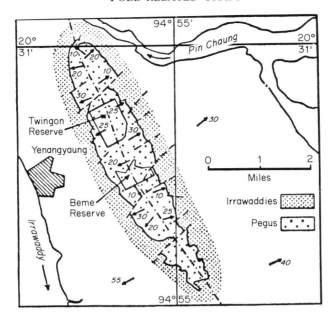

Fig. 3.15. Plan view of the Yenangyaung oilfield, Burma. The structure, a slightly asymmetrical anticline, is extensively faulted across the axis in the dip direction, but has no significant strike faulting. Reproduced by permission of Oxford University Press (from H. R. Tainsh, in V. C. Illing (ED.), *The Science of Petroleum*, Vol. vi, *The World's Oilfields*, Part I, *The Eastern Hemisphere*, 1953)

axial direction. Production has been from a series of lenticular Upper Pegu sandstones from the surface down to depths of about 5000 feet. The field was first drilled on in 1889, and is still producing (in 1980, 3112 bbl/day).

Singu (formerly Chauk) Field, on trend with the Yenangyaung structure in Burma, is an elongate, reverse-faulted dome with an almost vertical (80°) east limb (Fig. 3.16). The Padaung stage includes some 15 sandstone reservoirs, the oil traps (many with gas caps) being mainly in the western fold limb. A description is given by Tainsh (1953).

3.4.3 Simple Anticline: Complex Overburden

It is a common geological experience to encounter a situation where the near-surface structure is simple, but the deeper structure is complex. The converse is less common. An unusual example of a simple compressional anticline that has complex overburden deformation associated with it is the Kirkuk Field in northern Iraq.

This large field (the productive structure measures some 60 miles in length by 2 miles in width) was discovered in 1927, and is listed by Tiratsoo (1984) as the seventh largest 'giant' oilfield in the world. It has EUR (estimated ultimately recoverable reserves) of 16 Bbbl of oil, with about 9 Bbbl having been produced up to 1984.

Lees (1953) describes the structure as an elongate dome, slightly asymmetrical, flat-topped, with flanks dipping up to 50° and locally normally faulted, striking NW–SE. The anticline is a simple, compressional fold up to the top of the Main Limestone (which is the reservoir)

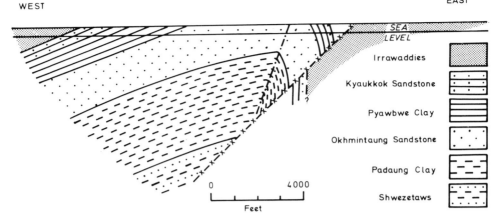

Fig. 3.16. Singu (formerly Chauk) oilfield, Burma, is an elongate, reverse-faulted dome with an almost vertical east limb, having 15 sand reservoirs in the Padaung Stage rocks. The structure is remarkably similar to the Assam examples—particularly to Digboi—in cross section. Reproduced by permission of Oxford University Press (from H. R. Tainsh, in V. C. Illing (Ed.), *The Science of Petroleum*, Vol. iv, *The World's Oilfields*, Part I, *The Eastern Hemisphere*, 1953)

of Middle Eocene to Lower Miocene age. The reservoir has both intergranular and fracture porosity.

What makes the structure unusual is that above the Main Limestone and a thin 'Transitional Limestone' (probably the basal 'pre-evaporite' carbonate of the evaporite basin) there is a substantial salt interval of Miocene age that has become plastically mobile. The salt rock is overlain by beds of gypsum, anhydrite, siltstone, and limestone of the Lower Fars, and sands, silts, and shales of the Upper Fars. During the compressional folding of the main anticline, the salt has acted as a lubricant, or *detachment zone*, resulting in thrusting and rheomorphic fold deformation in the beds overlying the Main Limestone/Transitional Limestone. This can be seen in the line-drawn cross-section across the Kirkuk structural axis in Fig. 3.17.

The writer has a particular interest in the Kirkuk structure, having been with a seismic party in this part of Iraq in the mid-1950s, and having visited what was probably one of the last (if not *the* last) operational full-sized steam drilling rig. It was working on the Kirkuk anticline (one of about 20 rigs drilling full time on the structure at that period), and was a magnificent array of machinery. There was a long line of vertical boilers belching smoke, but the whole thing operated with uncanny quietness when compared with the diesel-driven rotary rig that replaced it.

With its 'high aspect ratio' structure (60 miles by 2 miles), and the thrust-faulting involved in its development, the Kirkuk reservoir might be expected to have suffered considerable compartmentalization. However, within months of the beginning of a water-injection programme, there were detectable rises in oil–water contacts in wells up to 20 miles away from the injectors, demonstrating excellent reservoir continuity.

3.5 Uplift Anticline

3.5.1 Introduction

It is of some moment, from the viewpoint of petroleum trapping mechanisms, to consider

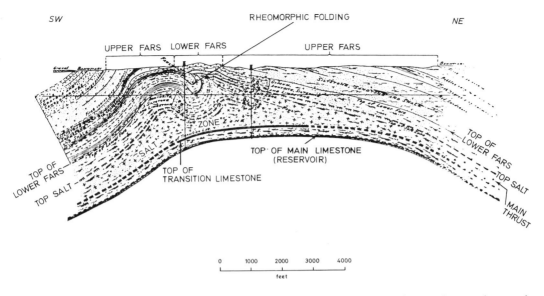

Fig. 3.17. Section across the Kirkuk oilfield structure, northern Iraq, based on surface geology and drilling data. Note the position of the main thrust on the north-east side of the structure, where the hanging wall has overridden the salt zone in the Lower Fars. Reproduced by permission of Oxford University Press (after G. M. Lees, in V. C. Illing (Ed.), *The Science of Petroleum*, Vol. iv, *The World's Oilfields*, Part I, *The Eastern Hemisphere*, 1953)

the differences between the flexural (compressional) folds discussed in the earlier part of the chapter, the uplift folds to be considered in the present part, and the drape folds that will be dealt with later. It is believed that this discussion will make it apparent why it is necessary to do all possible to determine the cause of any anticlinal fold under consideration as a potential trap.

In the sketches of Fig. 3.18, (a) and (b) represent respectively the profiles of a flexural and an uplift fold, with the stress zones indicated for a discrete part of the fold.

Resulting from (a), the fracture pattern shown in (c) will tend to develop progressively with increasing applied stress, the fractures in the zone of extension being inhibited from penetrating beyond a neutral surface separating the extensional stress from the compressional stress zone. Considerations of the thickness of stratal units involved, and the effect on fracture volume, were discussed in conjunction with Fig. 2.10.

On the other hand, in sketch (b), it is indicated that an uplift fold will typically develop only extensional stress (the whole fold being subject to this stress type, without the zonation met with in the flexural fold), and that any fractures (sketch (d)) will not be limited by stress zoning alone from penetrating inwards beyond the neutral surface towards the fold core.

The situation may not be as simple as this, however. In sketches (a) and (c), if stress increases progressively, the problem of room on the concave side of the fold element may become critical, leading to the breakdown of the rock fabric by crushing, thrusting, and dense packing. If at a later stage there is a change in the direction of maximum principal stress, the densely packed and deformed zone may undergo relief of stress, with the possible development of considerable fracture porosity.

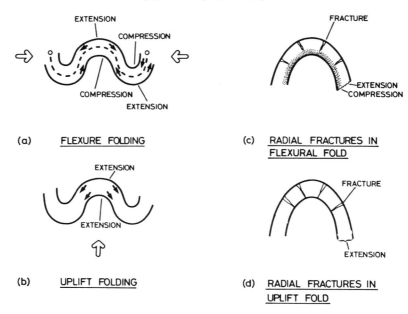

Fig. 3.18. Sketches showing in a simplified way the stresses applied during flexural (compressional) folding (a) and uplift folding (b), together with the different extent of fracturing associated with each type; (c) and (d) respectively. Reproduced by permission of Elsevier Applied Science Publishers Ltd (from Jenyon, 1986d)

As regards the difference between uplift and drape folds, an uplift fold may be either post-depositional or syndepositional. Whatever causes the uplift—whether it is a relative vertical movement resulting from tectonism, or an overburden uplift above a rising salt or shale body undergoing plastic deformation—if it occurs after the deposition and compaction of overburden sediments, then the latter will exhibit uniform thickness across the fold axis.

If, on the other hand, the relative vertical movement takes place syndepositionally, then there will be a thinning of the overburden stratal units over the fold axis. This thinning is mainly due to the winnowing and removal of sediments from the highs (in this case the developing fold axis) by bottom-water movements in the case of water-laid sediments, or wind in the case of terrigenous deposits, and their redeposition under gravity in the adjacent lows. Acting contrary to this effect will be differential compaction—greater in the thicker sediments of the lows—but this is usually quite insufficient to obscure the thinning across the crest owing to the winnowing/gravity processes. Thus, one type of supratenuous fold results.

Examples of both post-depositional and syndepositional uplift are visible in the seismic section of Fig. 3.19, which shows an overburden anticline above a Zechstein salt pillow P that has risen by plastic deformation and lateral flow of the salt rock within the interval marked B–T. The crest of the pillow has reached a level of about 1000 m above the top of the adjacent salt layer T. This salt movement was post-depositional with respect to the early Mesozoic (Lower Triassic) sediments in the interval marked T–X on the left of the section, which can be seen to remain quite uniform in thickness across the anticline and down the flanks.

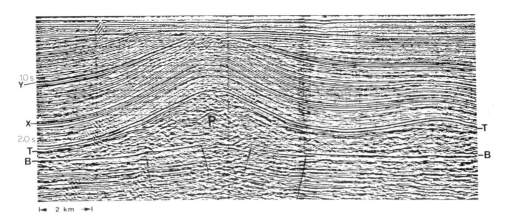

Fig. 3.19. Seismic example showing post-depositional salt movement effects (uniform overburden interval T–X related to salt pillow P uplift), and syndepositional effects (thinning of interval X–Y over the structural crest) in the same feature. Reproduced by permission of Seismograph Service (England) Ltd

However, the following interval of late Mesozoic (Upper Triassic to Cretaceous) sediments in the interval X–Y can be seen clearly to thin markedly over the anticlinal axis, thus showing that the uplift began and continued during the time interval represented, and such observations can be used to date the onset and pre-piercement movements of the salt pillow with reasonable precision. Sometimes zones of uniform thickness within the general zone of thinning indicate periods of quiescence with no movement of the salt, while other intervals showing excessive thinning, or even pinching-out, indicate periods of accelerated upward growth.

Inspection of the crestal region of the overburden anticline reveals irregularities in the data that are interpreted as crestal fracturing and faulting (parallel to the anticlinal axis, and normal to the paper) resulting from (in profile) extensional radial stress-zone faulting and incipient graben formation across the fold crest in the sediments.

Moving on to consideration of drape-folding, again there is more than one type of situation possible. Where drape results from deposition over an existing palaeohigh feature, a supratenuous fold will result from the concomitant differential compaction, but the regular stress patterns (extension/compression in the case of flexural folding, and overall extension in uplift folds) discussed in relation to previous examples will not be present. There may, however, be some fracturing and faulting present resulting from compaction, and dependent on the shape and amount of relief of the draped-high feature.

Other forms of drape structure will be mentioned in section 3.6.

3.5.2 Uplift Anticlines with Chalk Reservoirs

In the north-west European area there are several interesting uplift structures that have hydrocarbon accumulations in chalk reservoirs. In the North Sea, the Chalk of Cretaceous to Danian age is generally a fine-grained, effectively monomineralic rock of high porosity (30% or more is not unusual) but low permeability (often less than 1 mD). Clearly, for the Chalk to act as a satisfactory reservoir (when undeformed it is a very passable seal) it has to be stressed, fractured and faulted. Such conditions occur in the North Sea and adjacent areas where the Cretaceous and Danian Chalks overlie Zechstein salt rock that has become mobile.

The Ekofisk Field is a giant oilfield with reservoir in Maastrichtian (late Cretaceous) and Danian (early Tertiary) Chalks. The field is located in the Central Graben of the southern sector, Norwegian North Sea, and is notable for the high primary porosity retained by the reservoir rocks.

In the paper by van den Bark and Thomas (1981), the possible reasons for this retention of very high porosities (30–40%) are discussed. The main reasons put forward are: (i) overpressuring of the reservoir; (ii) presence of Mg-rich pore fluids; and (iii) early introduction of hydrocarbons. Overpressuring would reduce grain-contact stress; Mg-rich pore fluids may retard solution transfer; and early introduction of hydrocarbons—which is known generally to inhibit many diagenetic processes—may also have excluded or reduced pore waters and hence prevented solution transfer of calcite (chemical compaction). Samples taken from below the base of the hydrocarbon zone show destruction of almost all porosity by solution transfer effects.

The Ekofisk structure is a simple N–S-trending, slightly elongate, anticline. The seal above the fractured Chalk is an overpressured Palaeocene shale with an entry pressure that is capable of retaining the 1000-feet- (305 m) high hydrocarbon column in the reservoir. Vertical uplift by mobile Zechstein salt beneath the structure began in late Jurassic/early Cretaceous times and continued into the late Tertiary, forming the anticlinal trap and fracturing the now-lithified Chalk. Fracturing increases with depth. The N–S section along the structural axis in Fig. 3.20 shows a shaded zone of high porosity determined from seismic amplitude studies. Figure 3.21 shows an E–W section across the structure, with a low-velocity anomaly causing an apparent collapse left of centre. This is due to an extremely

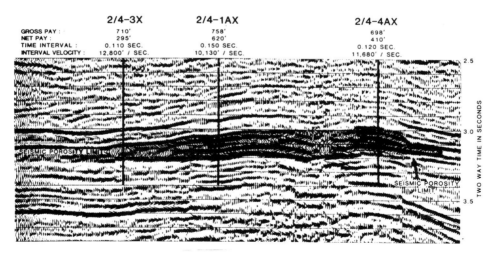

Fig. 3.20. A N–S seismic section across the Ekofisk oilfield. The shaded portion is a high-porosity zone determined from seismic amplitude studies that proved to agree quite well with the gross productive interval in the structure. From van den Bark and Thomas, 1986, reprinted by permission of American Association of Petroleum Geologists

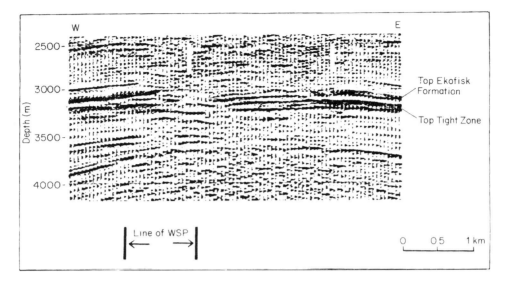

Fig. 3.21. An E-W seismic section across the Ekofisk structure, showing the pseudo-collapse zone (left of centre)—a seismic low-velocity anomaly caused by abnormally high pressure in this part of the reservoir. There is a 'gas chimney' above this anomaly in the section. Reproduced by permission of the publishers, Butterworth & Co. (Publishers) Ltd.©(from J. Brewster et al. (1986). Mar. Pet. Geol., 3, 139–169, Fig. 26)

low seismic velocity zone caused by the abnormally high pressure in this part of the reservoir. Gas leaks from a high-pressure part of the reservoir to create a 'gas chimney'. The seismic expression of this is a zone of irregular arrivals from the chimney, and an apparent collapse of the nearby sediments, which results from the unusually low seismic velocities associated with the high gas content of the porosities in the zone.

Figure 3.22 is a depth-structure contour map at Top Ekofisk level with the location of the seismic line in Fig. 3.21 marked.

The source rock of the hydrocarbons in Ekofisk is believed to be the Jurassic Kimmeridge Clay, as %Ro values of 0.93–1.16 have been measured for the Kimmeridge Clay in Ekofisk wells; this indicates the peak range of maturity for hydrocarbon generation. A possible alternative, the Palaeocene Lista Shale was found to have %Ro values of only 0.59–0.62, at the lower threshold of thermal maturity.

It must be noted here that the general interpretation of the Ekofisk hydrocarbon accumulation, as outlined here, has not gone unchallenged. There is a problem as regards the migration route from the Kimmeridge Clay up into the Chalk reservoir. There is a general acceptance of the proposition that this is by way of fault paths, although there seems to be little hard evidence. One dissenting view has been put forward in a recent paper by Moussa (1988), who suggests the interesting possibility that the Chalk itself may be the source rock. Coccolith ooze, the precursor of chalk, is said to include the faecal pellets of zooplankton (probably copepods) that contain dihydrophytol—a precursor of pristane, a precursor of some crude oil deposits. The matter will not be pursued further in this context, since we are mainly interested in the conditioning of the Chalk to form a reservoir.

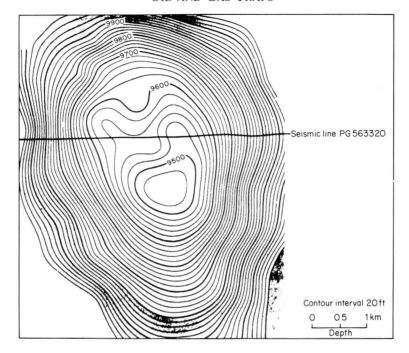

Fig. 3.22. Depth-structure contour map at Top Ekofisk Formation level (see the section in Fig. 3.20, the location of which is marked here. Reproduced by permission of the publishers, Butterworth & Co. (Publishers) Ltd.© (from J. Brewster et al. (1986). Mar. Pet. Geol., 3, 139–169, Fig. 27)

Brewster et al. (1986), in a waterflood appraisal of Ekofisk, divide the fracturing in the chalk into three classes: (i) healed fractures—early fractures filled with dense chalk matrix and not permeable; (ii) tectonic fractures—small-scale normal faults, often in parallel sets of conjugate fractures, and the most effective for enhancement of permeability; (iii) stylolite-associated fractures—as in (ii), these developed under a vertical principal stress system, and form parallel and adjacent to stylolite 'columns'; they have been interpreted as tension gashes. The incidence of stylolites and vertical stresses will occur again later in discussions of similar structures; it appears possible that the operative factor is the action of 'vice jaws' compression on the Chalk—the upward pressure of the mobile salt, and the downward pressure of the superincumbent overburden material. Stylolite development is thought to be a type of pressure-solution phenomenon that occurs in limestones, particularly at boundaries—at contacts with other lithologies, on bedding planes, and along fractures.

Another similar anticlinal trap in fractured chalk resulting from salt uplift is the Dan Field—also in the Central Graben of the North Sea, but in Danish waters. The reservoir is again Maastrichtian to Danian Chalk. Uplift by the Zechstein salt began in the late Jurassic and continued to early Tertiary times. Fracturing and stylolitization are common in the major part of the Maastrichtian, but are absent in the top Maastrichtian and in the Danian—see lithostratigraphic column in Fig. 3.23. Childs and Reed (1975) mention that high porosities (30–40%) and permeabilities (3–7 mD) occur at two levels—at the top of the Danian, and immediately below the Danian–Maastrichtian contact. Otherwise, most

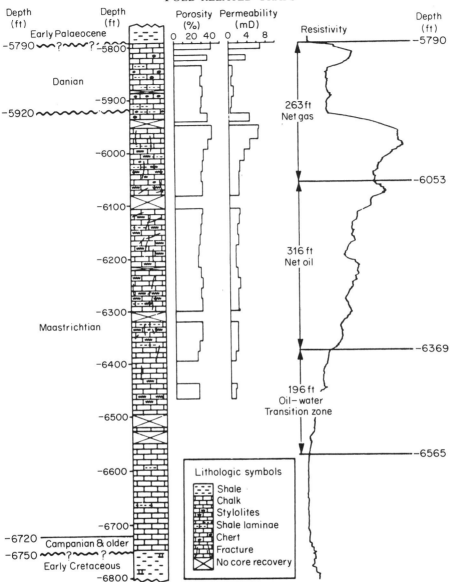

Fig. 3.23. Lithostratigraphic column through the Dan Field. High poroperm zones occur at Top Danian level, and immediately below the Danian–Maastrichtian boundary. Note the distribution of fractures and stylolites. Reproduced by permission of Elsevier Applied Science Publishers Ltd (from Childs and Reed, 1975)

of the Chalk section has 20–30% porosities and permeabilities of less than 2 mD. The reason for the two high poroperm zones is not certain, but a tentative suggestion made by Childs and Reed is that they represent secondary leaching at early and late Danian unconformities. A depth-structure contour map and schematic cross-section across the Dan Field are shown

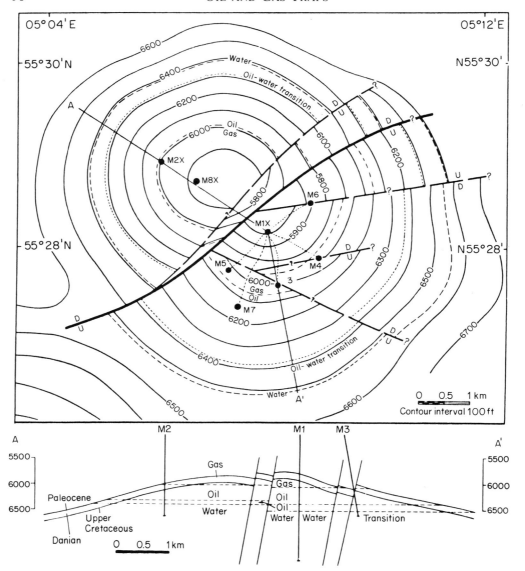

Fig. 3.24. Depth-structure contour map at Top Palaeocene level of the Dan Field (top), and a cross-section (bottom) located at A–A' on the contour map. Reproduced by permission of Elsevier Applied Science Publishers Ltd (from Childs and Reed, 1975)

in Fig. 3.24. In the column in Fig. 3.23, note the distribution of fracturing, and also of stylolites.

While discussing fractured chalk reservoirs, it is of interest to consider a study of a fractured chalk overburden rock uplifted to the surface by a salt diapir in Germany, thus giving a rare opportunity for the rock to be investigated at outcrop. Koestler and Ehrmann (1987) have carried out such a study on this exposed example of a possible hydrocarbon reservoir associated with a salt diapir at Laegerdorf, north-west Germany. The chalk has been uplifted some 1000 m through later sediments, and has reacted to the applied stress by brittle fracturing, with strain concentrated in distinct faults and fault zones arranged in complex conjugate sets (cf. Ekofisk). These sets subdivide the chalk into rhombohedroid volumes (with dimensions of tens to hundreds of metres) of low deformation bordered by large fault zones, and these form basic reservoir units.

Clay smearing, clay enrichment, and remineralization on and close to fracture surfaces reduce porosity. (Porosity within the undeformed rhombohedroids is as high as 40–50%, although permeability ranges only from 2 mD to 10 mD.) However, permeability is enhanced by microcracks on fault surfaces, and a high density of fractures close to fault zones, facilitating communication between undeformed chalk and fracture zones.

Deformation processes include: (i) pressure solution, leading to stylolitization (cf. previous examples) parallel to bedding and fractures, as well as concentration of clay minerals with sealing effects; (ii) compaction, leading to change in microtexture (coccolith collapse), volume decrease, and loss of porosity; (iii) remineralization (fibrous and blocky calcite, baryte), leading to porosity reduction; (iv) fracturing, causing brecciation, microfissuring, corrosion of particles, and increased permeability; and (v) faulting, causing clay smearing, sealed surfaces, and slickenside striations. Particle size of the chalk is reported as less than 5 μm.

The Valhall Field in the Central Graben, southern Norwegian North Sea, is an uplift feature that has been described by Hardman and Eynon (1977) and Munns (1985). The former authors categorize the field as a 'structural–stratigraphic' (i.e. combination) trap, which indeed it is. The principal trapping element, however, is uplift, and it is therefore included in this category, since it is another example of a chalk reservoir, although this time the uplift element is of tectonic origin rather than the result of salt movement, as in the previous examples.

The structure of Valhall is that of a domal uplift elongated in a NNW–SSE direction, the crestal region of which is modified by a shallow structural graben trending along strike, with other epi-anticlinal normal faults (throwing towards the structural axis on both limbs) paralleling the graben boundary faults.

Seismic interpretation of the structure was severely handicapped by gross distortion of the data over the crestal region. This has been found to be a consequence of the presence of gas-charged low-velocity silts in the Miocene section. The resulting lowering of the bulk density of the rocks produces a strong negative seismic velocity anomaly (a 'pull-down' effect), to which are added 'gas chimney' effects owing to upward leakage of gas in the geological section. See Fig. 3.25, which shows a seismic section across the structure, with the supracrestal distortion of data very obvious, and also, below, a geological interpretation of the seismic section.

In depth, the Valhall structure is bounded by a large normal fault—the western limit of the anticlinal Lindesnes Ridge, and the western bounding fault of a half-graben with a thick accumulation of Lower Cretaceous and Jurassic sediments. The inversion of this half-graben in late Cretaceous times is believed to be the cause of the uplift of the Valhall structure.

Valhall reservoirs occur in allochthonous chalks of the (Maastrichtian) Tor Formation, and in autochthonous chalks of the (Turonian–Coniacian) Hod Formation. They are noteworthy for the high primary porosity retained by the chalks; having regard to maximum depth of burial v. porosity for southern North Sea chalks, these should be only of the order of 10%, but range from 36% to 50% in the crestal zone of the structure. These high

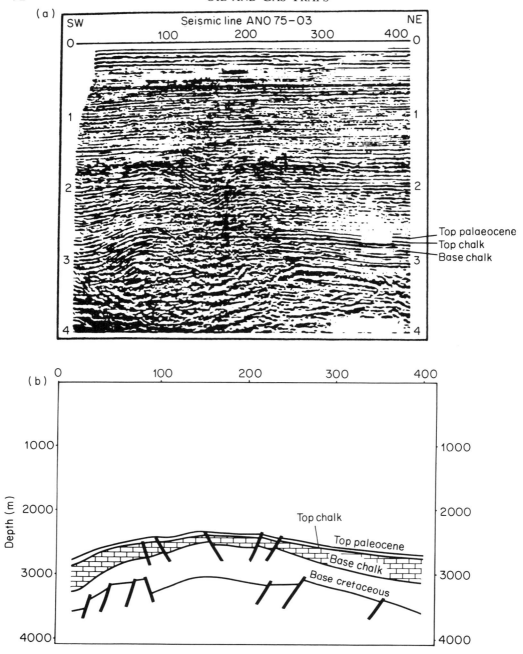

Fig. 3.25. Seismic section across Valhall Fields (a) with a geological interpretation (b). The serious distortion in the centre of the seismic section is due to a low-velocity anomaly, and gas chimney effects producing results similar to those seen in Ekofisk. Reproduced by permission of the publishers Butterworth & Co. (Publishers) Ltd.© (from Munns (1985). Mar. Pet. Geol., 2, 23–43, Fig. 17)

porosities, as in the case of other chalk reservoir traps, such as Ekofisk, are believed due to formation overpressure having inhibited compaction and porosity volume reduction in the chalks. The source rock for Valhall is the Jurassic Kimmeridge Clay, and the seal formation is the Palaeocene Rogaland Group of clays and silts. For a more detailed account of the field, the paper by Munns is recommended.

3.5.3 A Middle East Giant

A major field that is the result of gentle vertical uplift, which may be (but this has not been established) due to salt movement at depth, is the Burgan Field of Kuwait. Burgan is listed as the second largest 'giant' oilfield in the world by Tiratsoo (1984), with EUR of 72 Bbbl. The field is a subsurface dome, elongate in the N–S direction, covering an area of about 135 square miles. The immense size of the structure can be gauged by the fact that even though average dips are only 3°, the vertical closure at productive levels exceeds 12,000 feet. Depth to the top of the producing intervals is barely 3500 feet, and there are surface indications of the hydrocarbons—sulphur deposits and oil impregnation of Miocene sandstones.

A line-drawn section along the domal major axis is shown in Fig. 3.26, and an indication of the very gentle nature of the dips involved is given by the profile at Orbitolina Limestone level drawn at natural scale below the main cross-section. The location, and an idea of the great areal extent of the structure, is given in the sketch map of Fig. 3.27.

The main reservoirs in Burgan consist of several Middle Cretaceous sandstones of high permeability, named as the First, Second, Third, and Fourth Sand with increasing depth. There is also a deeper, Upper Jurassic limestone reservoir, the Minagish Oolite. Interbedded with the productive sands are impermeable sands and shales that constitute the seal formations. A somewhat unusual feature (see Chapter 1) of the field is that the specific gravity of the oils, as produced from the various horizons, *increases* with increasing depth.

The Orbitolina Limestone, present between the Second and Third Sands, acts as a good geological marker within the productive zone, since the sand and shale units are variable in thickness.

3.6 Drape Anticlines

3.6.1 Introduction

The features to be discussed and illustrated here are what might be termed the static or passive drape anticlines mentioned in section 3.5.1. Depending on the relief and shape of the older feature over which the drape occurs, these anticlinal traps vary from the subtle to the very obvious, the latter type usually involving sedimentary cover over a large palaeogeomorphic feature—perhaps a block fault or erosion-resistant cuesta on an unconformity surface. Sometimes this older, draped feature has moved again at several later stages, with vertical and rotational movements leading to quite complex depositional patterns in the draping sediments.

As previously noted, stress-zone fracturing and faulting are usually absent in this type of anticline, and therefore zones of possible fracture porosity must be sought elsewhere than over the anticlinal crest, in any favourable locations related to differential compaction.

94

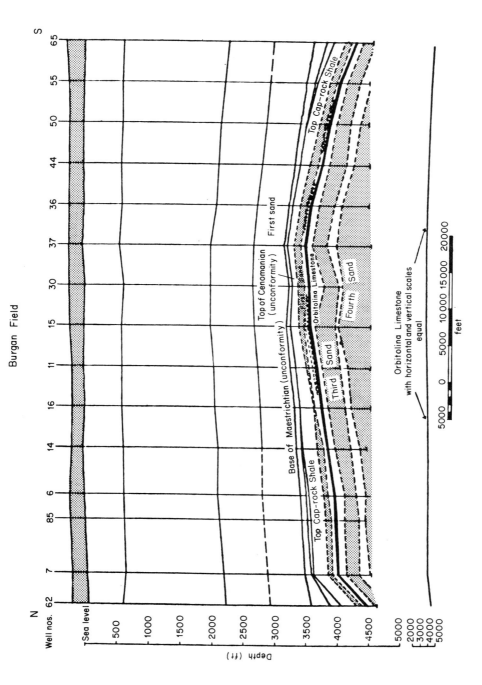

Fig. 3.26. Geological interpretation of the Burgan structure, with vertical exaggeration. The true-scale representation of the structure at the Orbitolina Limestone level, shown here beneath the main section, indicates the very gentle nature of the dips in this uplift anticline (mobile salt at depth being the possible cause). Reproduced by permission of Oxford University Press (from KOC Staff, in V. C. Illing (Ed.), *The Science of Petroleum*, Vol. iv, *The World's Oilfields*, Part I, *The Eastern Hemisphere*, 1953)

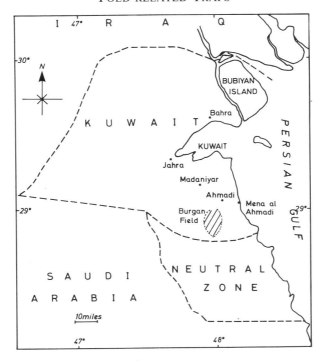

Fig. 3.27. Sketch map showing location (and giving an idea of the enormous size) of the Burgan Field, Kuwait—the second largest oilfield discovered in the world to date. Reproduced by permission of Oxford University Press (from KOC Staff, in V. C. Illing (Ed.), *The Science of Petroleum*, Vol. iv, *The World's Oilfields*, Part I, *The Eastern Hemisphere*, 1953)

Apart from these passive drape features, there are other, more dynamic structures that have such strong causative links with other phenomena that they have been included in other parts of the book. These include structures associated with some syndepositional fault movements, and also collapse drape, where an already compacted overburden subsides on to relict features (reefs, cuestas, erosional remnants) previously encased in a salt layer when the latter is removed by dissolution. A seismic example of two small drape anticlines resulting from such salt dissolution removal and overburden subsidence is shown in Fig. 3.28. Structures of this type will be discussed in detail in chapter 8.

3.6.2 North Sea Examples, British Sector

Three examples from the British sector of the North Sea have been selected for illustration and discussion. These are Montrose and Forties oilfields, and the Frigg gasfield (the latter straddling the British–Norwegian median line).

It is appropriate to commence with the Montrose Field, since this was the first oilfield to be discovered in the British sector, in December 1969.

The field is located 130 miles (209 km) east of Aberdeen in 300 feet (91 m) of water. The hydrocarbon reservoir is a Palaeocene (basal Tertiary) sandstone overlying Danian carbonates, and a schematic WNW–ESE structural cross-section is shown in Fig. 3.29.

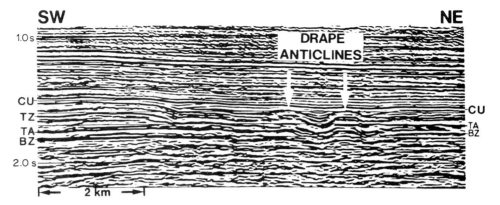

Fig. 3.28. A pair of drape anticlines formed in Mesozoic overburden that has subsided on to (probable) reef structures left as remnant positive features when the Zechstein salt interval that previously encased them was removed after dissolution of the salt. BZ = Base Zechstein: TA = top of the basal Zechstein carbonates; TZ = Top Zechstein salt on the left, which ends in a dissolution slope about one-third of the distance across the section. The Top Zechstein on the right is represented by TA); CU = Late Cimmerian Unconformity. Features of this type will be discussed in detail in Chapter 8. Reproduced by permission of Seismograph Service (England) Ltd

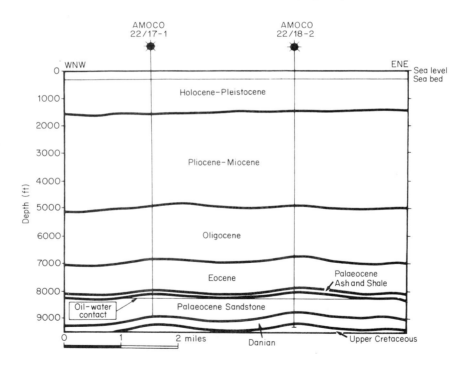

Fig. 3.29. A schematic section across the Montrose oilfield, the first oilfield discovered in the British sector of the North Sea, December 1969. Note the (approximate) position of the oil–water contact. Reproduced by permission of Elsevier Applied Science Publishers Ltd (from Fowler, 1975)

Montrose has been described by Fowler (1975), who mentions that the structure, which has three culminations, was probably initiated by Hercynian block-faulting in the late Palaeozoic, with some modification during the (Upper Jurassic) Late Cimmerian phase of movement. The area remained positive throughout the Lower Cretaceous, after which the Upper Cretaceous marine transgression occurred with deposition of a complete chalk sequence and afterwards a full Tertiary succession. The low-relief domal feature in the Palaeocene, draped over the deep-seated fault blocks, was developed fully during the Miocene Alpine movements, and the structural relief dies out upward in the Pliocene.

There have been problems in determining the oil–water contact level in Montrose, as it occurs at slightly different levels in the three separate culminations. The source rocks are not known with certainty, but are believed to be Jurassic organic shales. The maximum oil column is 190 gross feet (58 m), with the average net being about 60 feet (18 m) only. The water and oil zones are present in the reservoir, but so far no gas–oil contact has been observed in any well.

Core analysis indicates average porosity in the Palaeocene sand reservoir to be 22.8%. Horizontal permeability averaged from the combined core sets is 39.3 mD, and vertical permeability 1.21 mD only, the vertical to horizontal ratio thus being very low.

A N–S seismic section across the field is shown in Fig. 3.30(a), exhibiting one of the two culminations in the northern half of the field (Well 22/18-2), and the single culmination in the southern half (Well 22/18-1). The so-called 'ash marker'—a strong two-cycle seismic event well-known to seismic interpreters in this area—is seen in the section at the top of the Palaeocene. The drape structure is so gentle (recalling the vertical scale exaggeration on seismic sections of at least 2:1 usually) that there are no indications at the edges of the structure of faulting related to differential compaction. In fact, there are no signs of any significant faulting at the reservoir level. The rapid reduction of relief upwards in the section above the Palaeocene should also be noted—this is frequently the case in compaction-drape structures. Figure 3.30(b) is a depth-structure contour map on top of the Palaeocene sandstone reservoir, showing the three culminations of the trap, and also the position of the seismic section of Fig. 3.30(a).

Another drape anticlinal trap in Palaeocene sandstone, is present in the Forties Field, another early (1970) North Sea discovery. A schematic cross-section is shown in Fig. 3.31, with a small sketch location map. The structure, described by Walmsley (1975a) is due to an east-south-east-trending anticlinal nose crossing Block 21/10 and interrupting the regional east dip into the Tertiary Basin. It overlies a faulted high at the Base Cretaceous unconformity, which in turn overlies complex pre-Cretaceous block-faulting with associated thick volcanics.

The oil column in Forties, at maximum in the Palaeocene sandstone, is 155 m (500 ft) in height, and the reservoir covers an area of some 90 km^2, which makes this a major oilfield. The seal formation, as in Montrose, is a shaly mudstone, known as the Palaeocene Shale, which tends to be quite uniform in both thickness and lithology across the field. Core porosities in the massive reservoir sandstone range from 25% to 30%, and permeabilities vary up to 3900 mD.

It is salutary to note that in spite of the size of this largest of the UK North Sea oilfields (estimated recoverable reserves are about 2 billion barrels of oil), it does not appear on a list of the world's 36 largest 'giant' oilfields (Tiratsoo, 1984), which range from Umm Shaif, Abu Dhabi, with 5 billion barrels estimated recoverable oil at the 'small' end, to

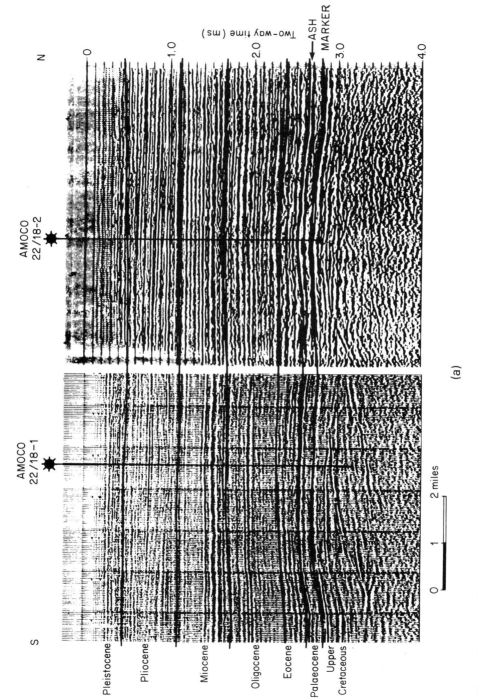

Fig. 3.30(a). A N–S seismic section across the Montrose Field, linking two of the three separate culminations of the reservoir (see also Fig. 3.30(b)). Note the 'ash marker', a prominent seismic marker event from a volcanic ash unit in this area. Reproduced by permission of Elsevier Applied Science Publishers Ltd (from Fowler, 1975)

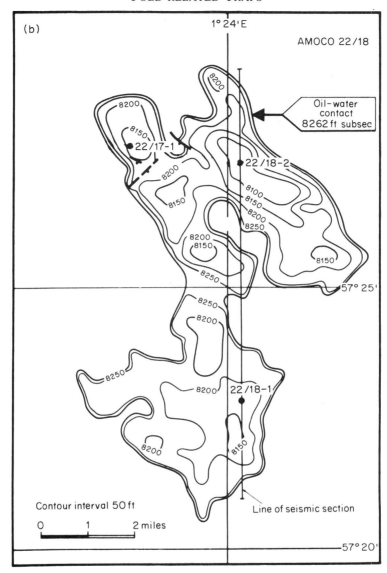

Fig. 3.30(b). Top Palaeocene depth-contour map of the Montrose Field, showing the three separate culminations of this drape structure, together with the location of the seismic section seen in (a).
Reproduced by permission of Elsevier Applied Science Publishers Ltd (from Fowler, 1975)

Ghawar Field in Saudi Arabia, with estimated recoverable reserves of 83 billion barrels of oil—over forty times the size of Forties.

Frigg gasfield is another structure resulting from drape and compaction above a tilted and rotated fault block. The field straddles the UK/Norwegian median line in UK Block 10/1 and Norwegian Block 25/1, with about one-third of its area in the UK Block. The structure has been described by Blair (1975) and Héritier et al. (1981).

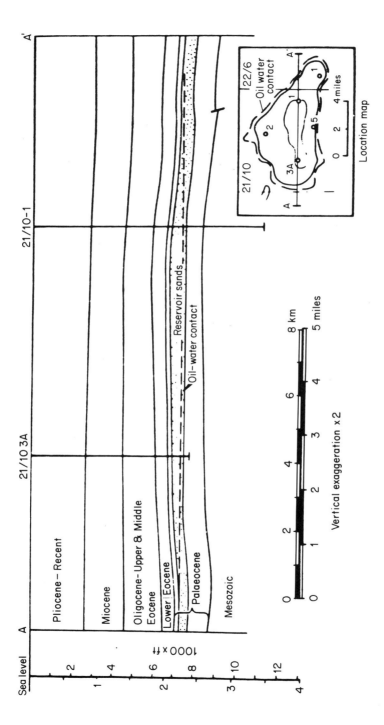

Fig. 3.31. Schematic cross-section of Forties Field, with small sketch location map. The section shows the Top Palaeocene sandstone interval that is the reservoir, with an indication of the OWC level. The relief of this drape structure is probably due to movements of deep-seated fault blocks that in turn affected a faulted high at the Base Cretaceous Unconformity below the anticline. Reproduced by permission of Elsevier Applied Science Publishers Ltd (from Walmsley, 1975a)

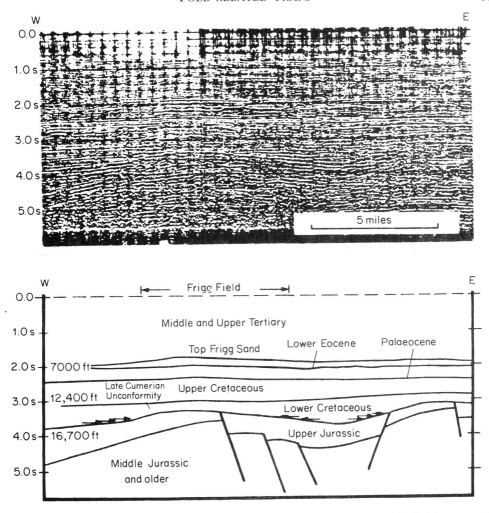

Fig. 3.32. Frigg gasfield—a seismic section from east to west, with (below) a geological interpretation. Another drape anticline above tilted and rotated fault block structure. Reproduced by permission of Elsevier Applied Science Publishers Ltd (from Blair, 1975)

The structure has been interpreted as a Lower Eocene deep-water submarine fan controlled by sediment supply from the Shetlands Platform to the west and south-west. The feature already had depositional mounded relief, which was later structurally enhanced by post-Eocene (probably Alpine) movements that rejuvenated Cimmerian faults present beneath.

In Fig. 3.32, an early E-W seismic section of rather poor quality together with a line-drawn interpretation are shown. It is possible to see in the seismic data the undulating Late Cimmerian Unconformity event (just above 4.0 s two-way time on the left), the steeply dipping Top Middle Jurassic horizon (4.8 s, left), and the strong band of events marking the Top Cretaceous (2.4 s, left). It is less easy to make out any detail in the Eocene reservoir interval on the section. A later, and better quality seismic section in a NW–SE direction

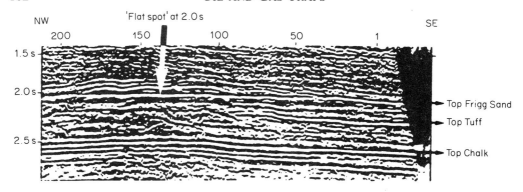

Fig. 3.33. Frigg gasfield—a seismic section from north-west to south-east across the field, showing an excellent 'flat spot' gas–liquid contact event (a 'DHI') at 2.0 s. Reproduced by permission of The Institute of Petroleum, London (from Héritier et al., 1981, Fig. 8, p. 387)

is seen in Fig. 3.33, in which the draped and mounded form of the Frigg Sand can be seen together with a good 'flat spot' gas–liquid contact event (a DHI) at 2.0 s. (It should be noted that the time zone covered by the seismic example in Fig. 3.33 is restricted compared with that seen in Fig. 3.32.) Again note the absence of significant faulting at the reservoir level, and the rapid reduction of relief above the draped structure (more apparent in Fig. 3.32 than Fig. 3.33 because of less limited time zone).

Features of this type will be considered in more detail, and from the depositional viewpoint, in Chapter 9.

For those wishing to know more of the complex depositional history of Frigg based on seismic and drilling results, the paper by Héritier et al. (1981) is strongly recommended. In particular, it is most interesting to note therein how, by differential compaction effects acting on varying lithologies, it is believed that a 'real' levee has been compacted out of existence (morphologically speaking) and a pseudo-levee developed in an adjacent location within the fan deposits.

As regards the relative influence of compaction and drape on the structure, Héritier et al. seem to believe that compaction had the major role, with drape playing a lesser part. However, Blair (1975) states that although the compaction factor cannot be determined, an estimate of the drape can be obtained. By making the reasonable assumption that the Top Cretaceous reflector (see Fig. 3.32) was flat at the time of Frigg Sand deposition, the amount of present-day reversal on the reflector can be determined as within the range of 200–500 feet (60–150 m). Therefore, at least this amount of drape enhancement of the Frigg reservoir closure can be assumed (remembering that the structure also probably had some depositional closure(s) owing to the mounded fan facies).

The Lower Eocene Frigg Sand reservoir is a clean, fine, unconsolidated sand with porosities ranging from 25% to 32%, and average permeabilities from 1200 mD to 1600 mD. The seal formation is a Lutetian (Upper Eocene) grey marine shale. Beneath the gas zone there is a 10-m-thick oil zone—however, the oil (estimated at 1200 MMbbl) is very heavy at 24° API, and non-commercial. Probable recoverable gas is estimated at 7.1 TCFG. Frigg Field is at present planned to shut down in 1993 (*European Continental Shelf Guide*, 1986–1987).

3.6.3 Libya—a Draped Carbonate Reservoir

The Zelten Field in the Sirte Basin, Libya, is a Lower Tertiary anticline draped over a pre-Cretaceous (? Ordovician) fault block. The field was discovered in 1959 about 175 km south of the port of Marsa Brega, and the discovery well was completed at a total depth (TD) of 5665 feet; on test, the well produced 17,500 BPD of 37° API gravity crude.

The reservoir of Zelten is a 350-feet interval of skeletal limestone of Lower Eocene age, with pockets of biostromal reef material distributed within it. The porosity varies from good to excellent over most of the field, with some limited areas of poor porosity in the southern part. The seal formation consists of a dense micritic limestone capping the producing interval, and being overlain in turn by a 250-feet shale section. Both of these formations contribute to sealing the reservoir (Fraser, 1967).

The pre-Cretaceous structure over which the field is draped is flanked on the west side by a normal fault system throwing down to the west. Most of the fault movement took place during the Upper Cretaceous and Palaeocene, with no evidence of any movements after the Lower Eocene. A line-drawn section across the field is shown in Fig. 3.34, and a depth-structure contour map at the level of the top of the reservoir forms Fig. 3.35, showing the three areas into which the field is divided. The separating lows between the culminations are due partly to structure, and partly to thickenings of the dense micritic limestone capping the reservoir. Note that faulting is absent at reservoir level.

As suggested by Fig. 3.34 (even allowing for the vertical exaggeration of the seismic section), dips over the structure are small except on the west flank where they steepen, probably owing to drape over the underlying faulting.

The general absence of faulting at reservoir level is, as seen in previous examples, frequently observed in relation to drape anticlines, which, apart from compaction, are relatively unstressed compared with compressional and uplift folds. Although not in itself diagnostic, this absence of associated faulting may, together with other evidence, help in determining the mode of origin of an anticlinal/domal structure.

3.6.4 Drape Traps in Ecuador

In some circumstances, it is not immediately obvious which of several factors is causative in the development of an anticlinal trap. For instance, Fig. 3.36 shows a schematic cross-section of what is described as a typical hydrocarbon trap in Oriente, Ecuador (De Righi and Bloomer, 1975). It is stated that 'where the fault-induced folds . . . (include) . . . zones of sufficient porosity and permeability in the Cretaceous sandstones . . . (of the Hollin and Napo Formations in this case) . . . accumulations of hydrocarbons occur. Some "roll-over" of the fold is encountered on the upthrow side of the faults'. It is not clear what is meant by 'roll-over' in this context—certainly, it cannot be taken to mean the roll-over anticline encountered on the downthrow side of a listric growth fault. In this case, assuming the line-drawn schematic is accurate, the trap occurs on the upthrow side of a feature that is certainly not a growth fault. There appears to be some fault drag represented, which is producing east dip into the fault plane on the upthrow side.

However, what the schematic seems to be indicating, with the trap situated above a Precambrian basement high, is that the hydrocarbon accumulations are present in a drape fold above the high, although there may be some element of fault drag in the east-dipping

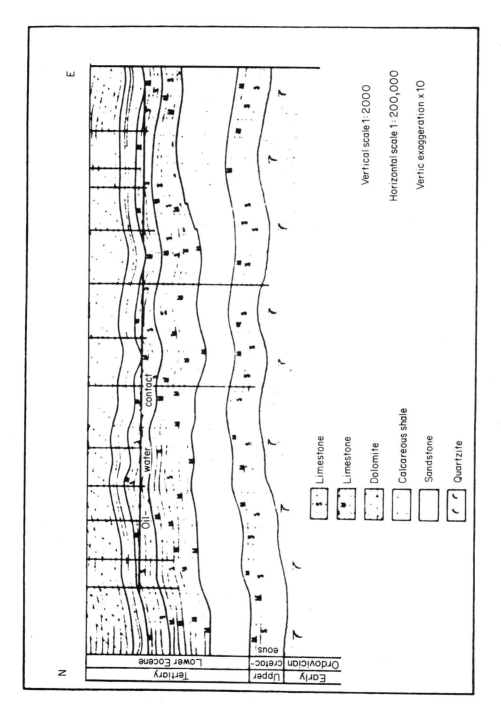

Fig. 3.34. Zelten Field, Sirte Basin, Libya. An oilfield with a draped carbonate reservoir (a skeletal limestone of Lower Eocene age). Reproduced by permission of Elsevier Applied Science Publishers Ltd (from Fraser, 1967)

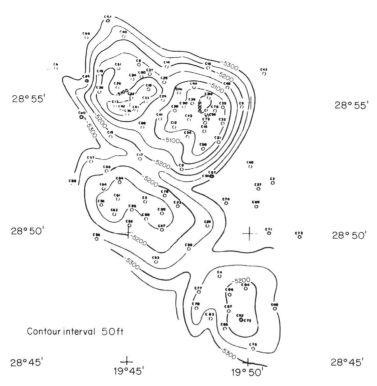

Fig. 3.35. Zelten Field: a depth-structure contour map at top reservoir level showing the multiple culminations of the structure. This seems to be a common feature of drape anticlines, where the reservoir sequence has been deposited over old relief. It can be used, with caution, as a diagnostic for drape structures. Reproduced by permission of Elsevier Applied Science Publishers Ltd (from Frazer, 1967)

limb of the overburden anticline in the reservoir beds. The short segment of a steeper west dip that is indicated just to the west of the structural crest seems to be clear evidence of drape here, as it is very difficult to suggest any alternative explanation. Any fault drag element must be small, since the fault throw is very small—a point particularly mentioned by De Righi and Bloomer.

3.7 Some Other Noted Anticlinal Traps

3.7.1 Hewett Gasfield

The gas accumulations of Hewett Field are of interest in that they occur at the effective southern limit of the underlying Zechstein salt, which has here become so thin that it no longer acts as a seal preventing gas from migrating up into the post-Permian section in the southern basin of the UK, North Sea.

The structure is a NW–SE-trending, elongate, doubly plunging anticline with associated faulting. Hewett is one of the few fields in this area where gas from the underlying Coal

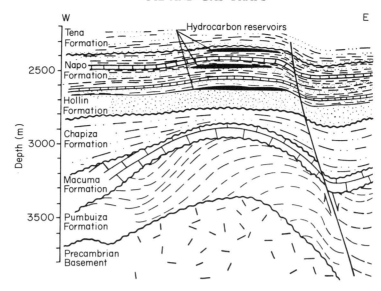

Fig. 3.36. Schematic section of a typical hydrocarbon trap as found in Oriente, Ecuador, with later stratigraphic units draped over a Precambrian basement high. Reproduced by permission of Elsevier Applied Science Publishers Ltd (from De Righi and Bloomer, 1975)

Measures (Upper Carboniferous) has migrated through the Upper Permian Zechstein marginal facies carbonates (with a vestigially thin Z2 salt), and been trapped in a Triassic structure above.

The gas has accumulated in two distinct reservoir zones. The lower zone, the (Lower Triassic) Hewett Sandstone, seems to be a purely local sand development immediately overlying the marginal facies Zechstein, and is sealed above by the Bunter Shales. The upper reservoir is the (Lower Triassic) Bunter Sandstone, which can be correlated with the German Middle Bunter Sandstone, and is sealed above by shales of the (Upper Triassic) Dowsing Formation. Average porosity and permeability of the Hewett Sandstone is 21.4% and 1310 mD, respectively, whereas for the Bunter Sandstone the figures are 25.7% and 474 mD (Cumming and Wyndham, 1975).

The sketch contour map at the Top Bunter Sandstone level in Fig. 3.37 shows the NW–SE elongation of the structure, and the major faults associated with it. It can be seen from the map, and from the NE–SW seismic section (with geological interpretation above) across the structure, shown in Fig. 3.38, that little of the vertical closure, if any, is attributable to the bounding faults. The doubly plunging anticlinal structure accounts for the major part of the closure, and this is also true at the level of the Hewett Sandstone reservoir unit.

Unlike structures further north, where the Zechstein salt is thick and has become mobile, the Hewett Field structure owes nothing to salt movement; there is here, as noted, only a vestigial few metres of Zechstein Z2 cycle salt rock in the section, insufficient to produce any important overburden deformation. The anticline is of tectonic origin, and probably related to compressional stresses developed during the Late Cimmerian (Upper Jurassic–Lower Cretaceous) movements. The NW–SE trend is commonly present in this part of the southern basin. The seismic section of Fig. 3.38 suggests some involvement of fault-drag

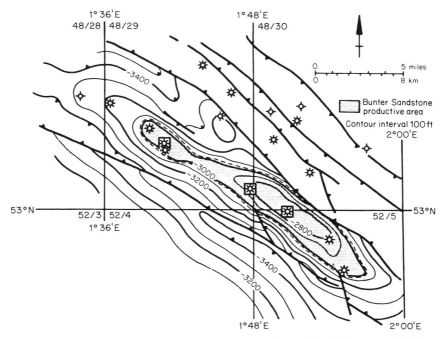

Fig. 3.37. Depth-structure contour map at Top Bunter Sandstone level, Hewett gasfield, southern UK North Sea. The area of the reservoir at this level is stippled, and shows that practically none of the vertical closure is dependent on the bounding faults. Reproduced by permission of Elsevier Applied Science Publishers Ltd (from Cumming and Wyndham, 1975)

in the structure, but this is not thought to be a major element in the main structure. It is, however, possible that the faulting has a strike-slip element, which is also common in the general area, and the compressional stresses involved may be those that can develop in a tectonic 'pull-apart' situation. Associated with the faulting at the south-west end of the seismic section example, there is an unusual upwarped segment of reflections (solid black arrowhead) that may represent a hangingwall compressional feature related to a lessening of dip in the lower part of the downward trajectory of the main fault plane here, which has been suggested in the marking of the fault trace on the section. This feature, which is believed to be real, is not indicated on the geological interpretation above, and could be an indication that the fault plane has more of a listric curve than is suggested. In the lower part of this upwarped feature, a partial diffraction hyperbola (see next chapter) originating at a truncated stratal event on the footwall (upthrow) side of the fault can be seen, terminating at the left-hand end of the Bunter Sandstone reservoir interval.

3.7.2 Wilmington Field, Southern California

With estimated total recoverable reserves approaching three billion barrels of oil, the Wilmington structure is one of many giant oilfields in the USA, and one of a number of important fields in this area. Located in the Los Angeles Basin, in the region of the Peninsular Ranges, the structure is a broad, asymmetrical elongate anticline that is compartmentalized

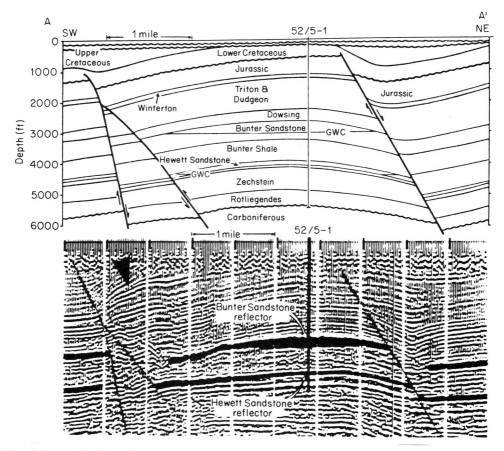

Fig. 3.38. A seismic section across the Hewett gasfield (the location of this line being through the northernmost well shown in Block 52/5 on the map seen in Fig. 3.37). Both Hewett and Bunter Sandstone reservoir intervals are marked in solid black between the two major faults. Reproduced by permission of Elsevier Applied Science Publishers Ltd (from Cumming and Wyndham, 1975)

longitudinally by a number of normal faults striking transversely across the axis. It lies within the zone of NW–SE-trending fault blocks that characterize the regional geology. In these fault blocks, elongate surface topographic highs are present that usually express underlying anticlinal trends often containing oil accumulations; Wilmington is one of these.

A depth-structure map contoured on the Top Ranger (Lower Pliocene) reservoir unit is shown in Fig. 3.39(a), with a line-drawn section (A–A') of the reservoir zone shown in Fig. 3.39(b). The location of this line is indicated on the contour map.

A description of the general geology of the region is given by Yerkes *et al.* (1965), and of the Wilmington Field by Mayuga (1970). The principal reservoir sequence consists of seven near-shore sands interbedded with mud and clay units that form seals. During the middle Miocene, compressional stresses in a N–S couple, folded the rocks and produced the NW–SE trend of the structure. Later in the middle Miocene, emergence and erosion

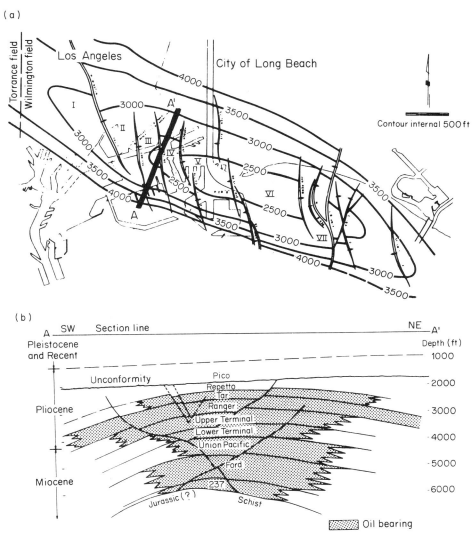

Fig. 3.39. (a) Depth-structure contour map at Top Ranger level in the reservoir interval of the Wilmington Field, USA. Note the many transverse normal faults cutting across the anticlinal axis. (b) Cross-section (location shown by line A–A' in (a)) showing the lateral 'edgewater' limits of producing zones in the 'old' part of the Wilmington Field. The 'new' part of the field is in the eastern half of the structure. From Mayuga, 1970, reprinted by permission of American Association of Petroleum Geologists

occurred, followed by subsidence and basinal deposition in the late Miocene and early Pliocene (Repetto) time. During this period, oscillatory vertical movements are believed to have produced overall uplift and arching (double plunging) of the Wilmington structure, which probably caused the development of the transverse faults seen crossing the structural axis in Fig. 3.39(a). From their position, trend, and nature (normal faults), these seem to be extensional stress-zone faults due to the arching, or perhaps tension-gash type fractures

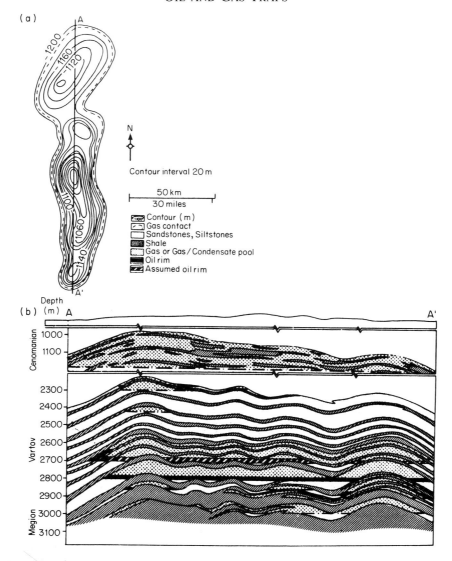

Fig. 3.40. (a) Depth-structure contour map on top of the Pokur Suite (Cenomanian). (b) Cross-section (location shown by line A–A′ in map (a)) approximately along the axis of the elongate Urengoy gasfield structure, northern part of West Siberia, USSR. From Grace and Hart, 1986, reprinted by permission of American Association of Petroleum Geologists

initiated during the activity of the N–S couple, and later moving as normal faults during the oscillatory vertical movement. The transverse faults have divided the reservoir zone laterally into a number of compartments.

North-westward along the Wilmington trend, a saddle feature separates the Wilmington structure from the Torrance Field. To the north-east, several highly productive subparallel trends exist, including such fields as Huntington Beach, Long Beach, Playa del Rey, Santa

Fe Springs, and Brea-Olinda, together with many others, making the Los Angelos Basin one of the most productive areas in continental USA.

3.7.3 The World's Largest Gas Trap

Described in 1986 as the world's largest gasfield (Grace and Hart, 1986), the Urengoy Field is in the northern part of West Siberia, south-east of the Ob Gulf, USSR. It is located on the central part of the Urengoy Megaswell, a projection from the basement surface that had a 2950-feet (900 m) vertical closure in early Jurassic times. Growth was reactivated in Neocomian times at unequal rates across the Megaswell feature.

The Fig. 3.40 illustrations show (a) a depth-structure contour map of the Urengoy structure at the level of the top Pokur 1 Suite (Cenomanian), seen in cross-section (along the structural axis) at (b) as the top of the prospective Cenomanian section. The Pokur Suite contains two gas reservoir sands with shale and siltstone lenses, and shale seals. The Neocomian Vartov and Megion Suites contain at least 12 reservoirs with gas and condensate, and about half of these contain non-commercial oil. A commercial oil accumulation was subsequently discovered below the Valanginian zone in the late 1970s.

Structurally, as seen in the map and section, the field is a drape anticlinal structure with multiple culminations, and has undergone some later growth as noted. The Cenomanian section, including the reservoirs in this zone, consists largely of thick, interfingering clastic marine and non-marine sands and clays, with coal and amber deposits found throughout, suggesting swamp–paralic–deltaic environments that would tend to produce gas-prone source rocks. The paper by Grace and Hart (1986) has a useful reference list containing some papers (by USSR scientists) that have English translations, for those interested in more detailed studies of hydrocarbons in this area.

During 1984, the Urengoy Field produced at the rate of 542.9 million m^3 of gas per day (19,170 MMCFGD), which was one-third of the total natural gas production in the USSR for that year (*Oil & Gas Journal*, 1984).

CHAPTER 4

Fault-related Traps

4.1 General Comments

As with most classes of hydrocarbon trap, it is possible to consider a number of different aspects of fault-related traps. The most basic aspect is that of the geometry of the faulting concerned. There may also be questions concerning: the effectiveness of the sealing properties of the fault or its characteristics as a 'carrier' feature for hydrocarbons migrating upwards through the geological section; the relationship of the faulting to specific types of structural or depositional setting; and details of the termination of the fault both vertically and laterally. Other questions may involve the shape of the fault plane/zone itself—whether it approximates a true plane, or whether there are departures of a regular nature (as in listric faulting) or irregular nature (as in some growth faults where the fault plane is irregular, resulting in compressional features in the hangingwall), and also whether or not the fault movements were syndepositional during a part or the whole of its history.

These and other related matters will first be considered in a general way before 'real Earth' examples of fault traps are discussed.

4.2 Introduction

4.2.1 Fault Types and Geometry

A fracture is a break in a rock owing to deformation by brittle failure, in which there has been no relative displacement between the two sides of the break.

A joint is a surface of fracture within a rock in which there has been no relative displacement between the two sides. Joints occur in sets, with numerous fractures orientated in the same direction, and systems, in which two or more sets intersect. They can be produced by, for example, depositional, tectonic, or thermal stresses in the rock.

A fault is a fracture in which there *has* been relative displacement between the two sides of the break. This is the type of fracture with which this chapter deals.

The break appears as a fault plane separating the two blocks that have moved relatively to each other.

The use of the word 'plane' in this context is unfortunate, in that it implies a mathematically flat surface of no thickness; usually, a fault 'plane' is not flat, and has (sometimes very substantial) thickness. Often the fault 'plane' is a complex zone of anastomosing fractures separating large numbers of fault-block slivers. The zone may be many tens, or hundreds, of metres wide between two major fault blocks. Because of the friction involved during movements of the fault, the fault plane/zone may be filled with cataclasites—fault gouge or fault breccia, which are mixtures of finely comminuted, loose-textured rock flour, clasts, pebbles, etc. At greater depths and higher temperatures–pressures, mylonitization and pseudotachylitization may take place.

As noted in Chapter 2, a seal may be formed in a 'seal fault', by this material within the fault zone having a capillary entry pressure so high as to exclude hydrocarbons in any reservoir unit in contact with it at the fault plane. On the other hand, a 'juxtaposition' fault may be present, in which the fault plane/zone is so thin that the reservoir bed in one fault block is juxtaposed directly with an impermeable bed in the other fault block (see Fig. 2.13).

Recalling the main types of fault based on the geometry, and the 'direction of displacement' criteria, these are as follows, referring to the sketches in Fig. 4.1.

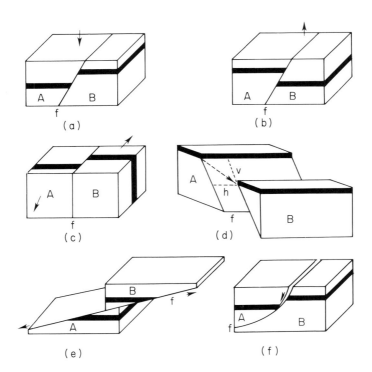

Fig. 4.1. Block diagrams showing the principal types of fault

4.2.2 Normal Fault

A normal fault is sketched in the block diagram of Fig. 4.1(a), being defined as a fault in which the fault plane dips towards the downthrown block (A), with the direction of relative movement shown by the arrow. Alternatively, it can be said that in this type of fault the hangingwall moves downwards, relative to the footwall (block A, *above* the angled fault plane, is referred to as the hangingwall; the block B *below* the angled fault plane, is the footwall).

Relative movement is mentioned as it is usually not possible to determine the details of absolute movement (i.e. did A move down, or B move up, or did both move in opposite directions, or by different amounts in the same direction?).

Normal faults are also referred to in some circumstances as extensional faults or gravity faults, these being descriptive of the cause of the applied stress that resulted in the fault. Gravity faults are typified by the bounding faults of graben features. The fault plane of a normal fault tends to be planar, with the fault trace in profile being straight, and the throw being constant with increasing depth. A seismic example is shown in Fig. 4.2, where a normal fault of small throw (about 10 ms two-way time down to the right) exhibits other characteristics that are typical of such faults in that the trace of the fault plane in the section is almost vertical, and the throw is approximately uniform with increasing depth (cf. later with growth faults). In this example, the fault appears to pass upward into a small flexure, which is a common mode of upper termination where the fault does not reach the surface.

As the seismic section is migrated, little is seen of the diffraction pattern often in evidence associated with faulting on an unmigrated section. This will be discussed later. The fault plane here seems to be very thin—approximating a true plane—and in the context of

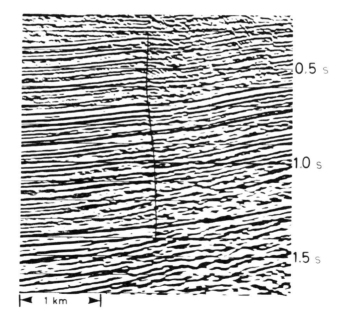

Fig. 4.2. Seismic section showing the trace of a normal fault of small throw. Reproduced by permission of Seismograph Service (England) Ltd

hydrocarbon trapping, we would here be looking for a 'juxtaposition' fault seal rather than a sealing fault with a thick 'fault plane' (Chapter 2).

Most planar normal faults have near-vertical fault planes. Some types—listric normal faults and growth faults—have fault planes that are steep in the shallow part of the section, but become progressively less steep, with listric (concave upward) or irregular curvature and approach the horizontal in depth; these usually flatten out (or *sole out*) at some detachment surface or interval, such as an unconformity, or an incompetent unit—a mobile salt or shale. Irregularities in the shape of the fault plane in these features may cause compression structures on the downthrow side, whilst regular listric curvature often results in antithetic and synthetic faulting in the downthrow of growth faults—points that will be discussed and illustrated in the part of this chapter devoted to listric faults.

In plan view, or in a borehole, a normal fault that has some dip on the fault plane shows a hiatus (bed cut-out) where the units across the plane have drawn apart horizontally.

A normal fault with a very shallow-dipping fault plane (that could be thought of as the extensional stress equivalent of a thrust) is known as a *lag*. Those faults are usually found only in areas of complex geology that include thrust faults, nappes, etc.—indeed, the terminology refers to the fact that the 'downthrow' side of a normal fault in the hangingwall of the thrust appears, relatively, to have 'lagged behind'.

The movement of a normal fault is described as 'dip-slip translational', since the relative movement of the two fault blocks is parallel to the dip of the fault plane, and all the parts of a fault block travel in the same direction (translation, as opposed to rotation). This description also applies to the movement of a reverse fault.

Fault-plane traces, as seen in vertical section (for both normal and reverse faults), appear to confirm the observation that the shape of a fault plane is influenced by the types of lithology through which it passes in the section. In practical terms, this seems to indicate that the downward trajectory of a fault plane alters (either shallowing or steepening) when passing, say, from a sand-dominated to a shale-dominated series. With increasingly high-resolution seismic data available, and studies of quarry faces, it should be possible to confirm or refute this interesting possibility. The effect must be related to the fact that the simple relationship between the horizontal and the orientation of the stress axes may hold good near the surface, but at greater depths, several factors can give rise to orientation variations in the stress axes, which in turn result in variations in the orientation of faults and the downward trajectory of the fault-plane traces in cross-section (Park, 1983). This adds another tool to the workshop of seismic stratigraphic analysis. There may be a slight effect of this nature in the example of Fig. 4.2 (note the very slight shallowing of the fault-trace dip, from 0.6 a down to about 1.0 s, below which it steepens again); this cannot be confirmed due to lack of lithostratigraphic information.

4.2.3 Reverse Fault

At a reverse fault, the relative movement of the fault blocks is in the opposite sense to the relative movement at a normal fault, although the movement type is still dip-slip translational. In the sketch illustration of Fig. 4.1(b), the hangingwall (A) is seen to have moved upward relative to the foot wall (B), and the fault plane dips towards the upthrown block. In general, reverse fault planes are near-vertical; conventionally, dipping at more than 45° (if less than 45°, the fault is a thrust). In plan view, if the fault plane is dipping, bed repetition is present due to the cross-over at the fault plane (i.e. if a vertical borehole is drilled at a favourable location, passing through the fault plane, the same section of the sequence will be encountered twice).

One of the problems attached to reverse faults is that although it is easy to visualize the movement of a normal fault, even with a near-vertical fault plane, under extensional stress, it is much more difficult to visualize the movement of a reverse fault with a near-vertical fault plane, under compression, especially where the throw is large and the translational movement that has taken place is substantial. The principal axes of stress are in the wrong direction. It is, therefore, an error (and a not uncommon one) to regard a reverse fault simply as the result of a compressive stress regime. It is most likely that many—perhaps most—reverse faults have come into being through causes other than simple compression. The reverse faults that are sometimes found bounding major crustal blocks give a clue that perhaps differential vertical movements are involved on some occasions, while in others, apparent reverse faults have been caused by strike-slip (wrench) faulting. It is only when the dip of the fault plane becomes small, and the fault is a thrust, that the action of simple compressional stress becomes credible.

In other cases, high-angle reverse faults are believed to have come into being through the application of stresses other than those of simple compression. They may be the result of a 'jostling together' of a series of fault blocks during a period of movement within a tectonically active zone, with some blocks finally coming to rest as horsts, bounded by high-angle 'reverse' faults. Such a 'jostling together' of blocks may be brought about by the injection or emplacement at depth of igneous material. Pinnate shears in a zone of monoclinal warping of the crustal rocks may also take the form of high-angle reverse faults (Sherbon Hills, 1972), while in some zones of compressional tectonics, a low-angle thrust may, with increasing distance from its origin, begin to steepen upwards and pass eventually into a high-angle reverse fault plane, somewhat analogous morphologically to the shape of a listric growth fault plane.

Although the convention is to regard 45° as the value of fault-plane dip separating high-angle reverse faults from low-angle (thrust) faults, the majority of reverse faults have fault-plane dips at angles that cluster about the near-vertical and near-horizontal. It is quite rare to come upon a reverse fault with a fault plane dipping at some intermediate angle—say between 40° and 50°.

An illustration of a feature falsely giving the appearance of a high-angle reverse fault is shown in Fig. 4.3. At the level of the upper marked seismic horizon (dashed line), there seems to be clear evidence of reverse faulting, with the 'upthrow' side to the right of the complex fault zone, which dips beneath the same side. However, at the level of the lower seismic marker the fault appears to be of normal type, with the fault plane (zone) dipping towards the downthrow side. (If the fault plane were vertical, it would seem that the throw was reversed in different time zones of the section). It is this apparent contradiction that marks the fault as of strike-slip type (to be discussed later), in which the fault movement has been largely horizontal, bringing into juxtaposition parts of the section—often of different thickness and/or with different dips—originally separated horizontally by some distance (often tens, or even hundreds, of kilometres). This is explained by the simple block diagram in Fig. 4.4, where a strike-slip fault (left-lateral, or sinistral) has displaced blocks 1 and 2 relative to each other. A thinning interval AB–A'B' is shown in block 2; after the fault displacement has occurred, it is separated from its original position (as indicated

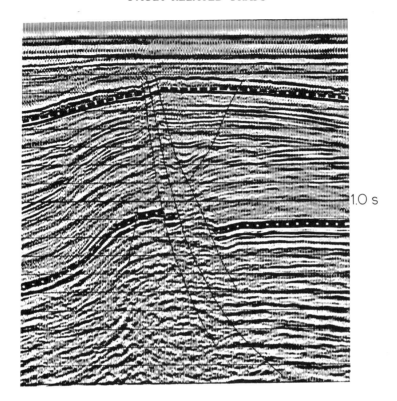

Fig. 4.3. Seismic example of a fault zone that appears to be reversed in its shallow part, but normal at depth. This is often characteristic of a strike-slip fault. Reproduced by permission of Seismograph Service (England) Ltd

by AB in block 1). If a seismic line were to be shot along the line X–Y, at the position Z on the cross-section there would be the large discrepancy seen in the interval thickness of AB across the fault plane. From the top of the interval AB it would appear that the downthrow side was in block 1, while from the base of the interval AB, the downthrow side would seem to be in block 2 (the fault plane being vertical). This may be further complicated when the shear stress causing the faulting is sufficiently large to produce compressional folding and buckling of the stratal units in the two blocks (tear faulting), in which case the apparent throw may vary in amount and direction along the fault strike. Anomalous throw directions, or variations in amount of throw along the fault strike, together with 'instant' apparent changes of bed thickness across a fault must always be regarded as evidence of possible strike-slip faulting, which will be considered next.

4.2.4 Strike-slip Fault

As noted in the previous part, a strike-slip fault (sometimes also referred to as a wrench, or transcurrent, fault) is a fault in which the relative movement between the fault blocks

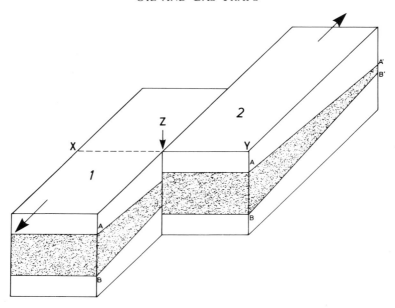

Fig. 4.4. Block diagram of a strike-slip fault showing sinistral (left-lateral) movement between blocks 1 and 2. This demonstrates why the shaded bed AB, thinning distally, would show an apparent 'instant' thickness change across the fault zone at Z on a seismic line shot along X–Y

is horizontal along the strike of the fault, as illustrated in Fig. 4.1(c). If an observer faces the fault plane (on either side), and the relative movement of the far fault block has been to the left, the feature is termed a left-lateral, or sinistral strike-slip fault. If the relative movement has been to the right, the feature is a right-lateral or dextral strike-slip fault.

With the increasing availability of high-resolution seismic data over the past two decades, it has been realized that the occurrence of this type of faulting is much more widespread than had hitherto been thought; in some areas, such as the southern North Sea, it can be the dominant fault type present, and can have important effects on the structure as well as complicating the stratigraphic relationships.

A geological situation in which strike-slip faulting is frequently recognized from overburden effects is where a competent and indurated basement series is overlain by a less-competent younger sedimentary series. If the basement rocks undergo strike-slip faulting, or older faults in the basement are reactivated in strike-slip movements, the less-compacted younger sedimentary series will show deformation patterns that are often recognizable as being the result of underlying strike-slip movements.

This situation was investigated experimentally by Riedel (1929), who used a slab of clay to represent a younger series overlying two wooden boards placed parallel and in contact to represent the basement series. Relative horizontal movement of the two boards, as shown in the sketch in Fig. 4.5(a), resulted in deformation in a zone overlying the plane between the two boards (representing a strike-slip fault plane). The deformation in such a zone is sketched in Fig. 4.5(b). Shear planes (often referred to as *Riedel shears*) in both sketches are lettered S, while tension gashes are lettered G in Fig. 4.5(a), and T in Fig. 4.5(b).

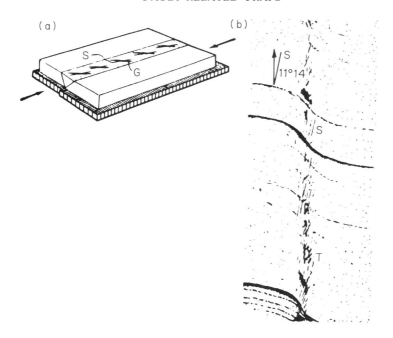

Fig. 4.5. (a) Block diagram showing Riedel shears 'S' and tension gashes 'G' produced in relatively undercompacted sediments by fault movements in a compacted basement series. (b) A plan view of the zone of deformation of (a) as seen in a model with Riedel shears 'S' and tension gashes 'T'. Both sketches from Sherbon Hills (1972), reproduced by permission of Chapman and Hall Ltd. The sketch (a) being after Riedel (1929) and sketch (b) after Cloos (1930)

The *en echelon* orientation of the resulting shear planes with respect to the deformation zone (and hence the underlying strike-slip fault plane) is important. In Fig. 4.6a(a), a portion of a seismic section is seen that shows a feature termed a 'flower structure' or 'tulip structure', consisting of two (or sometimes several) shallow branch faults diverging upwards from a single, deeper fault. In this instance, following the terminology used by Harding (1985) and others, the feature is referred to as a 'negative flower structure'. (For the moment, note that in this case, the reflections within the two diverging faults are low (as if downthrown in a graben) relative to those outside the branch faults. In positive flower structures, the reverse is the case, with the reflections within the upward-diverging fault system taking the configuration of an antiform, high shape relative to the adjacent stratal events outside the diverging fault system.)

In the sketch of Fig. 4.6a(b) the block diagram shows a basement fault that has undergone strike-slip movement in the sinistral sense, as indicated by the larger arrows. The stress applied to the less competent clastics above the basement (separated by the horizontal dividing line) has resulted in the development of a series of *en echelon* Riedel shears, with throw direction reversing on each side; three of these are shown in the plan view of Fig. 4.6a(c).

A profile across A–B in the plan view results in the section along the line A–B in the block diagram (b). This is a simple example—there can be many more upward-divergent

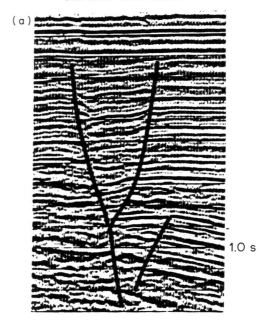
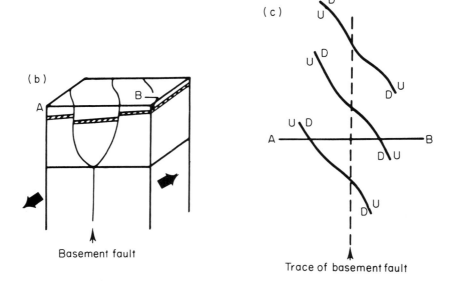

Fig. 4.6a. A negative 'flower structure' produced by Riedel shears *en echelon* in the deformed zone above a basement fault that has been initiated, or reactivated. (a) Seismic section; (b) block diagram, with the line of the seismic section in (a) along A–B; (c) plan view of the resulting series of Riedel shears, with throw directions reversing on each side (leading, in this case, to a *negative* flower structure). For an excellent discussion of such features, see Harding (1985)

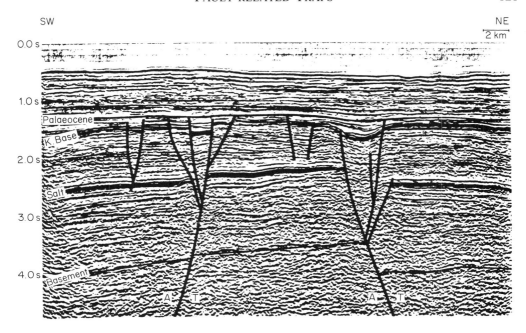

Fig. 4.6b. Seismic example showing several *negative* flower structures from offshore Norway. There is a great potential for fault traps in such complex situations. Reproduced with permission of Graham & Trotman Ltd (from Caselli, 1987)

shears present in a system than are shown here—but it indicates the nature of the faulting in the seismic section of Fig. 4.6a(a).

Excellent examples of 'negative flower structures' can be seen in the seismic section of Fig. 4.6b, from the Nordland Ridge area, offshore Norway. This is in a region of oblique-slip tectonics on the Mid-Norway Shelf, where 'basement' faults (the basement reflector is seen in the example at about 4.0 s at the left-hand end of the section) have been reactivated to produce the flower faulting. The late period of the reactivation is indicated by the effects on the shallow Palaeocene reflection marker above the faulting, particularly well seen above the right-hand major fault system. The potential for fault trap situations in these complex flower structures is clear; the potential for stratigraphic trapping in the shallow sediments owing to the reactivation should also be recognized.

It is worth noting that for the sake of simplicity, the discussion so far has involved only *translational* movements in faulting. *Rotational* movements may also take place either in conjunction with translation, or as the sole mode of movement, as a result of complex dynamic situations.

Some of the world's best-known faults are of strike-slip types. Although perhaps not quite 'household words', the San Andreas Fault of California, USA, and the Great Glen Fault of Scotland are two names that are familiar to many non-geologists. (The recently published *Seismic Atlas of Structural and Stratigraphic Features* (1988) of the European

Association of Exploration Geophysicists has an excellent seismic example of the latter fault at an offshore location.)

A combination of dip-slip and strike-slip movements, where there are elements of both types of translational movement, and/or of rotational movement, results in an *oblique-slip* fault, and an example is shown in the sketch of Fig. 4.1(d). The arrow on the dashed line running obliquely down the fault plane between blocks A and B indicates the resultant direction and amount (net slip) of relative movement between A and B; the related relative dip-slip (v) and strike-slip (h) movements are shown as dashed lines on the fault plane. This may be either an obliquely downward movement or an obliquely upward movement.

In reality, it is likely that most faults are oblique-slip types with v/h or h/v distance ratios being very small, in most cases nearly approaching true strike-slip, normal, or reverse faults.

4.2.5 Thrust Fault

A thrust fault, already described as a special case of reverse faulting with a very low angle of fault-plane dip, is illustrated in the sketch of Fig. 4.1(e). Thrusts are found typically in areas of compressional tectonics—in complex fold and thrust belts, such as in the Alps in Europe, the North-west Highlands of Scotland, and the southern Appalachian Mountains of the USA. In some places, where the thrusting is particularly intense, *imbricate structures* develop—a series of thrust sheets overlapping one another like roof tiles. Such an area does not appear, *prima facie*, to offer favourable trapping conditions for hydrocarbons because of the 'busy' nature of the tectonics, but examples of hydrocarbon traps in complex structural area of this type will be shown and discussed.

Some basic terminology and geometry of complex thrust faults are shown in the schematics of Fig. 4.7. In (a), sketches (i) and (ii) show two stages in the development of imbricate structures (or 'piggy-back thrusts') that evolve in the footwall. A similar type of faulting, in which the thrusts develop in the hangingwall, is known as overstep thrusting. In (b), the development of a *pop-up* structure (z) is shown; behind the frontal ramp (x) of the main thrust plane, an antithetic thrust or *backthrust* (y) develops. In (c), the basic terminology of thrust faulting is shown. Note how a fold occurs in the hangingwall due to the footwall ramp. See also Fig. 3.14.

Because of the low dip angle typical of the thrust plane—usually only a few degrees from the horizontal—it is often difficult to identify a thrust in seismic data with low-dipping or horizontal stratal events that are close to the dip of the thrust plane, and borehole information may be required to confirm a tentative identification.

A (fortunately) somewhat rare feature that can simulate a thrust, and may be impossible to detect seismically, is the special type of normal fault known as a *detachment fault*. This occurs where a mass of rock at a contemporary Earth surface becomes detached along a bedding plane, and slides under gravity for a large distance. The bedding plane slip surface usually coincides with an incompetent unit—an evaporite or argillaceous bed, for instance. Billings (1972) mentions an example at Heart Mountain, Wyoming, USA, which in places is a detachment fault, and elsewhere becomes transgressive. Where it is a detachment, net slip is estimated at 10 miles or more. There are numerous difficulties in identifying such a feature, even by field geological studies: there is generally no throw, the slip surface

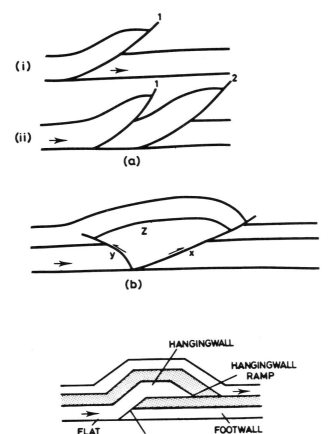

Fig. 4.7. Basic geometry and terminology of complex thrust faulting

coinciding as it does with a bedding surface, and the *breakaway* (the surface bounding the rocks that remained in place—the origin of initiation of the detachment) is usually difficult or impossible to identify after tectonism and erosion in an area.

Clearly, where such a feature is buried by later deposition, it would be very difficult to identify in seismic data, and if recognized as a feature, would normally be categorized either as a palaeogeomorphic surface, or a thrust.

4.2.6 Growth Fault

The term *listric* (or sometimes *lystric*) refers to the concave-upward curvature that is, in cross-sectional trace, the shape taken by (for instance) the slip surface in a terrestrial landslide; this shape is near to the vertical at the head (high point) of the slide, curving towards the horizontal at its toe. An example of the trace of a normal fault with listric curvature is shown schematically in Fig. 4.1(f).

Listric surfaces are encountered in various geological contexts. Previously it was noted that some thrust planes develop a concave-upward (listric) curvature distally as they approach the surface to terminate as a high-angle reverse fault. Palaeolandslip (or submarine slide) features occur, where the slip surface is a gravity controlled listric normal fault (although not a growth fault—syndepositional movement is not necessary in such features). Also, and most importantly for this discussion, normal fault planes may take on listric curvature during syndepositional movement, resulting in a listric normal *growth fault*.

The growth fault is of great importance in hydrocarbon exploration. It tends to occur not singly, but in sets of several main faults, and in areas of instability, either tectonic or otherwise. The most common depositional environments (although not the only ones) in which growth faulting is encountered are: (i) deltas; and (ii) evaporite basins with mobile salt rock, or where salt has been removed in solution. Both of these environments are closely associated in many parts of the world with the occurrence of hydrocarbon accumulations.

4.2.6.1 *Morphology and terminology of growth faulting*

Growth faulting tends to occur in unstable situations, where there is a high rate of deposition, and where subsidence is taking place.

In the case of a deltaic environment, the subsidence (at least initially) is of tectonic origin, and is relative—it may be actual subsidence, or apparent owing to rise of sea level. Later, the increasing weight of rapidly accumulating sediment may become an important factor, and perhaps the dominant factor, in the subsidence process, but it should not be forgotten that instability must be present initially in order to get the subsidence/deposition mechanism started.

In a salt basin, where rapid sediment deposition is occurring, instability and subsidence (although they may in part be due to tectonism) are promoted by lateral withdrawal of mobile salt, or its gradual removal by dissolution, accompanied by progressive overburden subsidence.

A schematic sketch cross-section of a growth fault as it might develop in a deltaic situation is shown in Fig. 4.8; later in this chapter, some seismic examples of actual growth faults

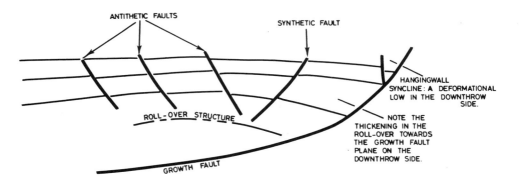

Fig. 4.8. Schematic section showing basic geometry and terminology of a growth fault

will be seen. In this figure, the listric curvature of the trace of the growth fault plane is seen; this plane often *soles-out* (i.e. flattens and merges at depth) at a detachment surface that may be an unconformity, an evaporite bed, or a mobile shale unit.

Frequently, the fault plane is less regular in shape than seen here, with antiform 'bulges' towards the hangingwall that produce compressional structures on that side of the fault.

One of the consequences of the listric, non-cylindrical shape of the fault plane is that after movement commences, the hangingwall tends to draw away from the footwall. The stresses due to gravity that are thus imposed on the hangingwall prevent an actual gap from developing at the fault plane; instead, the hangingwall 'settles down' on the footwall. The deformation involved includes ductile shear, brittle fracture, and faulting. Typically, the developing stresses are relieved by antithetic faults (faults in the hangingwall with fault planes dipping in the opposite direction to the main fault-plane dip). Synthetic faults in the hangingwall (dipping parallel to the main fault plane) are incipient growth faults that have not developed to their full extent and are usually inhibited from further evolution by the stress pattern resulting in antithetic faulting.

A brief explanation is desirable for the terms 'antithetic' and 'synthetic' with reference to subsidiary faults in a growth-fault system. The original usage of these terms refers to features with fault planes that dip in the same direction (synthetic faults) or in the opposite direction (antithetic faults) to the dip of the bedding of the strata in which they occur. Related to growth faults, however, the definitions given in the previous paragraph are generally accepted now.

Major growth faults often occur in sets of 'down-to-basin' type, as in the Niger Delta sediments, offshore Nigeria. In plan view they are frequently curved, with the concave side towards the basin, so that the whole fault-plane surface is spoon-shaped; this again is similar to a terrestrial landslip surface as observed in many examples.

The development of a growth fault involves syndepositional movement—although it is seldom either continuous or uniform. There are periods of deposition during which the fault does not move (indicated by uniformly thick intervals in the hangingwall) and others when movement takes place while deposition is under way (indicated by intervals in the hangingwall that thicken towards the main fault plane). The reason for this thickening is partly the downward movement of this fault block, and partly the pattern of sedimentation. Winnowing of sediments on the footwall (upthrow) depositional surface, with gravity and current effects, tends to cause the sweeping-over of sediments into the depositional trough on the hangingwall side.

This also seems to be accompanied by sediment redistribution due to current action from the higher areas of the depositional surface on the hangingwall side, tending to sweep further sediment down the shallow slope towards the fault-plane scarp. Such effects, together with the action of bottom contour currents, lead to a gradual interval thickening in the hangingwall (towards the fault) that is seen at some levels. This, together with the downward movement of the hangingwall block, results in an anticlinal feature in the latter that is known as a *roll-over structure*, and is one of the features typical of growth faulting that can be recognized in seismic data. Other such features are: (i) the increase in throw with depth usually seen; (ii) the listric shape of the fault-plane trace; and (iii) the frequent—although not invariable—presence of subsidiary antithetic and synthetic faulting in the hangingwall. Usually, the presence of some or all of these features makes the identification of growth faulting in the seismic section relatively easy.

It is obvious that the growth fault situation is one of great potential importance in hydrocarbon exploration. The regional environment often includes organic connections into the basin with marine shales of source-rock type, and shorewards with paralic and swamp-type sediments that are also potentially prolific sources of petroleum.

Within a delta sequence there are many favourable sedimentological features—distributary channel sands, point bars, beach bars, offshore barriers and other sand bodies depending on the type of delta—as well as silty argillaceous sealing units. It should also be remembered that in a set of growth faults, the footwall of one fault is the hangingwall of the next. The footwall should not be ignored from the sedimentological viewpoint; often it is the fines that are swept over the edge into the hangingwall trough, leaving coarser sand fractions behind.

Structurally, in addition to the possibilities proffered by the roll-over structure as a closed high feature, there are many potential traps related to the faulting, both major and subsidiary.

In a salt basin situation, a listric growth fault is shown in Fig. 4.9, with the base and top of the salt interval at B and T. The growth fault 'f' is seen soling-out at the base of the salt (solid black arrow). The situation has been caused by lateral flow of plastically mobile salt to the right, into a pillow that is off-section. This flow was probably initiated by strike-slip movement of a major deep fault at 'F'. Note that there is an interval of thickening towards the fault plane in the hangingwall from the level of the open arrow (on the right) to the 1.0 s time line. Between the 1.0 s time line and the upper seismic marker

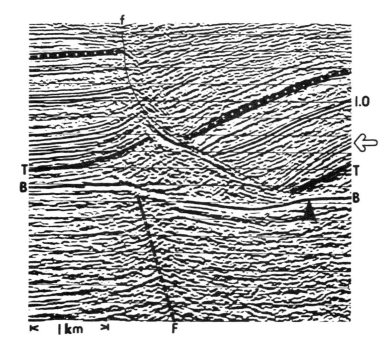

Fig. 4.9. Seismic section of a listric growth fault related to mobility in an underlying salt layer. B, T = base and top of salt interval. Reproduced by permission of Seismograph Service (England) Ltd

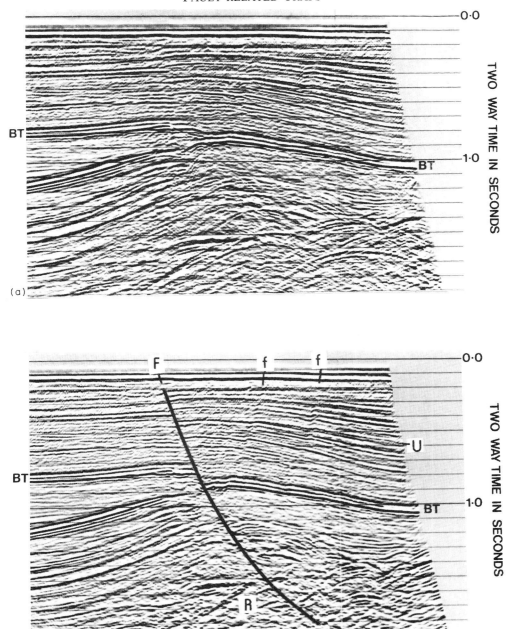

Fig. 4.10(a). Unmigrated seismic example of a listric growth fault, showing coherent noise events related to bedding truncations, fault-drag folds, etc. (b) Migrated version of the seismic section in (a). F = trace of listric growth-fault plane; f = antithetic fault traces; R = reflected refraction event related to the main fault plane F; BT = Base Tertiary; U = unconformity. Note that migration has removed much of the noise. Both (a) and (b) reproduced by permission of Seismograph Service (England) Ltd

(dashed line above 1.0 s), there is a relatively uniform interval above which the situation is not very clear, but would seem to include a thickening interval in the shallow part of the section. The depositional pattern shows that there was syndepositional movement over part of the fault's history, but significant periods of deposition when no fault movement took place.

The fault plane in this case is of irregular shape, and there are several unconformities/disconformities in the section that produce a number of gaps in the sequence. The possible reason for the marked salt uplift in the footwall side (left) will be discussed in a later treatment on the results of salt movement.

In Figs 4.10(a) and (b), a growth fault with a regular listric-fault-plane shape is seen; (a) being the stacked (unmigrated) version, and (b) the migrated version of the data. In the latter section the main fault plane is indicated at F, and there are several antithetic faults in the hangingwall side. The two most obvious of these are shown at f. The straight, steeply dipping event R in the bottom centre is a reflected refraction associated with the main fault plane. The vertical displacement of the Base Tertiary group of reflections (BT) is greater than that for shallower events up to and above the unconformity U, showing the typical increase of throw with depth that is common with growth faulting. The beginnings of a roll-over structure can be observed in the hangingwall, with thickening of some intervals towards the fault plane. The numerous fault diffraction and shape-response hyperbolas seen in the stacked section (Fig. 4.10(a)) are of interest, and should be looked at again after the later section that deals with these phenomena has been read.

A good general study of growth faulting has been written by Crans and Mandl (1980-81).

4.2.7 The Fault 'Plane' as a Seismic Reflector

Before continuing the consideration of various fault types, some thought will be given to the nature of the zone between fault blocks usually referred to as the fault 'plane', and the features of the fault plane that make it act as a reflector.

As noted previously, the term 'plane' is a misnomer, but by tradition continues to be used as the name for this feature. Here, it will also be used except where it is desirable to emphasize its non-planar nature, when it will be referred to as the fault 'zone' (not to be confused in this context with the alternative use—to refer to a zone in which faulting is present).

The nature of the fault plane must be of prime interest to the explorationist. Not only does it separate the stratigraphy in an area into compartments, but it also acts either as a seal or a conduit for migrating fluids, including petroleum.

When fault movement occurs, the massive nature of the forces involved, together with friction, cause the grinding and comminution of material of the bounding rock faces within the fault zone. The products of crushing, grooving, striation, and slickensiding vary from coarse fragments (breccia) to fine rock flour (gouge), forming *cataclasites* near the surface. At deeper levels, with increasing temperatures and pressures, recrystallization and flow of the materials may take place to form mylonite, ultramylonite, and pseudo-tachylite—glassy rock types. At even greater depths, the material may be incorporated in a zone of ductile shear.

In some instances, a type of deformation known as *fault drag* occurs, where the stratal units in one fault block, adjacent to the fault plane, are flexed downwards, while those in the other fault block are flexed upwards. This is normally attributed to incomplete restoration after the ductile shear phase that precedes brittle fracture in the rocks; however, the writer also believes that the frictional forces involved contribute to the drag effect in some instances—perhaps at deeper levels where the rocks are more ductile (Jenyon, 1987b).

The width of the fault plane/zone varies greatly. Some zones approximate a true plane, being only the width of a hairline crack in cross-section, while others (like that of the Great Glen Fault in Scotland) may be a mile or more wide zone of crushed and mylonitized rock material. Clearly, in seismic data, the width of the effects on the section produced by a wide fault zone will depend, *inter alia*, on the width of the 'real Earth' zone, and also on the *direction* of the seismic line across it.

The most common diagnostic feature of a fault in seismic data is the vertical displacement of stratal events (i.e. seismic events produced by the bedding-plane interfaces between the stratal units) across the fault zone. The next most common feature is probably the occurrence of diffractions at various levels up the fault-plane trace, produced at truncated bed-ends; these will be discussed later. Whether the fault plane/zone itself produces a recognizable seismic event is dependent on certain factors that will now be considered.

4.2.7.1 Dip of the fault plane

Consider the migrated seismic section examples of Fig. 4.11 and Fig. 4.12. In both of these figures, a relatively small fault is producing a clear fault-plane reflection at F. In the

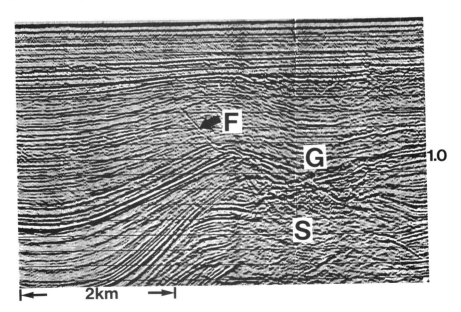

Fig. 4.11. A fault-plane reflection is seen at F. This fault is the result of dissolution and removal of salt from the crest of the salt structure at S, with concomitant formation of a collapse graben at G. The true dip of the fault plane must be less than about 45° (the dip, it must be remembered, is subject to approximately 2.5:1 vertical scale exaggeration in this, and most other, seismic examples).
Reproduced by permission of Seismograph Service (England) Ltd

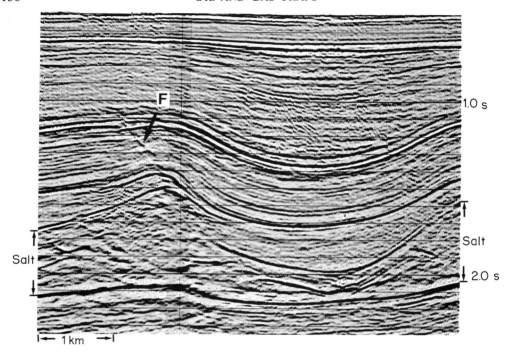

Fig. 4.12. A fault-plane reflection is seen at F, less continuous than that in Fig. 4.11. In this case, the fault is the result of subsidence of the overburden into the small salt-withdrawal basin present (the mobile salt interval being indicated). Reproduced by permission of Seismograph Service (England) Ltd

case of Fig. 4.11, the fault has been caused in the overburden at the margin of a collapse graben G due to dissolution removal of salt rock from the crest of a salt pillow at S. In Fig. 4.12, the fault occurs in the overburden at the margin of a small salt-withdrawal basin resulting from the lateral flow of salt in the interval marked on the section.

If Fig. 4.13 is now examined, with a clean version above at (a), and an interpreted version below at (b), it can be seen that the strong faulting in the zones marked A, C, and E does not in any instance produce a fault-plane reflection.

Even more striking is the observation in Fig. 4.14 that the major listric fault J shows no fault-plane event in its shallower part, but does show such an event as a strong reflection group at deeper levels (i.e. below horizon T—the top salt). At (b) is the interpreted version of the section at (a).

Why are there differences in the seismic response of these faults?

If a vertical feature in the geological section is considered—e.g. a fault plane, or a salt diapir flank—it is easy to understand that seismic signal energy travelling downwards from the surface and impinging on this vertical feature (angle of incidence = angle of reflection) would not be reflected back to the surface, but would rather be reflected downwards, so could never be recorded at the surface by seismic detectors. Depending on the geometry of the seismic detector spread at the Earth's surface, and the depth of the steep feature, this will still be true if the dip angle of the latter is reduced from the vertical to some lesser value—usually a limiting

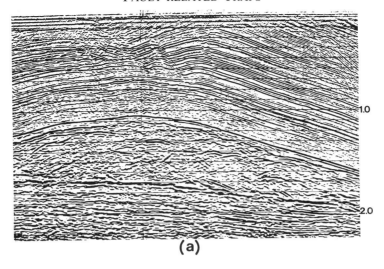

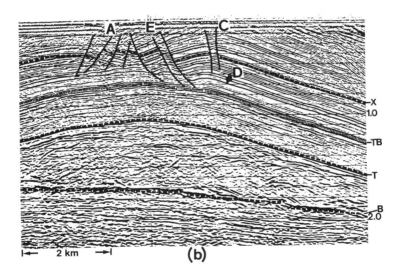

Fig. 4.13. Uninterpreted version of seismic section (a) and interpreted version (b) showing absence of fault-plane reflections associated with clear faults at A, C, and E. B, T = base and top of mobile salt interval; TB = top Bacton; X = Tertiary reflection marker; D = a Mesozoic marker. Reproduced by permission of Seismograph Service (England) Ltd

value around 45°, depending on the exact spread geometry, seismic velocities, seismic processing route, and depth of the feature involved. Thus the standard seismic section cannot be expected to show steeply dipping 'real Earth' features where the dip is greater than about 45°. Clearly this will rule out the majority of fault planes, which are vertical or near-vertical.

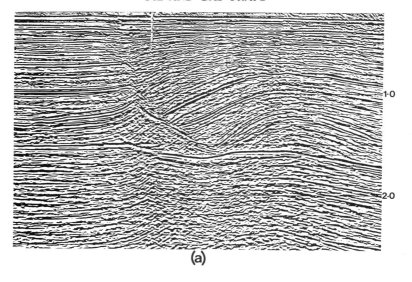

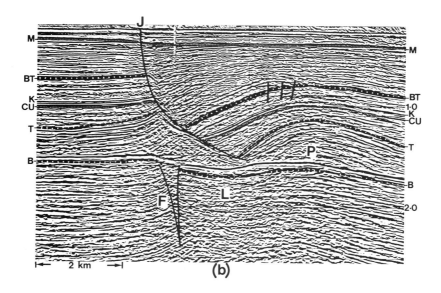

Fig. 4.14. Another seismic example showing the absence of a fault-plane event in the shallow part of the trajectory of the listric growth fault J, but with a strong fault-plane reflection group at deeper levels of the same trajectory. B, T = base and top of mobile salt interval; P = salt pillow; F = deep strike-slip fault; L = pull-down anomaly owing to salt withdrawal and the subsidence of low-velocity overburden; CU = Late Cimmerian Unconformity; K = top of Lower Cretaceous; BT = Base Tertiary; M = intra-Miocene Unconformity. Reproduced by permission of Seismograph Service (England) Ltd

FAULT-RELATED TRAPS 133

The seismic fault-plane events F in Figs 4.11 and 4.12 may appear to be dipping at more than 45°, but it must be remembered that there is scale distortion; this is usually about 2:1 or 2.5:1 vertical to horizontal. Thus, in the case of these two examples, the dips of the 'real Earth' fault planes must be less than some limiting value about 45°. In the case of Fig. 4.14, the fault-plane seismic event only becomes visible where the 'real Earth' dip becomes less than the limiting value (having regard to the vertical scale distortion). This is one of the reasons why fault plane events are relatively infrequently seen in seismic data.

4.2.7.2 Thickness of the fault 'plane'

It has been determined (see e.g. Widess, 1973) that the resolution of thickness of a stratal unit in seismic data depends on the bandwidth of frequencies in the basic seismic wavelet. Separate reflections from the top and base of a stratal unit, which are necessary for the resolution of an individual interval, can be separated on the section where the interval thickness is one-eighth (or more) the wavelength of the central (i.e. dominant) frequency of the seismic wavelet at that level in the seismic section.

A reflection group from a thin bed shows top and bottom reflections if the bed is sufficiently thick. For a thinner bed, the top and bottom reflections merge to give a broad pulse, which narrows if the bed gets thinner. At the 'critical thickness', the regular pulse width is attained. Thereafter, for still thinner beds, the pulse width remains the same but the amplitude changes; the thickness is said to be *amplitude-encoded*. If the top and bottom reflections have opposite signs (polarities), the amplitude decreases with thickness decrease. If these reflections are of the same sign, amplitude increases as thickness decreases (see e.g. Meckel and Nath, 1977).

Putting this another way, in order that a fault 'plane' should produce a discrete and recognizable reflection event, in addition to having a sufficiently large acoustic impedance contrast with units in both fault blocks, it must be a zone of thickness at least-one eighth of the wavelength of the dominant seismic frequency of the basic seismic wavelet in the data at that level in the cross-section (the *critical thickness*). Meckel and Nath (1977) quote critical thicknesses within a range of 50–80 feet (15–24 m) for typical seismic wavelets and representative subsurface seismic velocities, and one would probably not be too far out in assuming this range of thicknesses as being the minimum for a typical fault zone that is producing a good, discrete fault-plane reflection.

In the case of the fault-plane reflection in Fig. 4.12, which exhibits a strong but intermittent trace, it would seem likely that the acoustic impedance contrast of the material in the fault zone related to beds in one or both of the fault blocks is being reduced repeatedly, so destroying discrete segments of the reflection event. One could imagine, for instance, some form of cyclic sedimentary sequence being present in the fault blocks on either side of the fault plane/zone.

4.2.8 Diffractions and other Hyperbolas

An in-depth discussion of the generation of hyperbolic coherent noise events in seismic data is not within the terms of reference of this work, and must be sought elsewhere (e.g. Sheriff and Geldart, 1982; Jenyon and Fitch, 1985). However, some brief mention of these matters should be made here, since hyperbolas are effects often associated with faulting and other phenomena as seen in the seismic section, and may assist in diagnostic evaluations of some structures.

Hyperbolic events on the seismic section are caused by a variety of features that act as secondary energy sources, scattering back the propagating seismic energy in a coherent way. The causative features may be (in the case of marine seismic data) sea-bed irregularities including wrecks, truncated beds at a fault plane or diapiric structure, buried reefs, sharp flexures, monoclinal dip changes, and tight synform shapes, for example. Anywhere that a sudden, or rapid, change in the physical situation occurs at the surface, or in the subsurface, hyperbolic events may be a part of the seismic response.

In Fig. 4.15, which is a stacked (unmigrated) seismic section, a solitary hyperbola can be seen at A, related to a small graben that has developed in the overburden above a salt pillow (the base and top of the mobile salt interval are at B and T), due to extensional stress-zone faulting during uplift of the pillow.

Figure 4.16 is a migrated version of the same data. As can be seen, the migration process tends to eliminate the diffractions, attempting to collapse the hyperbolas back to their point of origin. This is often not perfectly achieved, owing to the model on which the migration process is based being insufficiently close to reality, and/or the seismic velocities used in the model not being very precisely determined.

The numerous fragmentary diffraction hyperbolas within the mobile salt interval in the stacked example of Fig. 4.15 are also noticeable. These are emanating from fragments of competent strata (carbonates and anhydrites) shattered and deformed by the salt movements, and 'floating' embedded in a mass of salt rock. As can be seen by comparing the two versions, the diffraction events appear to continue the true reflections, curving away and dying-out laterally. At C there is another (rather faint) family of hyperbolas that are

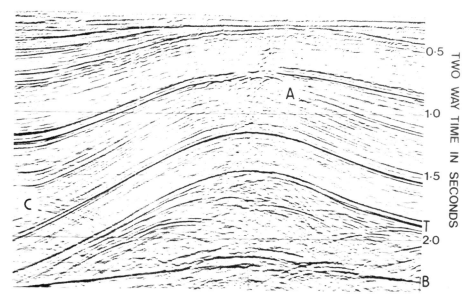

Fig. 4.15. Hyperbolic event (A) in unmigrated seismic data related to stress-zone faulting above a salt pillow. B, T = base and top of mobile salt interval. Reproduced by permission of Seismograph Service (England) Ltd

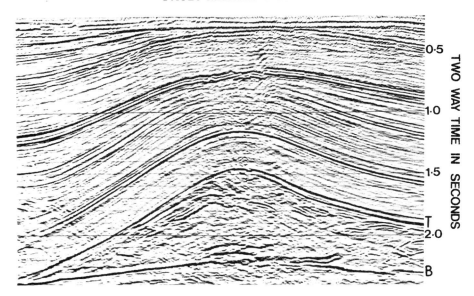

Fig. 4.16. Migrated version of Fig. 4.15 example, showing the resulting strong attenuation of the hyperbolic event A, and the small, tight synform shape that produced it in the unmigrated date. B, T = base and top of mobile salt interval. Reproduced by permission of Seismograph Service (England) Ltd

probably what might be called a 'shape-response' pattern produced by the rather tight synform shape in the overburden here; this is what is commonly referred to as the 'bow-tie' effect discussed again later. Shape-response hyperbolas are so-called here because they are generated in response to specific subsurface shapes—notably tight antiform/synform shapes, whether due to folding or to some other cause—rather than being the response to the truncated bed-ends at a fault plane, when fault diffraction hyperbolas are seen.

Shape-response hyperbolas are observable very frequently in the seismic section, and when associated with faulting, are often mistaken for diffraction hyperbolas. Frequently, tight antiform/synform shapes are associated with the fault zone—resulting from deformation such as fault drag, or fragments of bedded strata within fault slivers—and often these features result in shape-response hyperbolas. All the hyperbolas discussed so far have, in fact, been shape-response types.

In the upper example of Fig. 4.17, there is one quite extensive, and one partial, family of hyperbolas related to graben faulting, where segments of stratal units appear to have subsided—probably in this case due to dissolution and removal of thin evaporite beds in the underlying section, followed by subsidence of the overburden to form a graben. These hyperbola patterns are probably of the shape-response type, associated with synform-like configurations in the subsided material. Such extensive patterns are sometimes seen, whereas on other occasions only half the hyperbola is visible, involving only one or two individual cycles of energy, as in Fig. 4.17 (b). Two distinct 'legs' of one half of the pattern (A) are seen emanating from the upper strong seismic event; these are certainly true diffraction hyperbolas associated with specific reflecting horizons, and the result of either minor faulting, or sharp, very localized facies change.

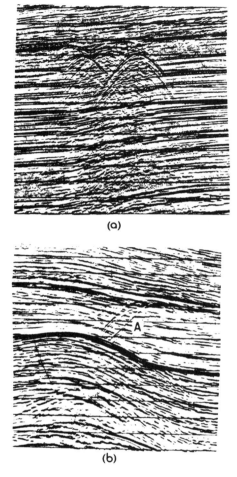

Fig. 4.17. Example (a) shows an unmigrated section with families of hyperbolic events produced by the response to synform shapes in the graben faulting present. Example (b) shows two separate 'legs' of what are probably true diffraction hyperbolas from bedding truncations at minor faults, or sharp, very localized facies changes, which can result in the same effects. Reproduced by permission of Seismograph Service (England) Ltd

In Figs. 4.18 and 4.19, the relationship of fault diffraction hyperbolas to the fault plane can be seen. The diffraction hyperbola is tangential to the generating reflector at the point where the reflection ends (truncated against the fault). The vertex of the hyperbola thus represents the true (migrated) position of the truncated end of the 'real Earth' units of which the reflector is the seismic expression. The reflection passes smoothly into the diffraction. This can be seen in unmigrated (Fig. 4.18) and migrated (Fig. 4.19) versions of the data, where a normal fault cuts a Base Tertiary group of reflectors (BT) and terminates upwards at an unconformity (U). Note that: (i) on the downthrow side of the fault in Fig. 4.19 there is some evidence of fault drag at the termination of the truncated reflectors against the fault plane; (ii) the trace of the normal fault zone in the section is dipping steeply

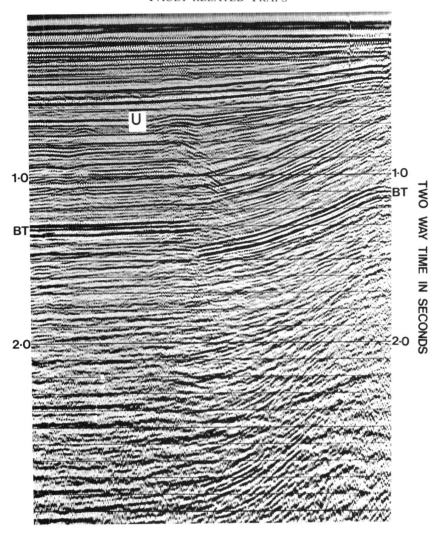

Fig. 4.18. Unmigrated seismic section showing truncation and shape-response hyperbolas associated with a normal fault. At U there is an unconformity surface, and at BT the strong group of events marking the Base Tertiary–Top Chalk contact. Reproduced by permission of Seismograph Service (England) Ltd

and indicates an approximately planar surface (both typical characteristics of an extensional normal fault); and (iii) the throw is roughly constant with increasing depth, again typical.

With fault diffractions, the phase of the hyperbola depends on velocity relationships, both vertical and lateral (across the fault plane, or with the material in the fault 'plane' if the latter is a thick zone). These relationships are discussed in Sheriff and Geldart (1982) and Jenyon and Fitch (1985), and also by Trorey (1970). In some cases a forward hyperbola (carrying the reflection beyond the truncation, and across the fault plane) is of positive polarity and the related backward hyperbola (if seen) is of negative polarity. In other

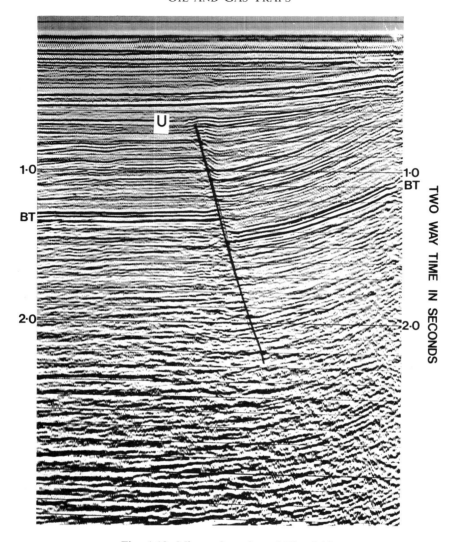

Fig. 4.19. Migrated version of Fig. 4.18

situations, the opposite is the case. Frequently, however, only half of the hyperbola (usually, but not invariably, the forward half of the curve) is seen, as in Fig. 4.17(b).

However, it is quite usual to see hyperbolas that show no change of phase, as in Figs. 4.15 and 4.17(a).

It is true to say that there is still much that is not understood about hyperbolas generated in seismic data, and many of the effects seen await a rigorous mathematical treatment. As regards the uniphase hyperbolas just mentioned, they sometimes appear to be associated with faulting, but also appear in relation to other phenomena. It is interesting to note the results of some experimental work carried out by means of ultrasonic sources with fault models using metal plates embedded in plexiglas, and other materials. Where wavelengths are not properly scaled, the effect of a slight burr on the end of the plate can be to generate centres of curvature, which would certainly produce hyperbolas with no phase change. 'In real-earth faults it is frequently the case that beds are pulled down or pulled up into a fault plane, thus creating sharp centres of curvature on steep truncated ends of beds ... (giving) rise to hyperbolas with no phase change.' (Jenyon and Fitch, 1985.) From the writer's experience, the 'burr on the plate' situation tends to arise in the 'real Earth'. This can happen, for instance, where a small graben drops some stratal unit down so that it is effectively surrounded (in cross-section) by the same lithology as is immediately beneath.

The effect can be seen by comparing unmigrated and migrated versions (Figs 4.15 and 4.16) in the location of the hyperbola at A.

An analogous situation arises where there is a sudden dip change, or a rather tight synform shape in the subsurface, as in Fig. 4.20. At the level of event A in the lower centre of the example, there is a sharp synform shape that has generated a family of partial hyperbolas, one of these being indicated by the arrows. Such a shape-response family of curves is known as the 'bow-tie' effect, for obvious reasons. It is seen on unmigrated sections where tight synform shapes are present in the subsurface, and is the result of data being displayed on this type of section vertically below the detector group receiving the returned signal, even though the latter may have come from a dipping reflector of substantial lateral displacement.

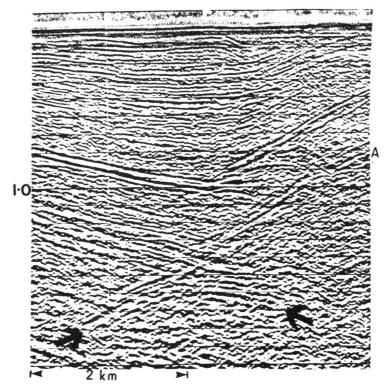

Fig. 4.20. Family of hyperbolas (arrowed) produced by a tight synform shape at the level of event A in the lower centre of the section. Reproduced by permission of Seismograph Service (England) Ltd

140 OIL AND GAS TRAPS

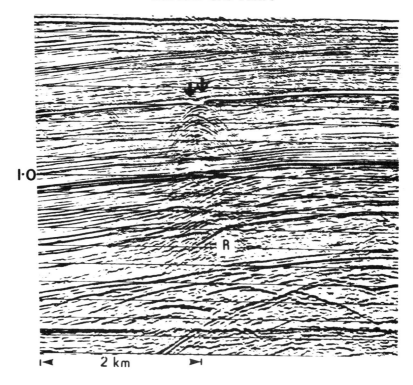

Fig. 4.21. Two alignments of hyperbolas (arrowed) related to the two bounding faults of a small graben structure. The many hyperbolic events in the lower part of the section emanate from truncated ends of competent bands within the salt interval that has been mobile, fracturing and disintegrating the competent bands. The steep, straight event R is a reflected refraction caused by either a minor fault, or a small collapse feature. Reproduced by permission of Seismograph Service (England) Ltd

Where a fault zone is wide, or the seismic line crosses a fault at an acute angle, each side of the fault can generate its own set of hyperbolas, so that two (or more, if the fault zone is complex and comprises a number of fault slivers) alignments of vertices of the hyperbolas can be seen. This is illustrated in Fig. 4.21, where the arrows indicate two such alignments related to the bounding faults of a small, narrow graben. Here also, in the lower part of the section, many hyperbolas can be seen associated with shattered fragments of competent units within a mobile salt interval. The straight, steeply dipping event marked R is not part of a diffraction pattern, but is a reflected refraction associated with a small fault at this level.

For any specific fault, the shape of the hyperbolas changes according to the depth at which they are generated, and also to the angle at which the seismic line crosses the fault zone. Where the line crosses at right angles, the resulting diffraction hyperbolas are of maximum curvature at any level, whereas if the line runs parallel to the fault zone, the resulting diffractions will be flat and horizontal (causing severe interference with the true reflections). At angles in between, the hyperbolas vary continuously in shape, becoming progressively flatter as the angle of intersection of the line with the fault zone becomes

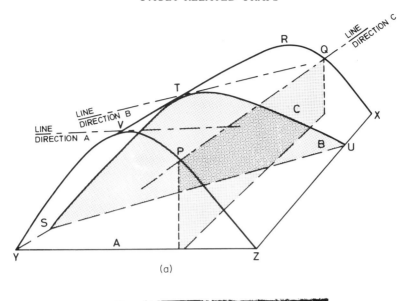

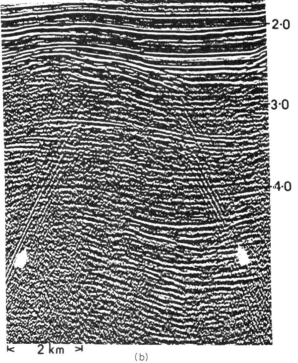

Fig. 4.22. (a) 'Hyperbolic prism' concept of the varying shapes of diffraction hyperbolas related to faulting. (b) Family of hyperbolic curved events (arrowed) of the type produced by sea-bed irregularities and other causes not related to subsurface structure. Reproduced by permission of Seismograph Service (England) Ltd

increasingly acute. A useful concept is to imagine that a hyperbolic prism lies along the fault strike, with its vertex (in cross-section) at the level of interest in the section, and that its axis is of infinite length. A vertical section through the prism along the seismic line will show the shape of the diffraction hyperbola.

The relationships are illustrated in Fig. 4.22(a). The line VTR, the crestal line of the hyperbolic prism, is the line of truncation of a reflecting horizon at a fault. Seismic line A cuts the crestal line at right angles, and the hyperbola YVZ shows the maximum curvature for a diffraction hyperbola from this feature. Seismic line C is parallel to the crestal line, and the hyperbola has degenerated to a straight line. Seismic line B is between these two directions, oblique to the crestal line, and the diffraction hyperbola is a flattish one, intermediate between the shapes seen at A and C.

In marine seismic data, families of hyperbolic curves are often seen that cannot readily be related to faulting or shapes in the cross-section. As noted, these are produced by a variety of back-scattering effects related to sea-bed irregularities, including wrecks, reefs, and outcrops, coastal promontories, other ships in the area (by reflection, and from propeller noise), etc. A typical example of this is seen in Fig. 4.22(b), where the family of curves indicated by the white arrows is produced by one of the features mentioned; it is not usually possible to identify the exact cause.

It should be clearly understood that the hyperbolic events under discussion are *not* the results of energy reflected back from plane surfaces, but of energy *repropagated* as if from a point source.

4.2.9 Graben and Block Faulting

Trough-like subsided blocks, bounded usually by normal faults, are known as graben structures. Typical examples are major rift valleys such as the East African Rift. 'Rift' implies extensional stress, and this is appropriate in many cases, but graben structures can also be the result of strike-slip faulting, as in the case of the Dead Sea feature, or to the vertical movement involved in subsidence, as in the case of salt dissolution–removal and overburden subsidence.

An example of the latter is seen in Fig. 4.23, where a mobile salt interval (H) has produced a small piercement structure above and to the right of H, which has subsequently undergone some dissolution resulting in overburden subsidence. A graben, bounded by the two gravitational normal faults F, has formed. Note the rapid decrease in throw upwards in the section of the two bounding faults, indicating syndepositional movement. The event R is a reflected refraction. Such structures, on any scale, clearly present many possible petroleum trap situations, both within the downthrown block and in the units truncated against the bounding faults.

Where normal faulting of a repetitive nature divides the crustal rocks into a series of large blocks, the system is referred to as *block faulting*. Often the stratal units within the blocks step down progressively in one direction, and this is known as *step faulting*. Tilted block faulting also occurs, as in Fig. 4.24 where the normal faults separating the blocks appear to be approximately planar surfaces, as seen by their traces in cross-section.

Rotated fault blocks are also seen in some circumstances—the fault movement having taken place on fault planes with listric curvature. In some areas of rifting, later reactivation has occurred, and this is the case in Fig. 4.25, where a series of rotated fault blocks (marked

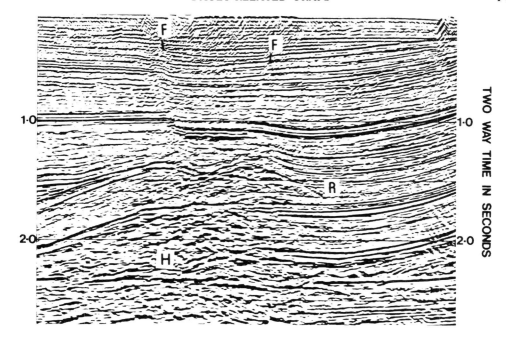

Fig. 4.23. A graben bounded by two faults (F) related to salt flow into a pillow (H) that later developed into a small piercement; the latter subsequently underwent some dissolution and removal of salt, accompanied by overburden subsidence and faulting. The event R is a reflected refraction related to the right-hand bounding fault of the graben. Reproduced by permission of Seismograph Service (England) Ltd

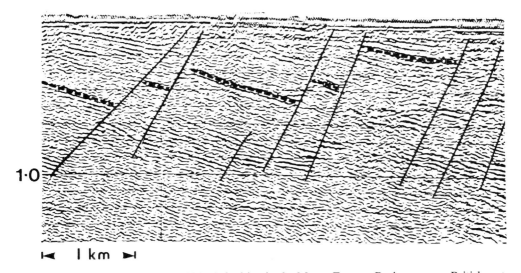

Fig. 4.24. Seismic example of tilted block faulting in the Manx–Furness Basin, western British waters. The fault traces in section show that the fault zones are planar, and the fault blocks have not undergone any rotation. The example shows a form of step-faulting. Reproduced by permission of Seismograph Service (England) Ltd

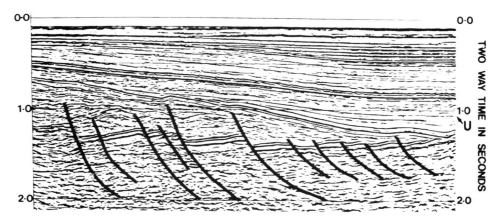

Fig. 4.25. Rotated block faulting. A series of apparent step faults that have undergone some rotation on listric fault planes. The overburden uplifts above some of the upper fault terminations indicate that reactivation of the faulting has taken place. In this instance, it is believed that there was at least an element of strike-slip movement in the reactivation. Reproduced by permission of Seismograph Service (England) Ltd

by the deepest strong reflector group) has subsequently undergone reactivation, with the movements causing uplift of later sediments, as seen in places in the strong reflection group starting at about 0.7 s on the left-hand side of the section. It is believed in this case that there was a strong element of strike-slip movement involved either in the initial fault phase, or in the reactivation phase, or in both. Evidence for rotation of the fault blocks is not very strong in this example, being more obvious in other data in the area (although the tilting of the blocks is obvious). Note that little or no reactivated movement took place after the time of the unconformity present at 1.05 s on the right-hand side of the section.

Tilted and rotated fault blocks, sometimes as combination traps, including buried topographic features, form an important class of trap, as in the many Brent-type fields of the North Sea.

4.3 Examples of Fault Traps

4.3.1 Beatrice—an Upthrow Fault Trap

The Beatrice oilfield, Inner Moray Firth, UK, North Sea has been described by Linsley *et al.* (1980). The reservoir is in an alluvial to marine Jurassic (Sinemurian–Callovian) sandstone–shale sequence, and the structure is an elongate fault-bounded and controlled anticlinal trap. A depth-structure contour map for the top of the reservoir sequence is shown in Fig. 4.26, and the seismic section shown crossing the contour map at X–X' is seen in Fig. 4.27.

Permo-Triassic clastics, mainly of continental origin derived from the Caledonian Highlands (either primarily, or secondarily from the thick underlying Devonian sequence of the Inner Moray Firth Basin) underlie the reservoirs, being capped by a Triassic cherty limestone, which is seen as the deepest of the three seismic horizons marked with heavy lines on the section in Fig. 4.27, indicating also the base of the reservoir sequence. There

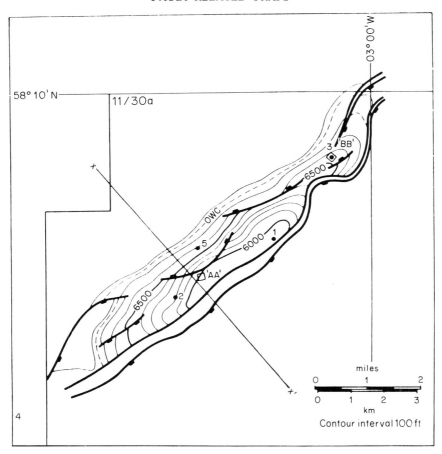

Fig. 4.26. Beatrice oilfield. Depth-structure contour map for the top of the reservoir sequence. From Linsley et al., 1980, reprinted by permission of American Association of Petroleum Geologists

is very little closure at the level of the shallowest seismic marker shown (the Base Lower Cretaceous). The dip reversal at the reservoir level may be due partly to drape over the underlying block, but appears mainly due to fault drag, which is visible on both upthrow and downthrow at the main fault plane. On the downthrow side (to the south-east, in the X' direction), there is some minor horst and graben faulting that probably dates from the same time as the main fault movement—i.e. middle to late Oxfordian.

Although the structure has some slight turnover, it is clearly a simple fault trap; most of the vertical closure is against the fault (although, of course, the closure along the fault strike is necessary for the trap). If the main fault were not present, or was not a sealing fault, there would be little or no oil accumulation.

The gross pay oil column in the Beatrice discovery well was 831 feet (253 m) and the well produced 6060 BPD of 38° API crude with a low GOR (gas/oil ratio). The crude has a high wax content (17%) suggesting a near-shore or paralic source-rock type, and a high pour point. Recent geochemical studies by Peters et al. (1989) have established that

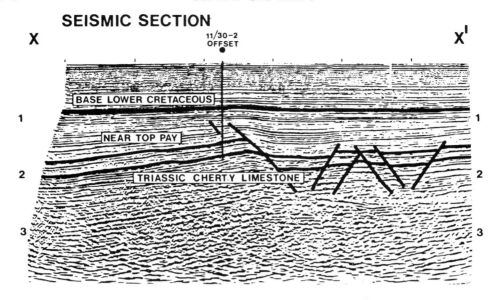

Fig. 4.27. Seismic section across the Beatrice structure along the line X–X' seen in Fig. 4.26. From Linsley et al., 1980, reprinted by permission of American Association of Petroleum Geologists

the oil here is a mixture from Devonian lacustrine and Middle Jurassic marine source rocks. Estimated total ultimately recoverable oil from this interesting field is 161 MMbbl using a 34% recovery factor.

4.3.2 Fahud—another Upthrow Trap

A similar basic situation to that of the Beatrice Field is found in the Fahud structure of Central Oman. Fahud lies between the foothills of the Oman Mountains to the north-east, and the Um as Samim (Arabic for 'mother of poison'—a great, dried-up salt lake on the fringes of the Empty Quarter) to the south-west. The writer remembers travelling in the Um as Samim in summer, when to be thrown against the inside of the Land Rover while crossing the rough mud-/salt-cracked surface meant receiving a severe burn that required treatment (shade temperatures, if there had been any shade, would have been hovering around the 130°F mark). It was a commonplace amongst seismic crew members that an egg could be fried anywhere it would stick on the superstructure of a Land Rover in these temperatures.

The Fahud structure was discovered by surface geology, and is clearly visible from the air, although the subsurface structure is more complex than is apparent at the surface. The field has been described by Tschopp (1967) and others, as being located in an asymmetrical anticline, elongate WNW–ESE, with oil accumulations in the reservoir sequence trapped in the upthrown block of a large normal fault striking parallel to the structural axis. The aggregate throw of this fault is approximately 4000 feet. The closure against the fault in the upthrown (northern) block has dimensions of some 27 km by 4 km, with vertical closure of about 3200 feet. Figure 4.28a shows a line-drawn cross-section of the field, with a depth-structure contour map at the level of the Top Wasia Limestone, the main reservoir unit, which is capped and sealed above by the Aruma Shale.

The Fahud carbonate reservoir consists of the Wasia Limestone Formation, a series of chalky argillaceous and dolomitic wackestones separated by shaly intercalations, with Campanian shales forming a sealing cap rock. Seven separate members are recognized, being designated 'a' through to 'g'. Net pay porosities vary from 25% in 'e' to 32% in 'a' and 'b'. Permeabilities are fair to good within a range of about 5 mD in 'f' and 'g' to 10,000–20,000 mD in 'a' and 'b'; paradoxically, some of the less porous intervals have the highest permeabilities.

The gross oil column (Tschopp, 1967) amounted to about 1500 feet, with a 360-feet gas cap (the GOC being at only 250 feet below mean sea level). There is some evidence that the reservoirs have a common free-water level, which suggests good communication between them. The API gravity of the oil is 33.6°, with sulphur content about 1.15%.

The depositional environment of the Wasia Limestone Formation is thought to have been a shallow carbonate shelf, not unlike that present in the Persian Gulf today. The source rock for the oil is believed to have been the Wasia 'b' member, which is a bituminous, mainly non-reservoir rock.

The Fahud structure is a residual gravity maximum within a larger Bouguer gravity minimum, suggesting, in the context of this area, that there is salt at depth. The presence of salt at nearly 12,000 feet was indeed confirmed in the Fahud-1 well. However, in this instance salt movement is not thought to have been the primary cause of the structure, although it may have had some secondary effects. It is believed that the structure is due to tectonic uplift during the deposition of the Middle Cretaceous Wasia Limestone Formation, and at the time of the major faulting. High porosities in the upthrown northern fault block (contoured in the lower sketch of Fig. 4.28a) compare with low porosities in the downthrown southern blocks. In the latter, insignificant calcite cementation indicates that the low porosities are not due to diagenetic effects. This, together with other evidence, indicates the age of the Fahud fault, and also suggests that the upthrown block was periodically lifted above sea level, where the concomitant sub-aerial leaching could account for the high porosities present in this block.

4.3.2.1 Basin margin faults

Faulting at the margin of a depositional basin is common, and often due to subsidence stresses at structural hinges. Two examples from the Porcupine Basin, offshore Ireland, are shown in Figs 4.28b and 4.28c. The first example, Fig. 4.28b shows a series of down-to-basin normal faults mainly affecting the Jurassic sediments present in this location in an off-shelf situation at the basin margin.

In Fig. 4.28c, shelf-edge faulting is present with a block of Carboniferous units seen in a rotated horst between a major basinward-hading fault (in a series of probable 'towards the upthrown block' faults), and a shelfward-hading fault that does not affect the post-Carboniferous section. Trap situations have been modelled in both examples (dashed intervals), which are from a paper by Croker and Shannon (1987).

4.3.2.2 An example from offshore Norway

The Askeladden gasfield, off-shore Norway, covers a large part of Norwegian North Sea Block 7120/8 in the Troms I area, about 100 km from the coast. Gas was discovered in sandstone sequences of early and middle Jurassic age (Westre, 1984).

The field is compartmentalized by faulting, and differing GWC have been found in various wells. A sketch depth-structure contour map at top Middle Jurassic level is seen in Fig. 4.29, and N–S seismic section through the southernmost of the wells marked on the map (7120/8-2) is shown in Fig. 4.30.

The seismic section shows E–W faulting of an earlier phase (middle Jurassic) during which some slight low-relief doming also occurred, and a later phase of more

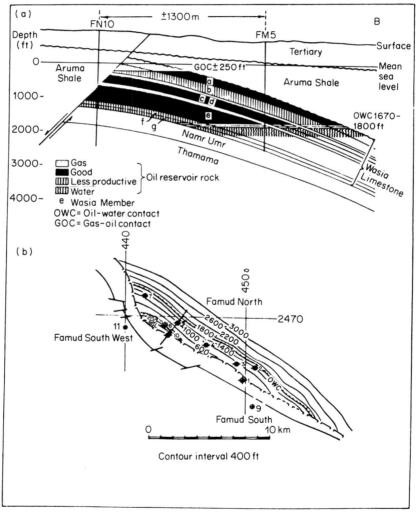

Fig. 4.28a. Fahud oilfield, Sultanate of Oman; (a) section across the field, showing reservoir details; (b) depth-structure contour map at top reservoir level. Reproduced by permission of Elsevier Applied Science Publishers Ltd (from Tschopp, 1967)

Fig. 4.28b. Seismic example of 'down-to-basin' normal faulting. Reproduced by permission of Graham & Trotman Ltd (from Croker and Shannon, 1987)

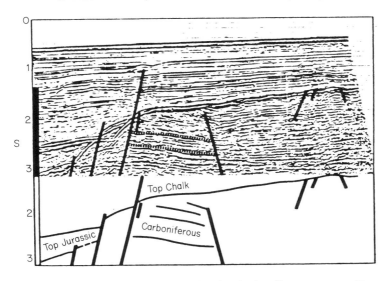

Fig. 4.28c. Seismic example of shelf-edge faulting in Carboniferous strata, Porcupine Basin. Reproduced by permission of Graham & Trotman Ltd (from Croker and Shannon, 1987)

extensive N–S and NNW–SSE faulting of Late Cimmerian (late Jurassic–Early Cretaceous) age. Some of the faults in both series terminate downwards in the Triassic, and upward through the Top Cretaceous reflector, and are associated in some locations with shows of 'heavy' hydrocarbons in the Tertiary section; from this, it can be assumed that some of the faults have been acting as fluid and gas conductors from deeper levels.

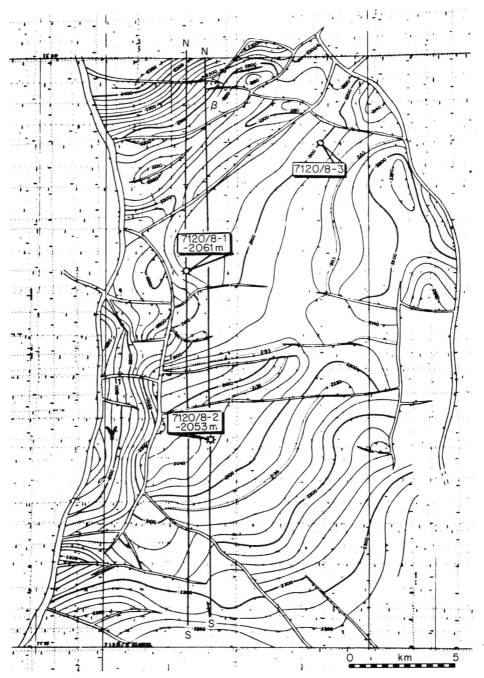

Fig. 4.29. Askeladden gasfield, offshore Norway; a depth-structure contour map at Top Middle Jurassic level. Note the fault pattern, and the position of the southernmost well (7120/8-2). Reproduced by permission of Graham & Trotman Ltd (from Westre, 1984)

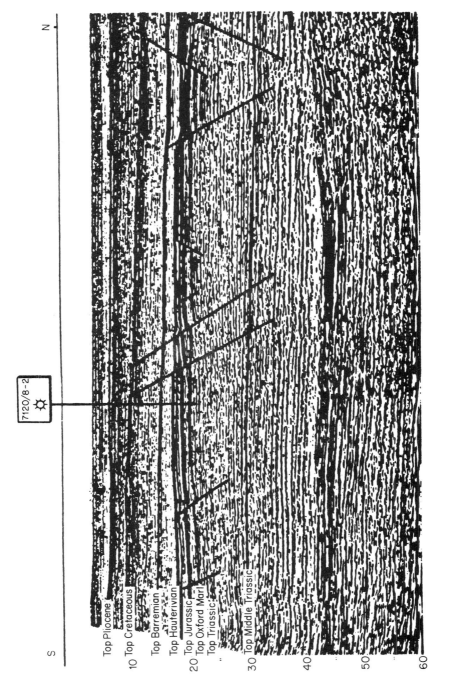

Fig. 4.30. Askeladden gasfield; a N–S section through Well 7120/8-2 (refer to Fig. 29). Reproduced by permission of Graham & Trotman Ltd (from Westre, 1984)

The Lower and Middle Jurassic reservoir sandstone sequences are interpreted (Westre, 1984) as a 'generally marine clastic coastal and inner shelf sequence'. The original reservoir properties in the general Troms I area must have been excellent, with clean, well-sorted sands of favourable grain sizes. However, these properties were impaired by several factors: (i) unfavourable (deep) burial history; (ii) lack of overpressure in the reservoir during burial owing to leakage along late-moving fault zones; and (iii) an absence of hydrocarbons in the structures owing to the preceding factors. Fortunately, at Askeladden the reservoirs had a more favourable burial history, and the net pay average porosity is approximately 15%.

As regards gas source rocks, there are Upper Jurassic shales with a high TOC increasing downwards, with mixed types II and III kerogen, and also post-Hauterivian claystones with fairly high TOC values and type III kerogen. Interbedded coals contain type III kerogen and are good potential gas source rocks. The exact source of the Askeladden gas has not yet been determined with certainty, but is not thought to be the Upper Jurassic shales, which are oil-prone. The answer may lie in either or both of the claystones and the coals. Initial gas migration into reservoir sequences probably took place in late Cretaceous to early Tertiary times.

The seismic example indicates the upward and downward terminations of the faults along the line of section. As noted, the evidence is that at least some of the larger faults have been 'leaky', perhaps during late movements only. It seems possible that they are of the 'juxtaposition' seal type that may have leaked at some period, and then resealed with later movements, retaining much of the gas in the reservoirs. Certainly, much of the vertical closure here depends on the trapping and sealing situation created by the faulting, and if the latter had continued to leak badly over along period, the large gas accumulation found is hardly likely to have survived.

The graben faulting at the north end of the seismic section indicates the extensional nature of the later (Late Cimmerian) stresses in this area.

4.3.3 Planar Step Faults

The examples shown in Figs 4.24 and 4.28b earlier in the chapter have a series of normal step faults with (from the evidence of the fault-plane traces in cross-section) approximately planar fault zones. An example of such features acting as closure control for petroleum traps is found in the Asl oilfield of Western Sinai. The line drawing of Fig. 4.31 shows a section across part of the field, with a series of approximately planar normal step faults, the line being in the dip direction for the tilted blocks separated by the fault planes. Two of the many wells drilled are seen, producing oil from two Miocene sandstone, and one Eocene limestone, reservoirs (Thiébaud and Robson, 1981).

Surface mapping carried out in 1937 indicated the presence of 18 beds of gypsum alternating with shales, marls, and sandy beds that were gently flexured with very little faulting. Subsequent information from nearly 30 wells, however, revealed that the section consists of numerous tilted fault blocks with the faulting dying out before reaching the surface. Only occasional minor flexures and scarps at the surface give some hint to the unexpectedly more complex underlying geology.

Most of the information regarding the field has been derived from the drilled wells. Seismic data for the area is of poor quality owing to the numerous subsurface faults overlain by a thick evaporite sequence. Thus in this case, as in others, the absence of an illustrative

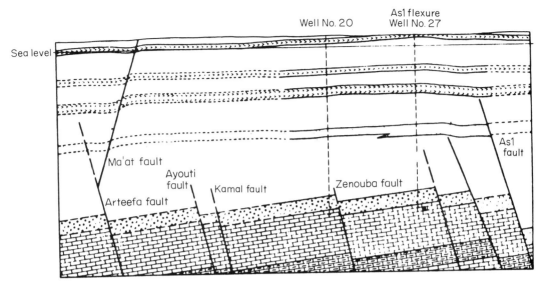

Fig. 4.31. The Asl oilfield, Western Sinai; section showing a series of planar step faults that die out upwards as gentle flexures. Reproduced by permission of the *Journal of Petroleum Geology* (from Thiébaud and Robson, 1981)

seismic section does not imply confidentiality, or a lack of seismic data, but simply the absence of *usable* seismic data.

The minor relief shown by the overlying evaporite sequence indicates gentle late reactivation of the underlying faults. The implication should also be noted that the deeper, faulted section is compacted and indurated to the extent that brittle fracture occurred to produce the faulting; the bedding meets the fault planes approximately normally, with no sign of ductile shear flexuring. In contrast, the shallower overburden shows signs of ductile shear and flexuring above the fault plane locations, indicating relative undercompaction.

The tectonics involved in the Asl oilfield are closely associated with those in the Clysmic Rift (the extended Gulf of Suez in Miocene times), and have features in common with the structural style in the Viking Graben of the North Sea—basic tensional stress with step faulting 'towards the upthrown block', followed by drape deposition (as, for instance, in Brent). However, there are differences—notably, a lack of symmetry in the pattern of block faulting across the Clysmic Rift.

It should be noted from the section in Fig. 4.31 that gentle flexuring of the objective section, mentioned earlier, was necessary to produce the required geometry for traps in the direction normal to the section, since there is effectively no cross-faulting, and closure has to be three-dimensional.

Thiébaud and Robson (1981) warn that interpretation of deep structure from surface trends alone (in the absence of good seismic data, or reliable drilling information of sufficient density), although it may be tempting, as in the case of the Asl Field, is a rash undertaking—a warning not unfamiliar to any geologist.

This message is brought home again by another, older Egyptian oilfield, Hurghada, that is also illustrated by a line-drawn section based on drilling results in Fig. 4.32. If anything,

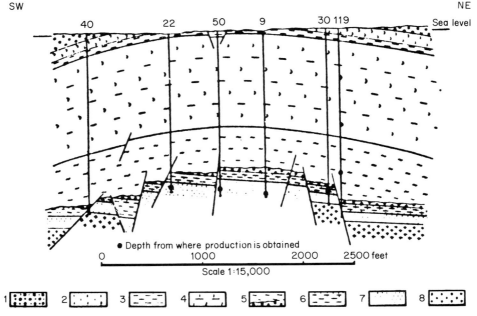

Fig. 4.32. Hurghada oilfield, Egypt; section across the field, showing the great difference between the geological structure at the level of the objective section, and surface trends and features. From van der Ploeg, 1953. Reproduced by permission of Oxford University Press (from P. van der Ploeg, in V. C. Illing (Ed.), *The Science of Petroleum*, Vol. iv, *The World's Oilfields*, Part I, *The Eastern Hemisphere*, 1953)

this is a more spectacular demonstration of the dangers of inferring the geology at depth from surface trends and features.

The deep structure at Hurghada consists of block faulting of Precambrian, Palaeozoic, and Mesozoic rocks. As in the previous example, the Miocene overburden reacted to the faulting mainly by ductile flexuring with little faulting, so that at the surface, what appears to be a simple anticlinal structure is seen, giving no hint of the underlying faulting.

In the example, key formations are numbered as (1) Plio-Pleistocene, (2–5) Miocene, (6 and 7) are Mesozoic and Palaeozoic (Nubian Series), with garnet- and staurolite-bearing metasediments included, above (8) the Precambrian igneous and metamorphic basement.

Oil was found in some of the Nubian Series rocks (7) that range in age from Devonian to Cretaceous, and also in dolomites of Miocene age. Seal formations are shales and marls. This field is no longer producing, having been discovered before 1914, and reaching peak production in the early 1930s, when the annual production was near to 2,000,000 bbl from over 100 productive wells (van der Ploeg, 1953).

4.3.4 A Note on Block Fault and Complex Fold–fault Traps

It may be wondered why rotated block-fault trap situations like that in the Brent and similar fields are not included in this chapter. The reason is that typical Brent-type fields are combination structural–stratigraphic traps, including the effects of faulting with those of buried topography and unconformities. It is considered that these are best treated in the

FAULT-RELATED TRAPS   155

later part of the book (Chapter 10) dealing with combination traps, as well as complex thrust-belt traps. The present chapter continues with a discussion of the important subject of growth-fault traps.

4.4 Examples of Growth-fault Traps

4.4.1 The Hibernia Oilfield

This field was discovered in 1979 in the Grand Banks area some 325 km east of St John's, Newfoundland, and initial calculations have indicated maximum recoverable reserves in excess of 1 Bbbl of oil. The structure has been described by Arthur *et al.* (1982); see also Tankard and Welsink (1987).

The time-structure sketch map in Fig. 4.33 shows the feature to be a north-north-east-trending roll-over anticline associated with a major growth-fault system to the north-west, and cut by a number of transverse, mainly north-west-trending, normal faults.

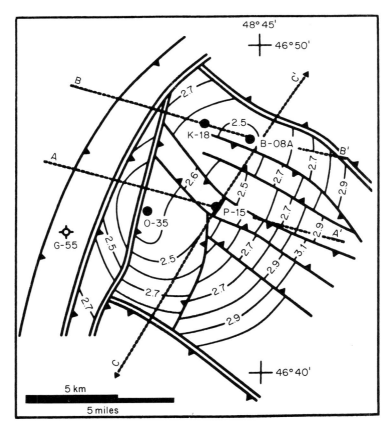

Fig. 4.33. Hibernia oilfield, offshore Newfoundland; a time-structure contour map on the Lower Cretaceous Limestone marker that delineates the shape of the structure. From Arthur *et al.*, 1982, reprinted by permission of American Association of Petroleum Geologists

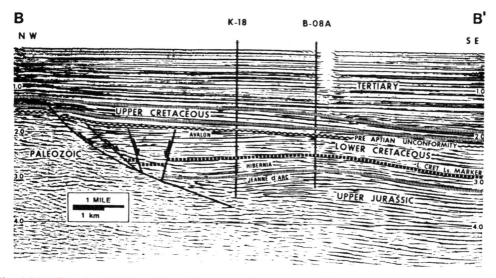

Fig. 4.34. Hibernia oilfield; a dip-section across the structure, showing all principal characteristics of growth faulting (line B–B' in Fig. 4.33). From Arthur et al., 1982, reprinted by permission of American Association of Petroleum Geologists

The seismic section of Fig. 4.34 is an excellent example to begin with, as it shows all the most important characteristics of growth faulting, and is shown on the sketch map as the line B–B'. The major listric normal fault with subsidiary synthetic and antithetic faulting is seen at the north-west end of the sections; the downward trajectory of the main fault plane becomes less steep with increasing depth, perhaps to sole-out (not shown) somewhere in the Jurassic section. The main fault shows evidence of decreasing throw with decreasing depth, and there is a strongly marked roll-over anticlinal structure in the handingwall. The well-seen thickening to the north-west of the section occurs particularly in the Lower Cretaceous interval, and indicates the syndepositional nature of the fault movement during this period.

The Lower Cretaceous Limestone marker seen on the section just above 3.0 s is some 335 m above the Hibernia member, the lowest of three Lower Cretaceous reservoir sandstones of probable deltaic origin. As noted before, growth faulting is a typical feature of unstable environments such as deltas or evaporite basins with mobile salt. A delta subsiding under a massive weight of sediment, with high depositional rates often leading to overpressured units, and located usually on the margin of a deep marine basin, is perhaps one of the most unstable depositional environments.

Porosities in the Lower Cretaceous sand reservoirs average 15–18%, with a maximum of about 30%. An Upper Jurassic sandstone and conglomerate reservoir is present in some parts of the structure, but seems less potentially productive than do the Lower Cretaceous intervals, from which each well so far drilled has produced over 20,000 BPD. The areal closure of the field is 155 km², and the vertical closure is 760 m. Northern and southern structural limits are imposed by the transverse faulting.

In the seismic section, note the productive Hibernia interval through which the wells pass, the relatively undisturbed look of the stratal events of the Tertiary section, and the way these are onlapping up the Tertiary–Upper Cretaceous unconformity surface from the south-east towards the north-west. Note also the hump-shaped roll-over anticline; in some growth-fault structures, extensional stress builds up over the crest of this feature, owing to subsidence of the beds between the crest and the main fault plane resulting from both fault movement and overburden loading compaction. This can lead to the development of a complex faulted collapse at the crest that can produce a proliferation of potential fault trap situations. A feature of this type will be illustrated later.

4.4.2 Eugene Island Block 330 Field

This example enables a closer look to be taken at the steep upper segment of the trajectory of a growth-fault trace in seismic data, and incidentally to see some excellent *DHI* events in the roll-over anticlinal trap.

The Block 330 Field is in a roll-over anticlinal structure in the hangingwall side of a major north-west-trending growth fault, which again involves deltaic deposits, doubly unstable here in that they are located in a salt tectonic setting. The field has been described in an interesting paper by Holland *et al.* (1980).

Eugene Island is located on the Texas–Louisiana shelf area of the Gulf of Mexico, with the Gulf Coast Basin Axis to the north, and the Texas–Louisiana Slope to the south. The field's reservoir units lie within a Pliocene–Pleistocene delta complex that prograded basinwards above a series of Jurassic to Miocene shelf and deep-marine sediments. Increasing depositional thicknesses were built up, probably contemporaneously with salt withdrawal movements in the underlying Louann Salt (Jurassic). The situation here is seen by the writer as being analogous to that in the Carolina Trough, offshore east coast USA, and, as in that area, the resulting instability led to major growth-fault development.

The reservoir sequence consists of 10 sandstone units that are further subdivided into 25 producing zones over a 6500-feet interval within the age range of late Pliocene to late Pleistocene. Analysis of reflection patterns shows that the predominant seismic facies patterns are complex sigmoid-oblique and shingled reflection configurations typical of the prograding delta-front depositional environment, with subsidiary patterns indicating delta-plain and delta-fringe facies. (These patterns are not obvious in the seismic section example in Fig. 4.35, which was processed for relative amplitude to bring out the DHI events.) Trapping mechanisms present include fault and roll-over anticline closures, plus facies-change traps—another fairly common feature in deltaic environments.

The migration of hydrocarbons into the structure is believed to have occurred in geologically recent times (within the last 500,000 years), during or after the late Pleistocene period: in fact, the reservoir sandstones themselves are less that 2.5 Ma old. The great majority of the reservoirs are abnormally pressured—again a typical feature of the rapid deposition in the delta environment.

The source rocks of the hydrocarbons in the Pleistocene (the oil being light and waxy, typical of a fresh or brackish water swamp-type source rock) have not been identified, and the hydrocarbons are thought to have migrated up the growth fault from depth. The Pleistocene shales present in the structure have been analysed for source rock potential—TOC, vitrinite reflectance, and TAI studies of the kerogen have been carried out. The kerogen is of type III, with average %Ro being in the range 0.31–0.47, and the shales cannot be regarded as a possible source for the hydrocarbons in this field.

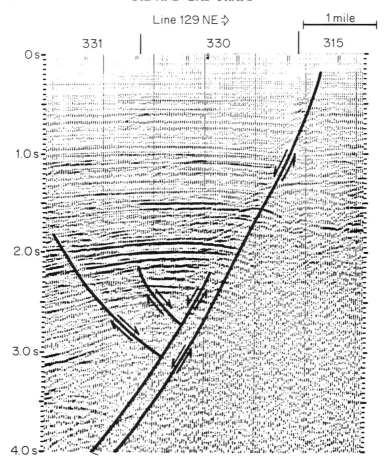

Fig. 4.35. Eugene Island Block 330 Field, Gulf of Mexico; a seismic section (relative amplitude) across steep (shallow) part of trajectory of a growth fault, with the field structure in the hangingwall of the fault. From Holland *et al.*, 1980, reprinted by permission of American Association of Petroleum Geologists

The relative-amplitude processed seismic section in Fig. 4.35 is of considerable interest apart from the structural indications of the faulting that it shows. Present are numerous DHI 'bright-' and 'flat-spot' events, indicating the presence of hydrocarbon-filled porosity. For instance, there is a prominent flat-spot event at about 1.5 s, located above the crest of the rollover anticline, and this event coincides with the OWC in one of the reservoirs (the HB reservoir). The overall reservoir sequence lies between about 1.35 s and 2.3 s in this section, and in the lower part of this interval particularly, there are many strong bright-spot events that appear in response to the hydrocarbons in the lower reservoirs. Also, other typical effects are seen, such as apparent amplitude broadening (due to absorption of higher frequencies by gas-filled porosity), time 'sag', and shadow zones, beneath the bright-spot events.

Down the trajectory of the trace of the main growth fault, diffraction hyperbolas—some quite faint—can be seen marking the route of the fault; at this level, the listric curvature of the fault trace is not strongly marked in this example. Note the directions of fault movement indicated for the synthetic and two antithetic faults present in the hangingwall.

The detailed history and geochemistry of this field cannot be dealt with in this brief note, and anyone wishing to go into more detail is recommended to read the paper by Holland *et al.* (1980). The background geology for this field can be studied in Martin (1978) and Martin and Bouma (1978).

4.4.3 Two Malaysian Fields

Two major growth-fault-related oilfields in Eastern Malaysia, interesting from different aspects, are the Baronia and Samarang oilfields.

The Baronia Field is illustrated by a seismic section, depth-structure contour map, and line diagram in Fig. 4.36, which is taken from an account of the field by Scherer (1980), who also provides some details as follows. The structure is a simple domal anticline formed by the roll-over structure between two major E–W-trending growth faults, the fault-plane traces of which are seen in the seismic example.

The field is located in the Baram delta off northern Sarawak (Borneo), 50 km north-west of the town of Miri. The main reservoir sequence consists of late Miocene sandstones with interbedded siltstones and clays forming seals for the ten separate reservoir sands (including five with large gas caps).

The oil is light and waxy with low sulphur, derived from land plants, and source rocks that were deposited in fresh water in common with most of the oils found in north-west Borneo. Chromatographic oil analyses showing high pristane/phytane ratios support an interpretation of source material deposited in a peat swamp environment (Lijmbach, 1975).

Looking at the seismic example in relation to the sketch depth-structure contour map, it is seen that the field lies between the two major growth faults, where there is a clear roll-over structure below about 1.0 s a two-way time, and some typical apparent interval thickening into the right-hand fault plane between 1.0 s and 1.7 s. Although there are antithetic (and one minor synthetic) faults in the hangingwall of the left-hand growth fault, there is no analogous faulting related to the right-hand growth fault, in the field location. This caused some surprise at the time of drilling (no sign of any faults were found after 32 wells had been drilled in the field location). However, it is not an uncommon phenomenon. Absence of hangingwall faulting, as in this case, may be dependent on the timing of the build-up of stresses, and the development of other growth faults and antithetic faults downdip, as here, that may act as a stress-relief mechanism relating to the hangingwall of the right-hand fault. The latter clearly finished moving earlier, from the level of its upper termination, which is deeper (older) than that of the left-hand growth fault. In most sequences of growth faults, the order of development follows a 'down-to-basement' pattern, with the basinward faults evolving later than the marginward faults.

It is interesting to compare this type of fault sequence with the 'towards the upthrown block' order of a series of planar step faults, such as those seen on the western margin of the Viking Graben in the northern North Sea, where the faults towards the margin develop later than the faults that are deeper in the basin.

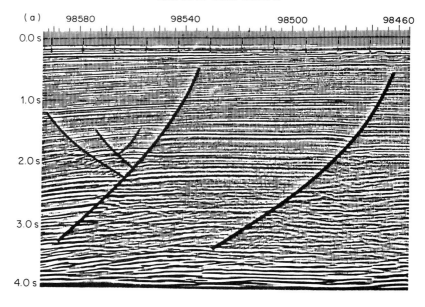

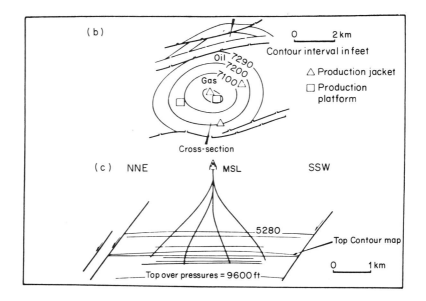

Fig. 4.36. Baronia oilfield, eastern Malaysia; a seismic section (a) across the field, with a sketch depth-structure contour map (b) and line-drawn interpretation (c) of the objective section. Reproduced by permission of F. C. Scherer and American Association of Petroleum Geologists (from Scherer, 1980), with acknowledgements also to Shell and PETRONAS

Additionally (although not in this case), subsidiary faulting may be present in locations in the block between two growth faults in some locations away from the seismic section, in directions normal to the section (i.e. faults that appear and die out laterally off-section).

This is a particular matter of fault dynamics that has yet to be exhaustively investigated. However, a detailed study of growth faulting is the subject of a series of papers by Crans and Mandl (1980–81).

The structural pattern of the Samarang oilfield is quite different from that in Baronia, although also in a growth-fault setting. A NW–SE seismic section across the field and the faulting in the dip direction are seen in Fig. 4.37. As with the previous example, the illustration and brief details are drawn from Scherer (1980). The structure is located in southern Sabah, some 45 km north-west of Labuan Island, in shallow water (9–45 m).

The structure of the Samarang Field is related to a major growth-fault system of Miocene–Pliocene age that was modified in late Pliocene times by tectonism involving basement strike-slip faulting and possibly some mobile overpressured clay/shale movements. The growth faults, hading westwards, form a series on the eastern and south-eastern side of the structure, with the fault planes appearing to sole-out at or close to a shallow regional unconformity surface (indicated by the broken arrow to the left of the section, below 2.0 s). The roll-over is marked by a major collapse structure at its crest, formed in a complex zone of synthetic and antithetic faults. The bulk of the commercial hydrocarbon accumulations are found in the relatively unfaulted western flank of the roll-over (the left-hand end of the seismic section in Fig. 4.37), with traps against the antithetic faults. The main hydrocarbon interval is defined by the vertical black bar on the section.

Most of the oil in the Samarang Field is of the light, waxy, low-sulphur type also found in Baronia, although in Samarang there is also some heavy, non-waxy oil in shallow reservoirs in the heavily faulted core of the crestal collapse. This oil has lost its wax content by bacterial degradation in the reservoir.

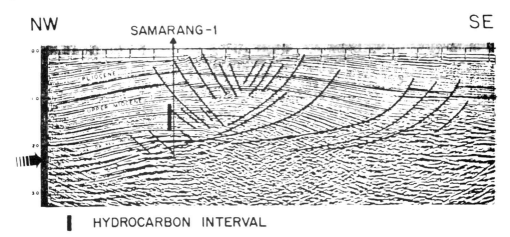

Fig. 4.37. Seismic section across the Samarang oilfield, eastern Malaysia, showing a complex growth-fault system with a roll-over anticline exhibiting crestal collapse with synthetic and antithetic faulting. This latter faulting is probably not related purely to the growth faulting—there may have been some 'external' tectonism involved. Reproduced by permission of F.C. Scherer and American Association of Petroleum Geologists (from Scherer, 1980), with acknowledgements also to Shell and PETRONAS

The fault system seen has been active throughout the whole depositional history of the field, with growth being observable across several of the major antithetic faults at various stratigraphic levels (Scherer, 1980).

The crestal collapse seen in the seismic section is not unusual in structures of this form. They are related to stresses that develop as a result of both the hangingwall movement of the growth faults themselves, and extensional stress developed over the crest of the roll-over owing to irregularity (in this case, an antiform high) in the shape of the growth-fault plane(s), or the surface at which the plane soles-out. In the steeper parts of the trajectory of the trace of the growth-fault plane (nearer to the surface), departures from pure listric curvature can result in compressional or extensional features (hangingwall anticlines or synclines) that may be of significance as trap structures in their own right.

The immense thickening of the section in the roll-over from left to right in this example is clear, and such thickening to some degree is characteristic. Also characteristic, but often less obvious (since it depends on detailed stratigraphic correlations across the growth-fault zone, which has not been carried out in this example), is the increase in throw—often not regular or progressive, but occurring over certain intervals only, as the fault has moved—with increasing depth. It is frequently very difficult to carry seismic markers across from footwall to hangingwall in a growth fault. Major thickening in the hangingwall just mentioned gives one clue as to the reason. Relatively short depositional time intervals in the hangingwall can produce large depositional thicknesses, but only small corresponding thicknesses (non-depositional hiatus or sediment starvation) in the footwall, resulting from the very different depositional patterns on the two sides of the fault scarp created by the developing growth fault. There may have to be considerable geological/drilling/logging input to the seismic data before reliable correlations can be made across the fault.

4.4.4 Reservoir Details in a Roll-over Structure

It is instructive to examine the types of structural trap that may be present in a roll-over structure that has been subjected to complex synthetic and antithetic faulting similar to the faulting in the Samarang Field structure just considered, and again to compare the detail of multiple reservoirs and traps that can be constructed from a combination of seismic and drilling data with the relative simplicity of the structure seen in a seismic section alone. This is a salutary exercise for both geologist and geophysicist, strongly suggesting as it does the absolute necessity for collaboration and combined studies.

For this purpose, an excellent example is afforded by the Nembe Creek oilfield of the Niger Delta, Nigeria, which has an important reservoir interval developed in a Middle Miocene deltaic sandstone–shale sequence. The hydrocarbons present include oil and gas, the latter in both associated and unassociated form, over a depth range of 7000 feet (2130 m) to 12,000 feet (3650 m). Values of API for the oils range from 16° to 41° with the heavier oils being found in the shallower reservoirs. Average recoverable oil reserves in the field, as at 1977, with 30 wells drilled, reached 645 MMbbl. Nelson (1980) gives an account of the field, and the illustrations used here are from Nelson's paper.

The Nembe structure is a complex roll-over anticline about 15 km long and 6 km wide at its widest point. The structure is dissected by numerous south-dipping, E–W-striking growth faults together with associated north-dipping antithetic faults, and oblique

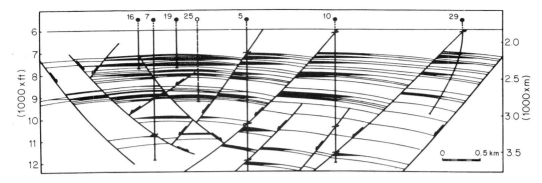

Fig. 4.38. Nembe Creek oilfield, Niger Delta, Nigeria; a true-scale section is seen here, drawn N–S through the field, showing the multiple reservoirs, and the types of trap present. From Nelson, 1980, reprinted by permission of American Association of Petroleum Geologists

cross-faults apparently compensating for lateral inequalities of displacement along the main series of E–W growth faults.

A true-scale N–S cross-section in the dip direction, approximately through the centre of the structure, is seen in Fig. 4.38, with the right-hand end being to the north. The steep upper trajectories of several south-dipping growth-fault planes can be seen, together with some minor synthetic faults and several major north-dipping antithetic faults on the south (left) flank of the roll-over anticline. Well locations (some projected on the cross-section from offset positions) are shown, numbered from 29 in the north (a deviated well) to 16 in the south. The main part of the prospective section ranges from just over 2.0 s two-way time (the time-scale is on the left) to about 2.7 s, in the centre of the section.

A seismic section that is approximately parallel to, and in close proximity to part of the constructed line of Fig. 4.38 is shown in Fig. 4.39, again with some wells projected on to the line from locations as shown (Wells 5, 13, 7, and 16). The well numbers that are common to both figures allow a comparison of locations to be made with the features in Fig. 4.38. The prospective section can be seen as the band of strong reflection events between 2.0 s and 2.7 s, with the (shallow) crest of the roll-over at about Well 13, north of which north dip is seen.

Reverting to Fig. 4.38, reservoir intervals are shown with oil represented as solid black, non-associated gas as dots, and associated gas as patches of dots within the oil symbol. Note that traps occur: (i) in high closures (within the plane of section) in the crestal zone and in the shallow part of the prospective section between Wells 16 and 25; (ii) as fault traps in north-dipping reservoirs on the footwall sides of growth faults, drilled by Wells 29, 10, and 5; (iii) as fault traps in south-dipping reservoirs on the upthrow sides of the antithetic faults south of the crest; and (iv) as fault traps between the antithetic and synthetic faults at the crest (just above TD in Well 25, with another trap below this).

On the seismic section in Fig. 4.39, some of the strong reflection events below 2.0 s are probably 'bright' owing to the presence of gas-filled porosity at some levels and locations (cf. Fig. 4.38). Note that there is a particularly bright, planar (dipping slightly to the north) reflection group extending northwards from a point just above the TD of Well 5, and there are several others to be seen. The observant will note that the TDs of Wells 16, 7, and 5 of Fig. 4.38 are deeper than the corresponding values in Fig. 4.39; these wells have been

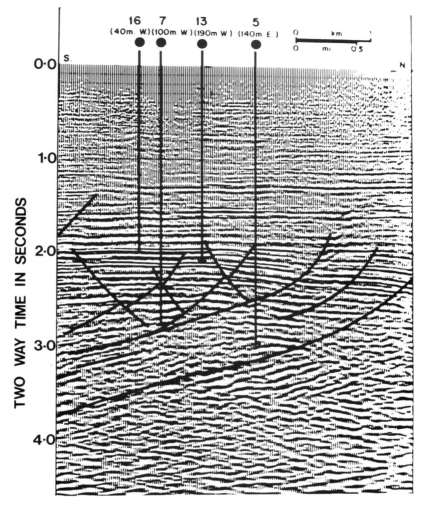

Fig. 4.39. Nembe Creek oilfield; seismic section across part of the field, showing the objective section over part of the Fig. 4.38 display (refer to well numbers at the top of the section for comparative locations). Note the difference in detail between the seismic section, and the true-scale section in the previous figure based on detailed drilling information. From Nelson, 1980, reprinted by permission of American Association of Petroleum Geologists

Deepened between the time of production of the seismic section, and the time of construction of the Fig. 4.38 section.

4.4.5 Generalized Model of Growth-Fault-related Traps

The seismic example in Fig. 4.40 shows in a general way the main types of trap associated with growth faulting, with model traps shown in solid black, and numbered 1–6. Trap type 1 is the thinning wedge trap that develops due to the thickening towards the main fault in a roll-over structure, whereas type 2 is the antiform trap that could develop at any level

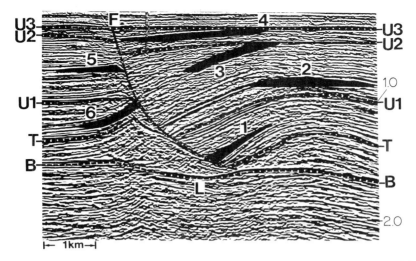

Fig. 4.40. A large growth fault F in a seismic section, showing two-dimensional models of the principal trap types associated with such features. Model traps are shown in black, and are numbered for discussion in the text. U = unconformity; B, T = base and top. Reproduced by permission of Elsevier Applied Science Publishers Ltd (from Jenyon, 1986d)

in the crestal zone of the roll-over. Types 3 and 4 are truncation traps related to two of the three major unconformities (U1, U2, U3) present above the salt interval B–T. Types 5 and 6 are footwall fault traps, both in varying degrees influenced structurally by the Parker–McDowell-type salt uplift on the upthrow side of the main fault. In this instance the growth fault is the result of salt withdrawal by flow into the pillow on the right, and the uplift on the left, of the main fault plane; the almost complete withdrawal of the salt, and its replacement by relatively low-velocity overburden clastics has produced the velocity 'pull-down' anomaly indicated at L.

The trap types shown are not the only ones to be found associated with growth faulting—there are often traps related to antithetic faulting, or collapse graben faulting in the shallow section over the roll-over crest, neither of which features is present here—but are the main general types to be seen in this situation. The geometry of growth faults varies greatly, and there are many different reasons for their occurrence, but wherever present, they are an indicator of structural instability in some form.

CHAPTER 5

Traps Related to Plastic Deformation of Salt and Shale

5.1 Introduction

The greater part of this chapter will be devoted to petroleum traps that result from the effects of plastic deformation in salt rock, since this is the most important and widespread type of material to undergo plastic deformation within the shallower levels of the crust (leaving aside igneous and metamorphic rocks).

Another class of salt-related trap involves overburden deformation by subsidence resulting from dissolution and removal of underlying salt rock by circulating formation water/brine. This is most conveniently dealt with as a separate subject, and will be discussed in detail in Chapter 8.

The subjects of salt geology and salt tectonics cover a great deal of territory and have their own large literature. Here, no more than a brief account will be given, covering basic aspects of salt geology and plastic deformation, before discussing some 'real Earth' traps related to salt movement. A briefer account of traps produced by other plastic materials, such as shale, will be included.

In recent years, many of the older concepts held regarding salt-rock mobility, and the causes and effects of salt tectonics have been questioned in the light of new evidence that has emerged from laboratory work, field studies, and the analysis of high-resolution seismic data. This questioning has led, in turn, to fresh hypotheses regarding the physical, chemical, and geological conditions required for the initiation and development of plastic mobility in salt rock. These new ideas will be included in the discussion that follows.

5.2 Physical Properties of Salt Rock

Salt rock consists mainly (often more than 98%) of the mineral *halite* (NaCl, common name—rock salt), density about 2.17 g/cm^3 and hardness 2.5 in its pure form. The small remaining percentage of other materials is taken up by impurities—other evaporites, such as anhydrite ($CaSO_4$) or potash minerals, together with minute amounts of insoluble detrital material. It is thus understandable that the terms 'halite' and 'salt rock' are used synonymously, and when the characteristics of salt rock are in question, it is really the characteristics of the mineral halite that are important.

Halite is a mineral with some unusual and interesting properties. Under certain physicochemical conditions, it becomes plastic and mobile. If there are any doubts on the potential for mobility of salt rock, these should be dispelled by examining Fig. 5.1. The example shows part of a seismic section in which a once uniformly-thick salt-rock layer (interval B–T) has become plastically mobile, and has flowed away in both directions (in the plane of the section) from the central 8 km of the line. In the direction normal to the section, this zone of complete salt withdrawal is over 20 km long. The complexity of the overburden above the salt, with its folding, faulting, and unconformities, is due entirely to the various stages of salt withdrawal movement, which appears to have been episodic. Complete salt withdrawal first occurred in the centre of the section, where the overburden in a large graben has subsided on to the erstwhile base of the salt interval, a sequence of carbonates and anhydrites. It is not difficult to imagine the potential for hydrocarbon traps held by such a situation.

Halite is also highly soluble, and can be dissolved and removed by circulating undersaturated waters. An actual example of this phenomenon is illustrated in Fig. 5.2, where a 900-feet-thick salt layer (interval B–T), which previously extended further up the basinal dip (to the left), has been progressively dissolved away by circulating intraformational water moving downdip. The original depositional edge of the salt lay to the left of the white arrow, but the salt now terminates at an edge-dissolution slope (arrow). Mesozoic sediments (between the top salt T and the unconformity U) have progressively subsided as the salt was removed, reaching a maximum of subsidence at the location of the salt slope. Further subsidence effects can be seen above the remnant salt layer as far downdip as the small fault marked by the black arrow. Although most of the dissolution and subsidence took place before the close of the erosional phase represented by the unconformity U, there

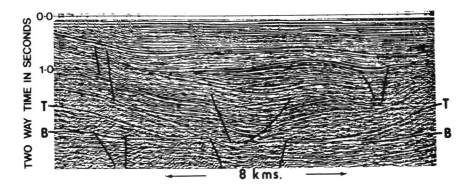

Fig. 5.1. Seismic example showing salt depletion by lateral flow. B,T = base and top of a mobile salt interval. Within the plane of section, a zone 8 km wide is seen where there is no longer any salt. Normal to the section, this zone extends for about 20 km. Reproduced by permission of Seismograph Service (England) Ltd

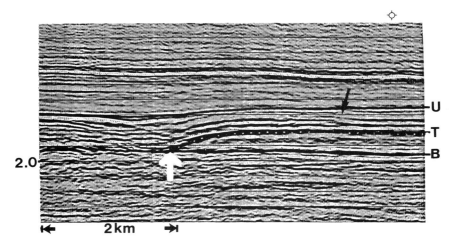

Fig. 5.2. Seismic example of a salt interval that has undergone dissolution and removal in part. Original salt thickness is shown marked B–T on the right. White arrow shows the extremity of the salt, terminating in a salt edge dissolution slope. Solvent water came down the basinal dip intraformationally from the left; water may also have percolated through the unconformity U. To the left of the white arrow, the section between level B and the unconformity consists largely of collapsed Mesozoic sediments. The small black arrow indicates a minor fault that probably marks the basinward limit of collapse effects in the overburden. Reproduced by permission of Seismograph Service (England) Ltd

was also some minor continuing effect later, as can be seen from the shape of the unconformity surface trace in the section. Again, it is not hard to see the potential here for hydrocarbon trapping. As noted, dissolution and subsidence effects of this kind will be discussed at greater length in Chapter 9, included with the subjects of evaporites, carbonates, and reefs.

Another important point about halite is that its density value falls about half-way between the density values of undercompacted and fully compacted clastic sediments, so that at shallow depths it may be denser than adjacent elastics, while at greater depth it may be less dense (or more 'buoyant') than the encasing sediments. This can lead in the case of a salt piercement structure (a diapir) of large vertical extent, to the shallow part of the salt body producing a positive gravity anomaly at the surface, while the deeper part of the same body produces a negative anomaly. (A dense cap rock may complicate the situation.)

The most common extensive natural occurrence of salt rock is as a member of a basinal evaporite sequence (see e.g. Jenyon, 1986d). An idealized evaporite cycle from the base upwards would probably include a clastic member (a product of the initial transgression), followed by 'pre-evaporite' carbonates (limestones, dolomites), sulphates (gypsum/anhydrite), chloride (halite), and potash minerals—the last 'bitterns'. Sometimes a 'regressive' anhydrite caps the cycle. In reality, there are usually omissions, and even reversals, in this order of increasingly soluble (and decreasingly dense) materials. This can be illustrated by the simplified stratigraphic columns for Zechstein sequences in basinal- and shelf-facies environments as shown in Fig. 5.3. In both cases, evaporite cycles Z1, Z2, and Z3 are marked, while the capping anhydrite is probably a 'regressive' anhydrite

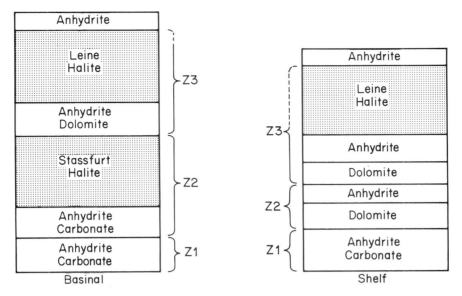

Fig. 5.3. Simplified stratigraphic column showing typical Zechstein salt basin and shelf-facies sequences in the North Sea southern basin area.

of cycle Z4. No distinct bittern (potash) stage is shown, but frequently potash-rich zones occur in the upper parts of the halite intervals. Note the compressed nature of the shelf (marginal) sequence, with a predominance of (non-evaporite) carbonates, and anhydrite.

A reversal of the normal evaporite sequence can occur if there is a sudden influx of fresher water (less concentrated brine, for instance) into the restricted environment of the evaporite basin, lowering the concentration, delaying, or completely preventing the precipitation of 'later', more soluble evaporite members. This is one of the reasons why, in some basins, there are great thicknesses of sulphate minerals (gypsum, anhydrite), but a complete absence of halite.

5.2.1 Deformation of Halite

Halite, which is a (relatively) hard and brittle mineral under standard temperature and pressure, is known to be thermoplastic, and the value of its yield stress is lowered progressively by increasing temperature, as it is also by differential pressure (although changes in confining pressure have negligible effects on the yield strength).

The most important discovery that has been made in recent years is that the presence of even very small quantities of free water associated with the mineral has a dramatic effect on its yield strength, causing it to become mobile under low stress conditions. When water is present—even the minute quantities of 'inherent' brine in fluid inclusions in the halite crystals—a polycrystalline halite mass will deform plastically following quite small changes in temperature and differential pressure. Low-stress deformation of halite has been demonstrated in the laboratory by Spiers et al. (1982, 1986), and in the field by, for instance, Talbot et al. (1982) under a relatively small overburden load in the Boulby Potash Mine, UK.

This work confirms the ideas of Kent (1979) and others as regards the flow of salt at the surface under ambient temperatures and pressures in the 'salt glaciers' of Hormuz, Iran.

Until quite recently, it was thought that high stress conditions (high temperatures, and massive overburden loading) were necessary to cause halite to pass from the elastic to the plastic state—for instance, temperatures in the range 200–350°C were, it has been stated, necessary to bring about the change under constant, normal pressures in the laboratory. Recent work shows that these experimental results on anhydrous halite do not represent the behaviour of naturally occurring halite where some free water is present. There is a range of low-stress-temperature environments in which halite is plastic.

The plastic deformation of halite takes place in two ways. The first is by dislocations and gliding within the crystal lattice (processes that are aided by the presence of water molecules), and the second is by diffusive mass transfer (pressure solution), which takes place by material being transferred by dissolution from points of high stress and reprecipitation at points of lower stress within an aqueous film that can withstand shear stress. In practice, both types of process will lead to a block of material under large vertical pressure becoming more extensive laterally, and thinner along the vertical axis (i.e. being flattened). Thus, a polycrystalline halite mass will actually change shape under stress, rather than undergoing translational movement, although the term 'flow' is often employed as a useful, if non-rigorous descriptive term.

It now seems probable that the effects of small quantities of free water on the processes just mentioned is the main cause of much apparently anomalous behaviour that has been noted in different salt bodies in the field. Primary halite, for instance of the 'chevron' type, has myriads of minute brine inclusions within the crystals. Secondary, recrystallized halite may have much less, or no free water in this form at all. At depth, free water may become available in the vicinity of a salt-rock interval by, for instance, clay mineral reactions (e.g. the montmorillonite–illite transformation), or the conversion of gypsum to anhydrite with increasing burial depth. Thus, some salt-rock units at depth will have, or develop contact with, free water, and others will not; this will materially affect the behaviour of the different salt-rock units when various forms of stress are applied. It is known, for instance, that in some German salt mines, the flow behaviour of different Zechstein salt units differs markedly, even when they are separated by only a few metres depth at total depths of hundreds of metres. Attempts have been made to explain this, yet they have all failed (such characteristics as colour, hardness, chemical purity, etc. have been compared). It seems probable now that a difference in inherent free-water content (never tested) is the reason for the behavioural difference.

5.2.2 Salt Rock and Petroleum

Many of the world's most prolific oil and gas areas are associated directly or indirectly with evaporite sequences, and there seems to be a particular affinity with those that include salt intervals. So far, definitive reasons for this frequent association have not been forthcoming; there are numerous aspects to be taken into account.

As noted, the early stages of deposition in an evaporite basin include 'pre-evaporite' carbonates—limestones and dolomites (the latter usually being secondary by dolomitization of original limestones). These basinal carbonates have usually formed in anoxic conditions of high brine concentrations at the floor of a restricted basin where the near-surface waters are still relatively fresh, and the warm, light conditions support a wealth of biota that thrive until the brine concentrations in the shallow waters become too great for life to continue.

As regards the ability of life forms to withstand increasing salinity, Kirkland and Evans (1981) have suggested, based on a study of present-day saline environments such as the Rann of Kutch, India, that with increasing salinity, the number of viable species may decrease, but the number of individuals in the surviving species would greatly proliferate (owing to the reduction of interspecies competition) as salinity increases, until bittern concentrations are almost reached, when no life form could withstand the dehydrating effect of the brine.

If such is the case, basinal carbonates in an evaporite environment could be organically rich, and prime source rocks for later petroleum development. In the Middle East, great thicknesses of marine carbonate rocks occur, often associated with evaporites, and it seems more than likely that the large accumulations of petroleum in the region are derived from such rocks.

Apart from this possible direct link between the evaporite environment and petroleum, there are many other less direct factors that favour the relationship. When salt rock becomes mobile, for instance, it causes deformation in the overburden rocks, leading to the formation of many potential structural and stratigraphic traps. These may be viable provided they include reservoir and seal units, together with migration routes into the traps from the source rocks, and provided that the migration of any petroleum in the area does not pre-date the development of the traps.

Also, salt rock is probably the best seal formation for traps, both as a capping seal and as a plastic sealant of faults and fractures. In this same capacity, it may also be preventing petroleum from migrating upwards into traps in the shallower section in some areas (as, for instance, in parts of the southern North Sea).

From the viewpoint of hydrocarbon exploration, it should by now be clear why it is important to understand what is known of the behaviour of mobile salt and its effects on surrounding rocks. Apart from initial discoveries, the future direction of some exploration plays in a salt area often depends on insight into the mechanisms of salt tectonics.

5.3 The Initiation of Salt Movement

It is believed, but not established, that a three-dimensional plot against axes of temperature, pressure differential, and moisture content would present a surface above which salt is plastic, and below which it is *not* plastic. A salt body will be potentially mobile when conditions rise above the surface, and cease to be so when they fall below this surface.

Geological history can produce complicated changes in the three parameters of the surface. One step of particular interest is a sharp rise in pressure differential, which can move conditions for a salt body from just below the critical surface to just above it, and so initiate salt movement.

The answer to the question 'What initiates mobility in salt rock?' must be, simply, 'We do not know for certain'. The movement of mobile salt is not observable at depth, and in any case is so slow that except in special situations, like that of the emergent salt diapirs of Hormuz, Iran, which are at the surface, would not be observable within a normal human lifespan. Only the results of the movements in deformation of salt and overburden can be seen in seismic data and from drilling information.

As to what initiates movement, there have been several suggestions made, none of which is generally accepted. Trusheim (1960) was non-committal, stating that initiatory causes were not known, but suggesting that an earthquake (perhaps due to major faulting) or a 'tectonic pulse'—also favoured by Sanneman (1968)—might start salt into motion. It is interesting to note that by implication, Trusheim seems to be saying that the salt is probably already in a potentially mobile state, but requires a transitory stress to be applied in order to start it into motion. Others have opined that basinal salt would begin to move spontaneously once the average overburden density exceeded the density of the salt (i.e. once a density inversion is present).

The writer cannot agree with this view, as there are too many examples of areas where a salt interval with thick, dense overburden exists, but where no plastic salt movement has taken place. Similarly, there are cases known of salt moving beneath overburden that has an average density less than that of salt (see Jenyon (1986d) for examples of both types of area). There are other questions that arise, such as exactly when does the salt start to move—at precisely the moment when the average overburden density equals the salt density, or when the density differential exceeds some specific value? Also, where do the factors of thickness and weight of overburden come into the reckoning? A density inversion of a 1-cm anhydrite band overlying a 1000-m-thick salt interval cannot be expected to generate major salt structures.

The idea of a 'tectonic pulse', as mentioned by Trusheim and Sanneman, deserves special consideration. Attention has been drawn elsewhere (Jenyon, 1986d) to the strange fact that salt rock in many basins down the eastern coasts of North and South America, and the western coasts of Europe and Africa, began to move, and had times of maximum movement (diapirism) over the same period (mid-Jurassic to mid-Cretaceous) in different basins. This is true too often to be mere coincidence, and implies some general factor affecting many basins at about the same time period—and this, together with the actual period mentioned as being involved, points inexorably to a relationship with lithospheric plate movements, and perhaps, further, to 'pulsation tectonics'. This involves periods of acceleration or retardation of plate movement, which could propagate pulses of compression or rarefaction that could be the 'tectonic pulses' mentioned by Trusheim and Sanneman.

Stretching an analogy with the 'elastic rebound' theory of earthquakes related to fault movement, the writer believes that initiation of salt movement in a basin begins suddenly, triggered perhaps by a tectonic pulse or fault-generated earthquake, when the basinal salt (as implied by Trusheim) is already in an unstable, or metastable state due to heavy overburden load pressures, and high temperatures in the deepest parts of the salt basin. The load pressure differential between basin centre and margin may be important. Also, from evidence of the effects of small quantities of free water on halite, it is believed that in some situations, relatively low-stress conditions, with small pressure differentials and temperatures may prevail at the initiation and during the later development of salt mobility.

The presence or absence of free water associated with polycrystalline halite masses may provide an explanation for areas where salt rock should have moved but did not (according to previous—and perhaps still current—ideas on salt mobility), and for others where salt should not have moved, but *did*. Both types of areas exist, and remain as 'skeletons in the cupboard' to be explained only with great difficulty by the older concepts of large overburden loads and high temperatures as being necessary for plastic deformation of salt rock.

A further possible initiator of salt movement is associated with faulting (either dip-slip or strike-slip, so long as there is a real, or apparent, upthrow and downthrow side to the fault). Where a salt interval is affected, but not completely severed across the fault, retaining at least a minimal connection, later planation of the depositional surface will cause a differential load stress to develop in the salt interval across the fault plane, leading to flow of salt across the fault zone from downthrow to upthrow side, with the formation of a salt 'high' in the upthrow side of the salt interval. This mechanism, which has been called the 'Parker–McDowell effect' (Jenyon, 1988f), is one way that salt features may be initiated and developed in association with faulting that does not require the operation of classical hypothetical processes, such as halokinesis or downbuilding (Barton, 1933). It will be mentioned again later, as will other possible salt/fault mechanisms that have been suggested (Jenyon, 1986a).

5.4 Salt Rock as a Seal

Halite is totally impermeable to liquid hydrocarbons, and effectively impermeable (although a negligible amount of lattice diffusion may occur) to gas. Given also that under certain conditions of temperature and differential pressure, halite can 'flow' plastically, it should obviously be, and is, a near-ideal sealing formation for petroleum traps. It can be particularly effective in some situations involving faulting, where the ability of the mineral to flow, seal, and 'heal' the fracture or fault provides a completely efficient sealing material, provided it is present in sufficient quantity.

As regards the durability of such a seal where water is present, there are different aspects to this question. As noted previously, it has been established that small quantities of free water (brine) inherent within, or in contact with, halite crystals, will facilitate passage of the halite from the elastic to the plastic state. However, such inherent free water is not *vital* to halite deformation, given the right physical conditions of temperature and differential pressure. Halite will undergo plastic deformation in quite anhydrous conditions, although the presence of the free water drastically reduces the yield strength of halite when it *is* present.

On the other hand, circulating undersaturated water/brine that comes into contact with a halite mass *can* dissolve and remove it provided the water has means of egress from the system as well as ingress. In other words, there must be the establishment of a circulatory system—by brine density flow under gravity control, by thermal convection, or by other means (see e.g. Anderson and Kirkland, 1980).

If such a circulatory system is not established, then *standing* water will not effect dissolution and removal of the salt: only ion exchange between brine and halite will take place. Under such circumstances, Baar (1977) states that a polycrystalline halite mass is effectively impermeable at burial depths (under a clastic overburden) greater than about 300 m.

Thus, a salt interval at depth can, under quite normal circumstances, provide an impermeable and ductile seal formation for a petroleum trap, and many examples are known from different parts of the world. In the North Sea, for instance, parts of the southern basin, where the Zechstein salt occurs in sufficient thickness to form a seal, are the locations for the so-called 'Rotliegendes play', with the salt forming a capping seal above the Lower Permian Rotliegendes sands. The salt here is particularly important and effective, since the basal Permian is much affected by faulting of strike-slip type, and the movements of the faults beneath the Zechstein have been largely accommodated by plastic deformation in the salt layer, which has prevented the fault movements from being transmitted upwards into the post-Zechstein section, and has maintained an efficient seal above the Rotliegendes

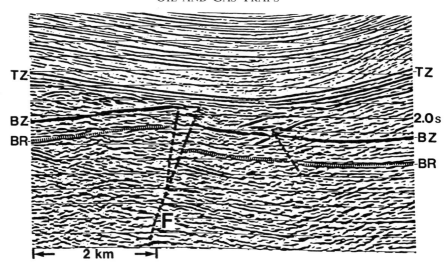

Fig. 5.4. An example showing how a mobile salt interval BZ–TZ (Base Zechstein, Top Zechstein) can accommodate movement of a major strike-slip fault (F), preventing its effects from being transmitted into the overburden above the salt. Note the thickness change in the BR–BZ interval across the fault zone. BR is the base of the (Lower Permian) Rotliegendes. Reproduced by permission of Seismograph Service (England) Ltd

gas accumulations. The amazing capacity of a mobile salt interval to accommodate strike-slip movements beneath it is illustrated in Fig. 5.4, which shows a fault system of this type at F, affecting the Lower Permian Rotliegendes sands (interval BR–BZ), and the Zechstein salt interval (BZ up to about 100 ms below TZ). Clearly, the uppermost 100 ms of the Zechstein interval, which is the strong band of reflections at TZ that is the response to carbonate and anhydrite units capping the Zechstein here, is quite unaffected by the major strike-slip fault F. The thickness difference in the BR–BZ interval across the fault, owing to the juxtaposition by horizontal displacement of two different zones of the truncated interval, is easily seen. The Zechstein salt interval here has completely accommodated the horizontal movements of the fault system.

5.4.1 Salt-sealed North Sea Traps

The UK southern basin of the North Sea provides a number of examples of traps that are sealed by Zechstein salt in what has come to be known as the 'Rotliegendes play'. The Permian in this area consists of Rotliegendes (Lower Permian), with red beds and sandy intervals that are of reservoir quality in some areas—especially where aeolian dune sands occur, overlain by Zechstein (Upper Permian) consisting of several cycles of evaporites, including massive salt-rock intervals that have become mobile, acting as a seal for gas accumulations in the underlying Rotliegendes sands. It is generally accepted that the gas source is in the Coal Measures (Upper Carboniferous) series, which is well developed in this basin.

The Zechstein salt acts not only as a capping seal for the Rotliegendes structures, but also 'heals' and seals many faults affecting the Rotliegendes/Base Zechstein zone, most of these being of strike-slip type due to reactivation during the Late Cimmerian (Upper

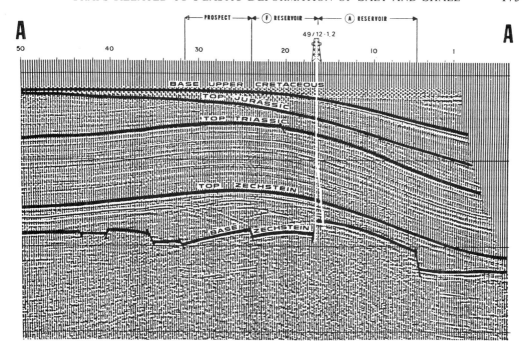

Fig. 5.5. Seismic section across the North Viking gasfield structure (see also the map in Fig. 5.6). The top of the Rotliegendes gas sand-pay can be regarded effectively as the Base Zechstein marker, above which the thick, mobile Zechstein salt forms a very efficient seal formation above the faulted reservoir. Reproduced by permission of Elsevier Applied Science Publishers Ltd (from Gray, 1975)

Jurassic–Lower Cretaceous) movements. The salt has accommodated the horizontal translational movements of these faults and prevented the effects of the faulting from penetrating above the salt into the upper section of Mesozoic and later rocks.

This is well illustrated by the seismic section in Fig. 5.5, which runs in a NE–SW direction across the North Viking gasfield. The line of the section is shown on the depth-structure contour map at Base Zechstein level seen in Fig. 5.6. The field has been described by Gray (1975) as one of a series of fault-bounded structures with separate GWCs. The main structures are two *en echelon* anticlines, one of which is the asymmetrical structure about 10 miles long and 3 miles wide, trending NW–SE, which forms the North Viking Field illustrated.

The Base Zechstein marker is mapped for the underlying Rotliegendes, since the latter interval is relatively thin and poorly resolved in the seismic data—particularly older data, as in the case of the Fig. 5.5 illustration, which gives a fairly clear and simple indication of the structure but lacks any significant detail. Since only 150–190 feet separates the Base Zechstein marker from the top of the Rotliegendes reservoir, it has been the normal procedure to map the marker, which always provides a strong, clear reflection group. Information on significant thickenings/thinnings of the Rotliegendes sand intervals has had to be obtained 'retrospectively' from drilling data. The seismic illustration shows the Zechstein salt pillow beneath which the trap lies, and the strongly faulted Base Zechstein/Rotliegendes zone with the faulting being accommodated by the mobility of the Zechstein salt.

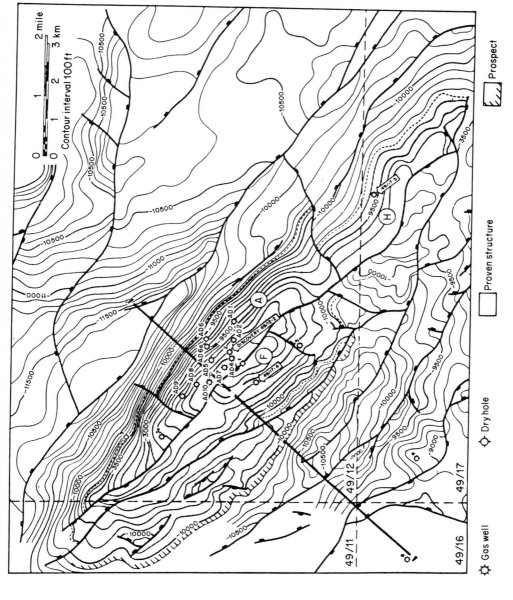

Fig. 5.6. Depth-structure contour map of the North Viking gasfield at Base Zechstein level, with the line of section of the Fig. 5.5 example marked A–A'. Reproduced by permission of Elsevier Applied Science Publishers Ltd (from Gray, 1975)

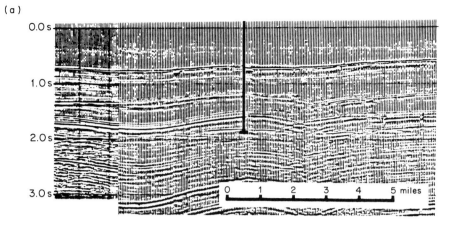

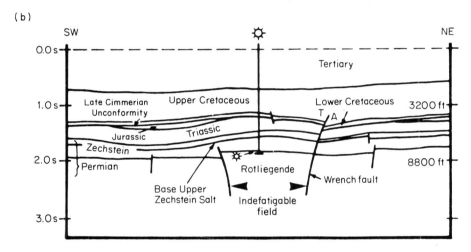

Fig. 5.7. Seismic section (a) and geological interpretation (b) of the Indefatigable gasfield, North Sea. Another Rotliegendes play, with Zechstein salt sealing the faulted reservoir. Reproduced by permission of Elsevier Applied Science Publishers Ltd (from Blair, 1975)

It is interesting to note that in the early days of exploration in the North Sea, seismic data quality was poor, and often locations were 'spudded' on vague seismic indications of structural highs in the section. In cases like that of the North Viking Field (although this field is not specifically being referred to in this context), provided the well was drilled deep enough to penetrate the Rotliegendes, a serendipitous discovery could be made.

This would also be the case with a structure like that of the Indefatigable gasfield illustrated in Fig. 5.7 with a seismic section from south-west to north-east, and with an accompanying geological interpretation in the line-drawn section beneath. The field has been described by Blair (1975) and others, and is a polyphase structure with a complex history. Drilling information indicates, by a thinning of the Rotliegendes interval within the zone of the strike-slip fault-bounded trap, that this location was structurally high in

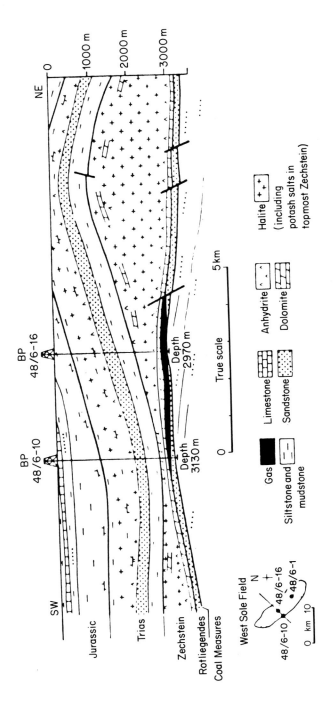

Fig. 5.8. Geological cross-section of the West Sole gasfield, based on seismic and drilling data. The problems faced in depth conversion of seismic time data can be imagined from the way the Rotliegendes structure is offset from beneath the thick Zechstein salt pillow (right), and lies beneath relatively thin but variable salt. Reproduced by permission of Elsevier Applied Science Publishers Ltd (from Butler, 1975)

Rotliegendes times. However, in the view of the complex structural history of the feature (see Blair (1975) for details), it is purely fortuitous that the Rotliegendes feature is also structurally high today, and lies below a clear Zechstein salt pillow.

The strike-slip fault south-west of the field is accommodated by salt mobility and does not affect the top of the Permian or the section above. However, the major strike-slip fault on the north-east side appears to be one of the exceptional faults of this type that has traversed the (relatively thick) salt interval, and it is interesting to note the evidence for the nature of the fault that lies in the sudden change of thickness of the Triassic section (obscured to some extent by the Late Cimmerian erosion) from thick, on the south-west side of the fault, to thin on the north-east side.

An unusual example of a salt-sealed Rotliegendes reservoir trap is provided by the West Sole gasfield, another Rotliegendes play feature in the UK Southern basin. A description of this field has been given by Butler (1975), and the illustration in Fig. 5.8 shows a geological interpretation based on seismic and drilling data, in a NE-SW section across the field.

The Rotliegendes structure is unusual as it is offset from beneath a large Zechstein salt pillow (as seen, to the north-east of the field), which results in the Rotliegendes trap being located largely beneath a structural low in the overburden.

In a fascinating account of the historical stages of the interpretation of the field from 1964 to 1970, Hornabrook (1975) describes the sophisticated methods of depth conversion of the seismic data developed over this period to deal with apparently intractable problems resulting from the configuration of the feature, velocity hysteresis effects related to a complex burial history, and the variable thickness of salt overlying the reservoir. The seismic section in Fig. 5.9, from Hornabrook's paper, clearly shows the Rotliegendes trap in cross-section (seismic data of 1969 vintage). The work was carried out during the early period of North Sea exploration, when the information base for seismic velocities and offshore geology was negligible.

The three fields just mentioned briefly are included purely to illustrate some structural aspects of the salt-sealed trap. For further details, the cited papers should be consulted.

Another interesting situation involving salt sealing is present in the area known as the Broad Fourteens Basin, in Dutch waters of the North Sea—mainly in K Block.

Following early Cretaceous uplifts in adjacent (flanking) areas, a Broad Fourteens inversion axis developed in late Cretaceous times, as shown in the sketch map and schematic sections of Fig. 5.10.

In Fig. 5.11 a seismic section across the inverted structure in K-17 is shown as an unmigrated section (a), and a migrated section (b); a geological interpretation is shown in (c). These figures are from the paper by Oele *et al.* (1981), which should be referred to for further details.

Under other circumstances, the faulting that resulted from deformation attendant on the inversion tectonics would almost certainly have breached any other type of seal except mobile salt rock. In this case, the mobile salt has accommodated the fault movements, and the faulting (lower series) has not penetrated upwards through the Zechstein salt interval. This may indicate a strong element of strike-slip movement in these faults, as the 'throw' on some of the faults is too large to have been accommodated by the salt were the faulting to be of dip-slip type.

The Zechstein salt, as usual, forms an excellent seal above the Rotliegendes sand reservoirs. It is thought that much of the gas present in the traps today probably remigrated

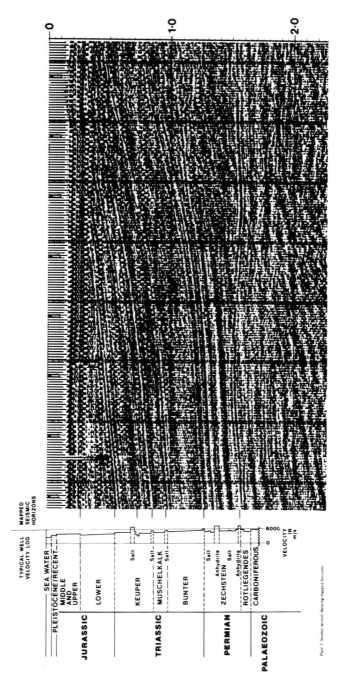

Fig. 5.9. A seismic example from the early phase of exploration of the West Sole gasfield, after some sophisticated processing methods had been applied. The strong dip reversal at Rotliegendes level, sealed by Zechstein salt, can be seen clearly. Reproduced by permission of J. T. Hornabrook (from Hornabrook, 1975) and the NGU

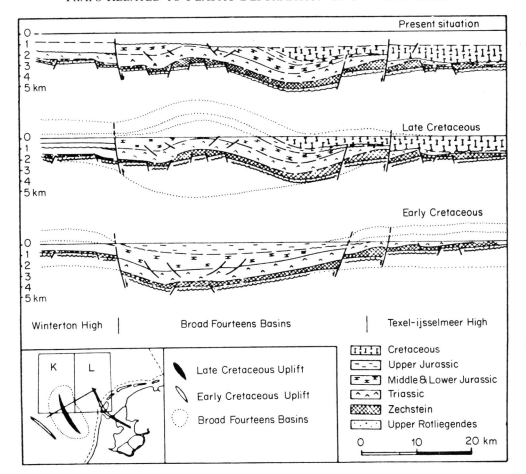

Fig. 5.10. Schematic sections across the Broad Fourteens Basin in Dutch waters of the North Sea show how the late Cretaceous inversion axis evolved from early Cretaceous times to the present. Reproduced by permission of The Institute of Petroleum, London (from Oele et al., 1981, Fig. 12, p. 293)

from earlier traps during the inversion tectonics period, and some gas may have escaped at this time. In the general area, some fourteen gasfields have been found, varying in size between 3×10^9 m^3 and 25×10^9 m^3 of recoverable gas.

5.5 Effects of Syndepositional Mobility

When a potentially mobile stratigraphic unit (e.g. salt, shale) begins to deform plastically beneath a clastic overburden, the effects of this movement are transmitted upwards through the overburden to the contemporary depositional surface. Depending on the rate of the plastic deformation together with other factors, such as the exact lithologies present in

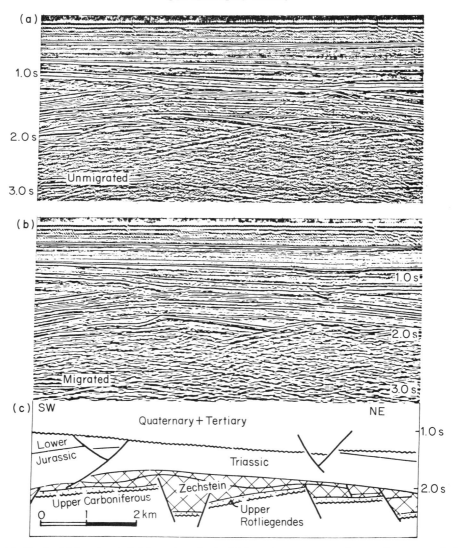

Fig. 5.11. Stacked (a), migrated (b), and geological (c) sections across an inverted structure in K-17 Block (Dutch waters of the North Sea) showing faulted Rotliegendes traps in the inverted section beneath Zechstein mobile salt. Reproduced by permission of The Institute of Petroleum, London (from Oele *et al.*, 1981, Fig. 15, p. 297)

the overburden, their state of compaction and competence, pore fluids present, etc., there will be a time lag between the movements of the plastic unit occurring, and the effects appearing at the depositional surface. Eventually, however, zones of subsidence and/or uplift will develop at the surface corresponding to the shapes evolved at the surface of the mobile unit.

Since in this chapter the 'mobile' unit involved will generally be salt rock, it will hereafter be referred to as such in most instances.

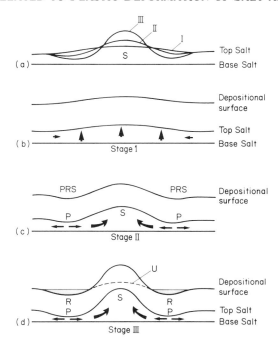

Fig. 5.12. Schematics showing the development of a salt uplift, and some early effects on the overburden (i) as regards deformation of strata already deposited before salt movement began, and (ii) in deposits formed during the salt movement period. See text for explanation

A series of sketches is shown in Fig. 5.12 that may be imagined as sections across either a subcircular or an elongate (in plan view) salt swell. Sketch (a) shows a salt interval that becomes plastically mobile, indicating stages I, II, and III in the progressive deformation of the top of the salt to form a salt swell/pillow at the location marked S. In the sketches (b), (c), and (d) following, the three stages of the salt surface deformation are shown, together with an imaginary clastic overburden layer (the interval TS–depositional surface). In sketch (b), a vertical salt uplift is beginning to develop (large arrow heads) due to lateral flow of salt in towards the uplift (small arrows) within the salt interval BS–TS. The depositional surface is also expressing this uplift. Note that in this discussion, the originating cause of the salt movement is not stated—but could, for instance, be a fault at the Base Salt level beneath location S.

In sketch (c) a distinct low (P) has developed in the top salt surface on each side of the swell S, due to flow of salt inwards towards S and outwards away from S. The development of low features, such as P, was studied from the viewpoint of fluid mechanics by Nettleton (1934, 1943) in model investigations using two immiscible liquids of differing specific gravity; the basic premise was that a mobile layer and a clastic overburden both behave under stress like viscous liquids. Nettleton termed the low that developed in the mobile-layer surface (P in Fig. 5.12) the *peripheral sink*, and demonstrated that it is an inevitable result of the pattern of 'flow' in a mobile layer beneath a more dense overburden layer, once an uplift has been initiated. The reason for the cause of the flow towards and away from the salt uplift S, as referred to the trough of the peripheral sink P, is the pattern of differential overburden load pressures that develop (see e.g. Nettleton 1934, 1943; Jenyon 1986d).

The peripheral sink (P) will be of annular form if the salt swell S is subcircular in plan, or in the form of two subparallel troughs if the salt swell S is elongate (normal to the paper).

The deepening of P is expressed at the contemporary depositional surface by the development of a subsidence trough (or troughs), such a feature being known as a *primary rim syncline*—PRS in sketch (c). Again, this feature may be of complete (or more usually, incomplete) annular plan view, or a flanking pair of subparallel troughs in plan, depending on the basic plan shape of the salt swell. It should be clearly understood that the PRS may be thousands of feet above the top of the salt surface, depending on the thickness of the clastic overburden. (Some interpreters inexperienced in salt geology tend to confuse the peripheral sink with the PRS, which is excusable, since in the past there has been considerable ambiguity in the use of these terms. However, the significance of these two quite distinct features should be clearly understood.) The overburden, it should be noted, will have uniform thickness owing to deposition before any salt movement takes place, unless affected by external factors, such as tectonism, differential erosion, etc. This uniformity of pre-salt movement overburden is an important factor in the dating of salt movement periods.

As the structure develops—sketch (d)—the uplifted overburden above S will tend to undergo erosion (either subaerial or submarine). The products of this erosion from above some unconformity surface U (dashed line) will be redeposited within the PRS. Other erosional products from the extremities of the sketch section may also be transported into the PRS, the basal surface of which is marked R in the sketch. The PRS deposits (horizontal

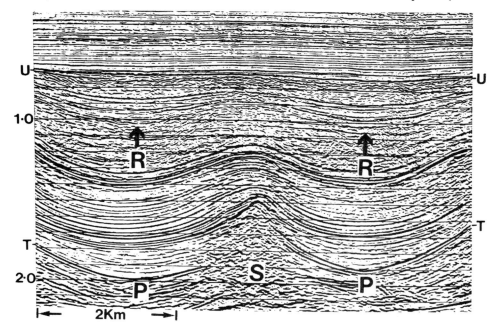

Fig. 5.13. Seismic example of primary rim syncline deposition (at R) related to a mobile salt uplift (S) during the pre-piercement stage of salt movement (although S has begun to pierce the overburden at a late stage). Top of the mobile salt interval is at T, and peripheral sinks in top salt surface are at P. U = unconformity. Reproduced by permission of Seismograph Service (England) Ltd

TRAPS RELATED TO PLASTIC DEFORMATION OF SALT AND SHALE 185

line shading) are thickest at the trough centre(s), thinning in both directions away from this; note particularly the thinning towards the salt swell S.

It should be clear that the presence of PRS sediments marks the onset and continuation of salt movement in the structure during this *pre-piercement* stage of development (i.e. before any diapiric piercement of the overburden by the salt has occurred).

A 'real Earth' seismic example of the situation developed in Fig. 5.12(d) is shown in Fig. 5.13, where a salt interval, the top of which is indicated at T, has become mobile with the development of a swell at S. (Note that S has become a piercement at a late stage, but this is not important in the context.) The trough points of the peripheral sink developments are shown at P. Above this, at R, the trough points of the PRS developments are seen beneath an unconformity surface U. Note the thinning of the stratal event intervals above R and below U—both inward towards a location above S and outward towards both extremities of the section. It will be seen how such interval thinning could lead to trap situations, and also how good stratigraphic identifications of events at the level of R can lead to precise dating of the onset of salt movement in structures of this type. Later, it will be shown how in some circumstances, the PRS sediment lens may undergo inversion in a 'turtle' structure, producing an even more attractive potential trap situation.

In many cases, similar situations to that seen in Figs. 5.12(d) and 5.13 exist, but where the unconformity U has bitten down deeper into the section, removing all evidence of the PRS sediments and leaving only uniform-thickness pre-salt-movement sediments between the top of the salt and the unconformity.

The results of the continuation of the salt uplift into the diapiric stage, with piercement of the overburden and the development of secondary (and in some cases, third-order) rim synclines will be discussed later in the chapter.

5.5.1 Traps Related to Pre-piercement Salt Structures

The structures produced by salt movement create many potential hydrocarbon traps. They can be grouped into types for convenience of description, although there is much overlap between types, and details vary infinitely. In Fig. 5.14, five of the most common trap situations associated with a pre-piercement salt structure (a swell, or as here, a pillow) are modelled (in solid black) in two dimensions within the plane of the seismic section, and numbered 1–5. It must be remembered that these two-dimensional models must, in the 'real Earth', be accompanied by suitable geometry in the third dimension in order to produce either a structural or stratigraphic trap.

Trap type (1) is the simplest of the models, representing a closed high feature resulting from uplift of the overburden by the rise of the salt pillow—the mobile salt interval being indicated at S. Three-dimensional closure must be present; this may exist because the structure is a domal pericline, or is an elongate anticline with double plunge in the two directions (towards and away from the observer) normal to the section, or is some combination of these with faulting. The hydrocarbon accumulation, represented as being in the reservoir in solid black, may have a flat base (OWC, GOC, or GWC), or two separate basal contacts, depending on the thickness-controlled geometry of the reservoir.

Trap type (2) is a stratigraphic feature formed by an interval that thins and feather-edges updip in the flank of the overburden fold, with the assumption that it is enclosed by seal formations. More will be said of this later in relation to trap type (5).

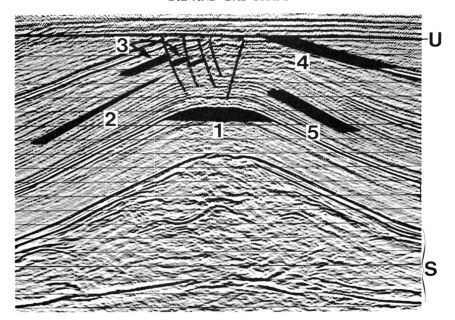

Fig. 5.14. Some possible hydrocarbon trap situations related to a pre-piercement salt structure (S). Trap types, numbered 1–5 are modelled in a seismic section. U = unconformity. Reproduced by permission of Elsevier Applied Science Publishers Ltd (from Jenyon, 1986d), with acknowledgements to Seismograph Service (England) Ltd

Trap type (3) is a series of possible traps against faulting—the latter in this case being present over the crest of the anticline due to extensional stress zoning produced by the salt uplift. Other types of faulting over the crest and down the flank may occur, such as gravity faults hading downdip as a result of slope instability in the flank sediments.

Trap type (4) is a truncation trap beneath an unconformity U. This is a common situation where there are unconformities present in the section and uplift has occurred before the erosional hiatus. The immediate post-unconformity sedimentary unit must be of seal quality, of course, with capillary entry pressure sufficiently high to prevent the escape of the hydrocarbons. This must also be true of the pre-unconformity units above the reservoir.

Trap type 5 is a stratigraphic trap due, for example, to tightening porosity updip related perhaps to diagenesis; alternatively, it may be the result of updip facies change. Some would also argue the possibility of hydrodynamic trapping, depending upon the shape of the reservoir unit.

The history of salt movement and overburden deposition determines whether trap types (2) and (5) are created.

Figure 5.15 is a seismic section across one flank of a large salt pillow (H), the top of the mobile salt layer being at T. A number of erosional hiatuses are present in the overburden representing the removal of considerable amounts of sediment at various times, including the crest of the pillow (just off-section to the left). The principal visible unconformities are marked as U1, U2, U3, and U4.

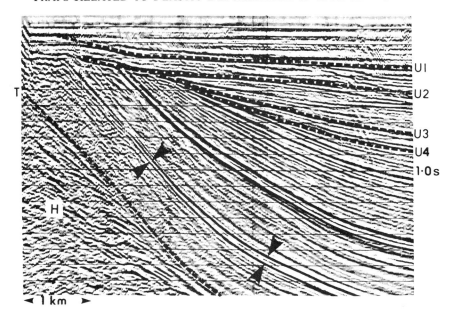

Fig. 5.15. Erosional hiatuses, and thinning intervals due to syndepositional movement of a large salt pillow (H), top of the mobile salt being at T. One of the many thinning intervals is arrowed. Four unconformities are marked U1–U4, but there are many other depositional hiatuses present. Reproduced by permission of Seismograph Service (England) Ltd

It is clear from the stratal reflection events above the top salt T, that growth of the salt pillow began after deposition of only 200–300 ms (two-way time) of the overburden sediments present above the salt. The evidence for this lies in the thinning and feather-edging of many of the seismic intervals—one of these being indicated by the arrowheads—upflank, demonstrating syndepositional growth of the salt pillow. The top of the uniformly thick layer of post-salt, pre-salt-movement sediments immediately overlying the salt is somewhere below the arrowed interval. The general thinning of the section above this and up to U4 shows that the salt pillow underwent relatively continuous growth during this period (sometimes, such growth is episodic, evidence by thinning intervals separated by uniformly thick intervals).

The Fig. 5.15 situation leads to the possibility of traps of types (2), (4), and (5) as seen in Fig. 5.14 (since there could also be porosity traps within the thinning intervals). This is one instance where the primary rim syncline sediments, as seen in Fig. 5.12(d) and Fig. 5.13, are absent, having been removed during the erosional period represented by the unconformity U4.

If Fig. 5.16 is now examined, it is seen that the seismic section shows the top of a mobile salt layer at T, and a slightly asymmetrical salt swell in the lower centre of the section. There are two principal visible unconformities (U) present in the overburden section.

It is immediately noticeable that in comparison with the section in Fig. 5.15, there is little evidence of any updip thinning and feather-edging of reflection intervals—i.e. there is no supratenuosity in the overburden above the salt swell, although at some levels there

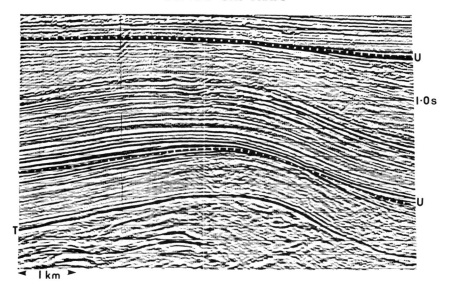

Fig. 5.16. Seismic example of a salt swell with salt top at T. In this case, there is no evidence of supratenuous thinning above the salt uplift. Stratal events in the interval between T and the lower unconformity (U) are uniform in thickness, and are clearly pre-movement deposits. Between the two unconformities, there is some slight general thinning from left to right across the section, which must also have pre-dated the salt movement. There is some minor extensional stress-zone faulting above the crest of the salt swell. Reproduced by permission of Seismograph Service (England) Ltd

is a slight tendency for thinning *across* the whole section from left to right, which clearly must pre-date the salt movement.

What is seen here is a substantial section of effectively uniform thickness above the salt—i.e. the uniform, pre-salt-movement deposits. Any post-movement deposits, including primary rim syncline sediments have been removed by erosion. Trap types possibly present are (1), (4), and (5). There is obviously little possibility of trap type (2) being present. On the other hand, trap type (3) may occur, as there is some minor stress-zone faulting present above the pillow crest between the two unconformities as indicated.

(It is interesting to note that the sedimentary pattern discussed can sometimes be seen in line-drawn sections based on drilling results, as seen in older works. The sketch in Fig. 5.17a shows a section across Oligocene sediments adjacent to the Port Neches Salt Dome, Orange County, Texas (Levorsen, 1967). The marked thinning towards the salt body shown by the Frio and other units categorize this as a primary rim syncline sequence, deposited during the pre-piercement 'pillow' stage rise of the salt.)

The trap types discussed have been models only; in the 'real-Earth', problems would be posed as regards hydrocarbon migration into these traps through the thick salt barrier, although such accumulations *do* occur in special situations. It was seen in section 5.4.1 how the salt seal can operate in favour of the explorationist; in the present situation, the opposite is often the case.

Thus, in an area of mobile salt geology, it can be of significance to the determination of trap-type potential to identify and differentiate between events in seismic data that

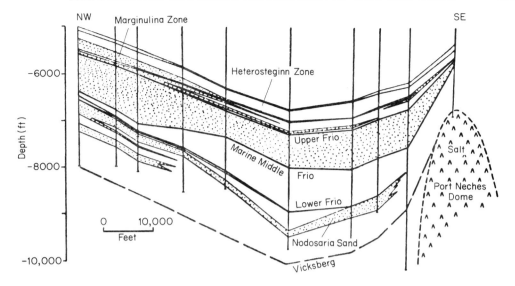

Fig. 5.17a. Geological section shows thinning towards the salt diapir—the Port Neches Dome, Texas. These thinning Oligocene sediments are primary rim syncline deposits that indicate the period of pre-piercement movement of the salt body. From GEOLOGY OF PETROLEUM, 2/E by A. I. Levorsen. Copyright 1954, © 1967 W. H. Freeman and Company. Reprinted with permission

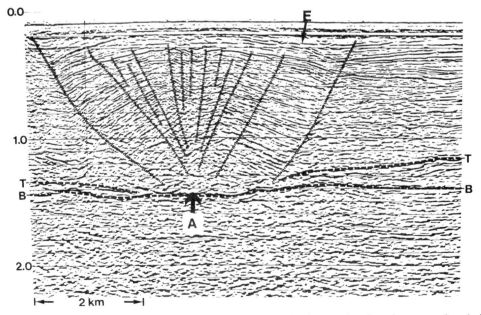

Fig. 5.17b. An unusual 'reversed-flow collapse structure' in the overburden, interpreted as being due to salt withdrawal (salt interval between B and T) from a developing pillow. Reproduced by permission of the *Journal of Petroleum Geology* (from Jenyon, 1988b)

represent pre-movement overburden sediments (of uniform thickness), and those that represent deposition contemporary with salt movement (i.e. those sediments that show supratenuous thinning above a salt uplift location) as well as to identify any erosional hiatuses in the overburden. It will be seen later that seismic differentiation can also be made between these deposits, which are pre-piercement overburden sediments, and those sediments (post-piercement) that are deposited after a diapiric piercement of the overburden by the rising salt mass. The latter type of sediments occurs in a depositional environment peculiar to a salt tectonics setting, and can result in another set of characteristic trap types.

It is also worth noting at this point that the depositional patterns discussed here may still be recognizable in the overburden in areas where, at later stages, the original mobile salt interval has been removed partly or wholly by dissolution or flow, or even where the source salt layer is too deep to be identifiable in the seismic data.

An interesting example is shown in Fig. 5.17b where the seismic data show what is believed to be a case where a salt pillow was originally developed owing to uplift and thickening within the salt interval B–T. This uplift resulted in extensive fracturing of the overburden. Later, it is believed that the flow of salt into the pillow was interrupted and *reversed*, with bilateral flow out of the pillow taking place, and the antiform overburden structure subsiding on to the base of the salt interval at A. During the process, the subsiding wedge developed a fan-shaped series of faults as shown, the earliest being in the centre above A, and the latest being the two large bounding faults (as determined by the upper terminations of the faults). The onlapping of the upper part of the wedge, as at E, indicates that the feature (which is elongate normal to the section) developed from mid-Cretaceous to late Tertiary times. If some of the faults are of sealing type, and reservoir/seal formations are present in the wedge, the complete withdrawal of salt at A could have led to very favourable trapping situations here. This structure has been referred to as 'reversed-flow collapse structure' (Jenyon, 1988b).

5.6 Diapirism and Post-piercement Salt Movement

Local uplifts of mobile salt that begin as non-piercement swell and pillow features may continue upward movement until the salt pierces the overburden rocks and penetrates towards the surface. Such a feature is known as a salt *diapir* or *piercement* structure (shales, clays, and other materials can also form diapirs), and may produce important deformation effects and new depositional patterns in the overburden. Often the possibility of hydrocarbon traps is enhanced by the operation of these processes.

In Fig. 5.18 a seismic section across an elongate salt-wall diapir (C) is shown. The diapir has developed from a source salt layer (A) between horizons BZ and TZ. The undulatory deformed event B is produced by a competent dolomite/anhydrite band (the Plattendolomit event, P), which is encased by the Z2 (below) and Z3 (above) salt intervals.

The deformation of P, seen at (B), is due to lateral flow of salt rock into the diapir C. A good idea of the thickness of the salt interval as it was originally is given by the position of TZ (the top of the salt) at the south end of the section. Before any salt movement took place, the top of the salt interval was approximately at this level across the whole section. As seen, marked subsidence has taken place on both sides of the diapir owing to flow of the salt from the BZ–TZ interval into the diapir from both sides. On the south side, the subsidence owing to salt withdrawal into the diapir has caused a major gravitational normal

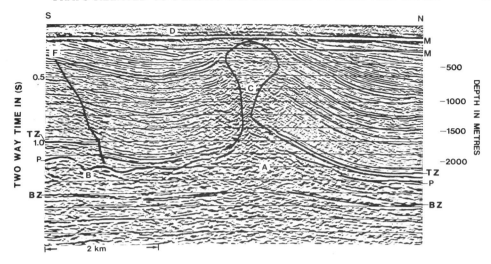

Fig. 5.18. Seismic section across an elongate salt wall diapir (long axis is normal to this section). C = salt diapir; A = mobile salt interval below the piercement; BZ = base (Zechstein) of salt interval; TZ = top (Zechstein) of salt interval; P = competent marker band (Plattendolomit) within salt—also indicated at B where it has undergone severe deformation owing to lateral salt flow out of the salt withdrawl basin adjacent to the diapir. M = multiple reflections. Also marked, at D within the section, is the shallower of the two multiple reflections. A large overburden fault is indicated at F, which is due to overburden subsidence into the salt withdrawal basin. Reproduced by permission of Seismograph Service (England) Ltd

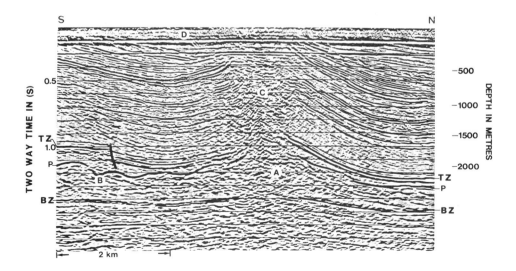

Fig. 5.19. Same example as in Fig. 5.18, but without interpretation marks

fault, F, to occur. In the shallow part of the section, the strong horizontal event marked D and the weaker parallel event 100 m below it are both multiple reflections (marked M on the north side of the section). The cross-sectional outline of the diapir C is approximate only.

An unmarked version of the section is provided in Fig. 5.19 for comparison.

Note the shape of the subsided overburden to the north and south of diapir C. The top of the original salt layer at TZ is forming a peripheral sink into which the overburden sediments have subsided in two trough-like developments often referred to as *salt withdrawal basins*. The uplift in the top salt surface as it approaches the diapir, and the upturn of the stratal events as they approach the diapir flanks and crest, indicate the upward thrust of the mobile salt mass through the overburden material.

The overburden sediments themselves seem for the greater part to consist of the 'pre-movement' deposits of uniform thickness discussed previously. Only in the shallowest part of the section, above about 0.4 s in the troughs of the salt withdrawal basins, is there some indication of thinning towards the diapir that would be evidence for the presence of primary rim syncline sediments; at this shallow location in the section, there is considerable obscuration of the primary data by horizontal shallow multiple reflections. As is often the case, depositional patterns produced by the pre- and post-piercement salt movements are either too shallow to be seen, or have been removed by erosion (there is a shallow unconformity that is obscured here, again by the multiple reflections).

However, the relatively simple structural situation shown by this example demonstrates the possibility of the existence of traps in the 'uniform' overburden by truncation against the diapir flanks and also against any shallow unconformity present above the diapir. Some shallow-angled truncations below the base of the salt interval BZ (acting as a seal) also hold the possibility of pre-salt traps in this situation. The gravity fault at F caused by the subsidence into the peripheral sink of the sediments in the salt withdrawal basin also offers truncation trap possibilities. Apart from the structural traps mentioned, there could also be porosity or facies-change situations present in the 'uniform' overburden units.

Note that there may be an important fault beneath the diapir at and below BZ level, which is obscured by the positive velocity 'pull-up' anomaly produced by the large overlying thickness of salt resulting from the vertical extent of the diapir. Faults in this type of location are often the initiatory cause of the salt piercement, especially where a later planation of the surface has caused an overburden load pressure differential to develop across the fault zone. This can lead to a salt uplift on the upthrow side, whence a diapir. The schematic diagrams in Fig. 5.20 show in a very simplified way the stages of the diapiric process, and the depositional patterns that develop during the syndepositional salt movements.

Diagram 1 represents an initial salt uplift that occurs for some unspecified reason, which could be for example, a fault movement at the base of the salt interval (cross symbol). The blank section above the salt up to the depositional surface represents a clastic overburden of uniform thickness overlying the salt layer.

Diagram 2 shows a developing salt pillow, with lateral flow of salt into the pillow leading to the peripheral sink development (PS)—either one annular depression or two parallel depressions in the salt surface, depending on whether the diapir is subcircular or elongate in plan. Eventually the deformation of the uniform clastic overburden above the salt leads to primary rim synclines (PRS) at the depositional surface that receives sediments from the surrounding positive areas, including the uplift above the salt body.

TRAPS RELATED TO PLASTIC DEFORMATION OF SALT AND SHALE

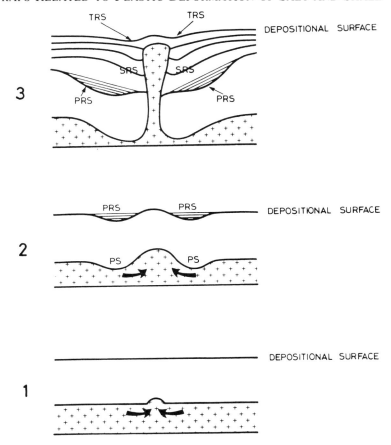

Fig. 5.20. Schematics showing the stages of diapiric process, and overburden depositional features, in very simplified form

In Diagram 3, the salt pillow has become a diapir, piercing the clastic overburden. There is an accelerated flow of salt into the diapir in the early stages (see e.g. Seni and Jackson, 1983), the salt being drawn from the source interval in the immediate vicinity of the diapir root—and sometimes only from the deeper part of the salt interval. Subsidence occurs rapidly into an annular sink about the diapir. (The horizontal cross-section of the diapir is usually relatively small compared with the area from which the salt is being drawn; because of this and the depth below the surface—not appreciated from these sketches owing to the scale—the actual uplift above the diapir is not noticeable in the subsidence pattern until a later stage when the diapir crest is approaching the surface.) The subsiding trough deepening towards the diapir—either all round it, or on both sides, depending on the plan view of the salt structure—is filled quite rapidly with sediments that thicken *towards* the diapir. Note the contrast with the PRS sediments that *thin* towards (and away from) the diapir from a trough zone within the rim syncline.

This trough of sediments that thickens towards the diapir is called the *secondary rim syncline* (SRS in the sketch) and should be compared with the PRS basin of deposition

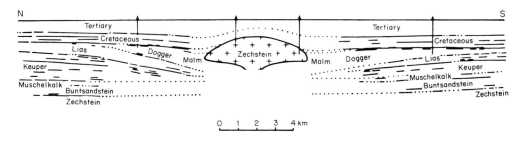

Fig. 5.21. Outline geological interpretation of a salt diapir in north-west Germany, with both primary and secondary rim syncline deposits present. From Trusheim, 1960, reprinted by permission of American Association of Petroleum Geologists

as just noted. Just as the onset of PRS sedimentation (effectively with supratenuous thinning in the overburden over the crest of the salt feature) signals the beginning of pre-piercement salt uplift, so the onset of SRS subsidence and sedimentation indicates the beginning of the diapiric piercement stage of salt movement.

The geological interpretation of an example that shows the two different types of sedimentary feature is seen in Fig. 5.21, where a Zechstein salt diapir in the north-west German area is illustrated from the classic paper by Trusheim (1960). It is clear that the pre-piercement salt uplift in this location was initiated at the beginning of the Keuper, and continued through Liassic times (with the pre-Keuper section, consisting of Bunter Sandstone and Muschelkalk sequences, forming the uniformly-thick pre-movement section). The Keuper and Lias sequences can be seen thinning towards the diapir on both sides. The onset of piercemcnt occurred at the beginning of (Jurassic) Dogger times, and continued through the Malm period, as is shown by the *thickening* towards the diapir of this interval on both sides of the section. A basal Cretaceous unconformity removes some of the later evidence of SRS sedimentation, and at the base of the Tertiary section, there is some indication of a minor, third-order (or 'tertiary' with a lower-case 't') rim syncline (TRS) formed at a late stage with the approach of the diapir to the surface, accompanied by small uplift/subsidence effects.

As Trusheim (1960) points out, the basal beds of the primary rim syncline (to some extent), and particularly those of the secondary rim syncline, mark a separation between very different depositional environments. They could be expected, therefore, to show up as strong and distinct events in seismic data. In general this is true of the base of the SRS in a seismic section, provided it is present. Sometimes, the source salt layer is at very great depth, and together with the PRS and SRS sedimentary stratal reflections is beyond the vertical range of high-resolution seismic data.

Often, on the other hand, these sediments have been removed during one or more erosional phases, perhaps together with some of the pre-movement 'uniform' sediments, and all that can be seen are the remnants of the latter, which may have been present in great thickness before any salt movement took place. In an unfamiliar salt area, it is therefore important from the viewpoint of deciding what trap types may be present, to determine the relationship of the overburden sediments that are present, to the source salt layer and any movements thereof.

TRAPS RELATED TO PLASTIC DEFORMATION OF SALT AND SHALE   195

5.6.1 Primary Rim Syncline Inversion, and 'Turtle Structures'

It is clear that the sediments of a primary rim syncline must be of interest to the explorationist. Generally, they have been close to the sediment source (the clastic sediment uplift above the diapir, and the immediately adjacent areas), and so frequently contain coarse lithological units that may retain good primary porosity, or have developed adequate secondary porosity. As suggested in the sketches of Fig. 5.20, the subsidence attendant on the SRS development may have caused at least the proximal half of the PRS sediments to tilt towards the diapir, and indeed may lead to structural inversion of this half of the PRS nearer to the diapir. Figure 5.22a shows again in simplified sketch form the type of structural process envisaged. Sketch (a) illustrates the situation at the time of development of the pre-piercement salt swell and the PRS deposition, whereas sketch (b) shows the evolution of the secondary rim syncline during the piercement stage, with the rapid

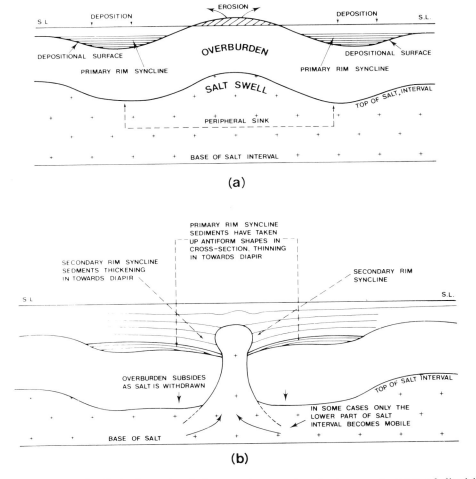

Fig. 5.22a. More detailed schematic of the pre-piercement and piercement stages of diapirism, indicating depositional and erosional patterns in the overburden

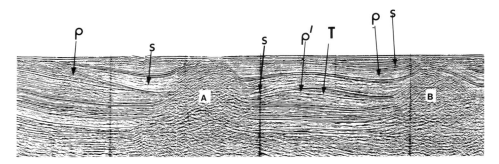

Fig. 5.22b. Seismic section showing two salt diapirs, A and B, with an asymmetrical 'half-turtle structure' between them. Asymmetry is due to the salt movement in diapir B having developed later than that in diapir A, as can be determined by the relative positions of the primary (P) and secondary (S) rim synclines. Primary rim syncline at P' largely removed by the erosional phase marked by the major unconformity immediately above. Reproduced by permission of Seismograph Service (England) Ltd

subsidence leading to structural inversion of the halves of the PRS troughs nearest to the diapir.

It can be appreciated from the foregoing that were there to be other diapirs located at the extremities of the line in sketch (b), and if these had developed *pari passu* with the diapir in the centre, then the PRS deposits would be common to the three diapirs. During subsidence related to the SRS deposition, the PRS inversion would be total *in the plane of the section*—that is, full antiform shapes would develop from the original synform lenses of sediments of the PRS troughs. These antiform shapes are known as *turtle structures*, and can be most important trap situations.

Note that the remark 'in the plane of the section' is meant to draw attention to the fact that a turtle structure seen within one seismic section may not necessarily have the correct trapping geometry in three dimensions (i.e. normal to the plane of section) unless other factors, such as faulting, or stratigraphic trapping situations, are present. Good turtle structures are most likely to have satisfactory three-dimensional geometry where salt structures occur as subcircular features spread over an area. A linear arrangement of such features may lack 'three-dimensional' trap capability. In Fig. 5.22b an asymmetrical 'half turtle structure' is seen, developed in this way because formation of salt diapir B was later than that of diapir A.

The reality of primary and secondary rim syncline sediments as seen in seismic data can be examined in some detail in the illustrations following. In Fig. 5.23, the base of a secondary rim syncline sequence dipping strongly to the right can be seen marked at S. Vertically beneath S, a series of uniformly-thick seismic intervals is present, representing the uniform 'pre-movement' strata. In the upper left part of the section, the letter P marks the base of a primary rim syncline sequence thinning strongly to the right; both these features are related to a salt diapir off-section to the right. The PRS sequence, originally a generally horizontal, synform lens-shaped deposit is now tilted to the right because of the piercement-related subsidence that has resulted in the SRS deposition. Although tilted in this example, the PRS deposits are not showing any signs of actual inversion.

TRAPS RELATED TO PLASTIC DEFORMATION OF SALT AND SHALE 197

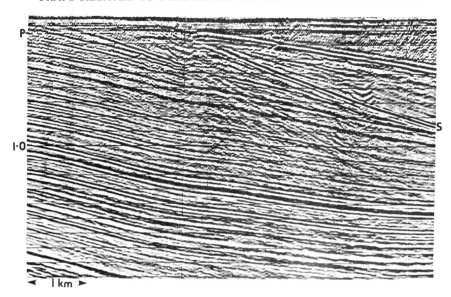

Fig. 5.23. Seismic example showing the base of a secondary rim syncline sequence at S related to a salt diapir (off section to the right), showing stratal thickening towards the diapir, while beneath it, in the interval P–S, primary rim syncline deposits (thinning towards the diapir) are seen. Reproduced by permission of Seismograph Service (England) Ltd

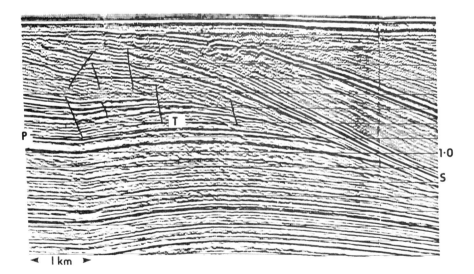

Fig. 5.24. A similar situation to that seen in Fig. 5.23, but in this case, continuing major subsidence and thickening towards the salt diapir of the secondary rim syncline (base at S) has led to structural inversion of the primary rim syncline sediments at T (base of the sequence is marked P). It is this type of deformation of the primary rim syncline deposits that produces 'turtle structures' in between salt diapirs that have developed approximately concurrently. Reproduced by permission of Seismograph Service (England) Ltd

198   OIL AND GAS TRAPS

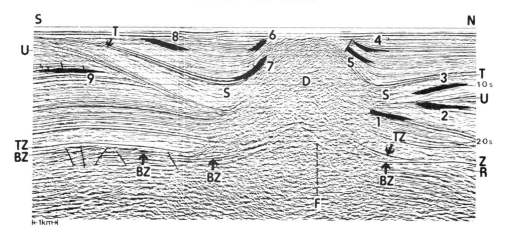

Fig. 5.25. Seismic section across an elongate salt-wall diapir, with hydrocarbon trap types numbered 1–9 modelled within the section. Trap types related to both pre-piercement and post-piercement stages of salt movement are shown, and are discussed in the text. D = diapir; BZ = Base (Zechstein) of salt; TZ = top (Zechstein) of salt; remnant basinal carbonates after total salt withdrawal are indicated at Z; R = Base Permian; S = secondary rim syncline deposits; T = Base Tertiary; U = Late Cimmerian Unconformity; F = fault. Published by permission of Elsevier Applied Science Publishers Ltd (from Jenyon, 1986d), with acknowledgements to Seismographic Service (England) Ltd

In Fig. 5.24, a similar situation is seen, but with the PRS sedimentary sequence (base at P) related to a diapir off-section to the right now showing structural inversion. The PRS deposits have now taken up an antiform shape centred at T, showing normal faults due to inversion, and folding from salt movement. Again, the steeply dipping and strongly marked basal reflection group of the SRS is seen at S. This example is a detail (see feature numbered 9) from the seismic section in Fig. 5.25 that is to be discussed in the next part of the chapter.

5.6.2 Trap Types Related to a Salt Diapir

In Fig. 5.25, a seismic section across an elongate salt diapir (long axis normal to the section) at D is shown. Some characteristic trap types are modelled and numbered 1–9 (shown in solid black). These do not represent all possible types of trap in this situation, but include the most frequently encountered types. They are related to both the pre-piercement and post-piercement phases of salt movement.

The base and top of the Zechstein salt interval are marked BZ and TZ, respectively, where they can be separated on this migrated section. Where complete salt withdrawal has occurred, the remnant Zechstein interval—consisting of carbonate anhydrite marginal facies units—is marked Z. Beneath the base of the Zechstein, Rotliegendes sands are truncating (R) at the north end of the section, with the possibility of traps against the overlying salt seal (not numbered). The letter F indicates the probable position of a fault that may have initiated the salt movement, affecting the basal- and pre-Zechstein sections. There is a major unconformity ('Late Cimmerian') at U, and a strong Base Tertiary group event at T. The letter S marks the deep trough of the secondary rim syncline (SRS) that is present. Primary rim syncline (PRS) stratal events are present immediately beneath U at the two extremities of the section.

Trap type 1 is a truncation trap in the dipping, pre-movement 'uniform' section units against either the salt diapir flank, or the unconformity U; such traps are potentially present both to the north and south of the diapir.

Trap type 2 could be either in 'uniform' section units, or basal PRS units, truncating against the unconformity U.

Trap type 3 is an SRS interval feather-edging updip. The base of the SRS in this example happens by chance to coincide with the unconformity U, on which the SRS events can be seen to be downlapping. The qualifier 'by chance' may not be strictly correct, as the erosional phase that produced the unconformity U may have had some influence on the initiation of at least some of the salt movement in this area, as did probably the tectonism that took place during the U interval.

Trap type 4 is formed in shallow units in the SRS upturned by the diapir, feather-edging to form a trap updip.

Trap type 5 is formed in shallow units in the SRS upturned by the diapir, and truncating against peripheral, or radial, faulting caused by the diapiric uplift.

Trap type 6 is formed in shallow units in the SRS upturned by the diapir and truncating against either: (i) a shallow unconformity; or (ii) a shallow salt overhang spreading from the crest of the diapir—not present in this case, but does occur elsewhere in this situation, and will be discussed later.

Trap type 7 is formed in SRS units upturned by, and truncating against, the diapir flank.

Trap type 8 is formed in SRS units truncated updip away from the diapir by the shallow unconformity.

Trap type 9 is formed in PRS sediments in a structurally inverted 'half-turtle' feature between the lower ('uniform') section, and the SRS subsidence above (see also Fig. 5.24).

The crest of the salt diapir in Fig. 5.25, together with the overburden above it, is not visible (and may have been removed by erosion). The more general case, where the diapiric crest is at some depth below the surface, is now illustrated by the schematic in Fig. 5.26a loosely adapted from Halbouty (1967) and others. Most of the trap types present were met with in Fig. 5.25, and are numbered in accordance with the latter illustration. Four trap types, lettered A, B, C, and D were not present, or not shown, in Fig. 5.25, and are as follows:

A—Simple anticlinal/domal trap due to uplift above the diapir (the latter indicated by the cross symbol).
B—Porosity traps with geometries related to the uplift.
C—Cap-rock traps, usually due to leaching and the presence of secondary porosity.
D—Pinchout traps due to SRS sediments thinning away from the diapir.

Of these traps, types A and C were not mentioned in connection with the diapir in Fig. 5.25, since the diapir crest and the overburden above it were not visible. However, type A was mentioned earlier related to pre-piercement salt movement, and an actual example related to a salt piercement structure is represented by the Gannet North structure seen in the seismic section of Fig. 5.26b. One of the group of five small oilfields on the Western Platform of the North Sea Central Graben, the field has dip closure at the levels of both reservoir sands (the Eocene Tay Formation, and the Palaeocene Forties Formation).

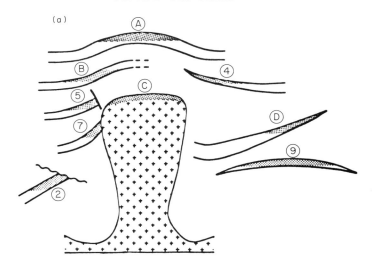

Fig. 5.26a. Schematic of hydrocarbon trap types associated with salt diapirs. Types A and C not present in Fig. 5.25. Based loosely on Halbouty (1967) and others

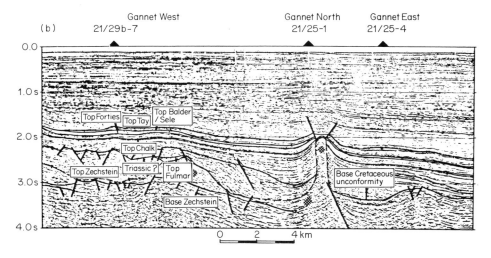

Fig. 5.26b. Seismic example showing three of the Gannet Group oilfield structures. Gannet North is discussed briefly in the text; depositional patterns in the overburden owing to the stages of Zechstein salt movement are clearly seen, as is the trap situation above the diapir crest. Reproduced by permission of Graham & Trotman Ltd (from Armstrong *et al.*, 1987)

The structure is seen in the seismic example with two of the other fields in the group, and a description is given by Armstrong *et al.* (1987).

The Gannet North diapir's period of salt uplift can be seen clearly indicated by the depositional patterns on the section. Strong thinning towards the diapir marks the primary rim syncline deposits during the Triassic, showing that the pre-piercement (pillow) stage of movement took place at this time. The Base Cretaceous Unconformity represents the basal surface of secondary rim syncline deposition thickening towards the diapir, indicating the period of piercement and rise of the salt towards the surface. Upward movement continued well into the early Tertiary period, but at variable rates expressed by the variation in bed thinning above the Top Chalk horizon.

5.6.3 Cap Rocks, Overhangs, and Diapiric Shale

The illustration in Fig. 5.27, from Levorsen (1967) is a sketch representing a typical Gulf Coast salt diapir, showing several of the characteristic features of salt piercement structures and associated petroleum traps (solid black) in this area. The traps shown schematically here are of types already discussed, except for those shown within the feature named as 'cap rock' on top of the salt body proper. This is shown as having several patches of petroleum reservoir rock within it, as well as a small overhang on the right-hand side.

Cap-rock sequences at the crests of (and sometimes sheathing the flanks of) salt diapirs are not uncommon—in fact, it may be that they are the rule rather than the exception, and where currently absent, may have existed at an earlier time on many salt bodies until destroyed by, for instance, dissolution of the salt beneath, and collapse of, the overlying sediments.

A typical vertical section through a cap-rock sequence from top to base shows an upper zone of calcite/limestone, a middle transitional zone consisting of calcite, gypsum, sulphur, anhydrite, and a basal zone of anhydrite (Martinez, 1980). Laterally, the geometry of the various mineral species present can become complex.

In seismic data, it is often rather easy to see the expression of a cap-rock event, since some of the minerals present (particularly the most common—anhydrite) produce a very

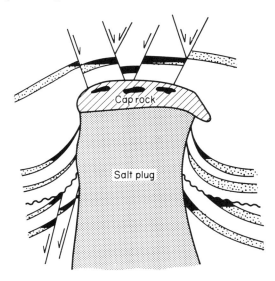

Fig. 5.27. Schematic of a typical Gulf Coast 'salt dome', indicating various trap types, including leached secondary porosity traps in a cap-rock development. From GEOLOGY OF PETROLEUM, 2/E by A. I. Levorsen, 1954 ©1967 W. H. Freeman and Company. Reprinted with permission

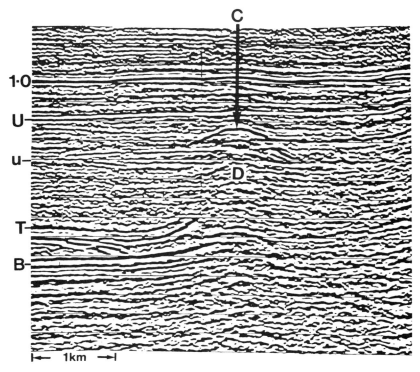

Fig. 5.28. Seismic section showing a cap-rock development (C) associated with a spreading overhang at the crest of a small salt diapir (D), which is not well seen on this noisy section. At least one other spreading overhang may have developed below C during the rise of the diapir. Base of the mobile salt interval is at B and top at T; u and U are two unconformities present in the section. Reproduced by permission of Seismograph Service (England) Ltd

strong acoustic impedance contrast when juxtaposed with salt rock or with overlying clastics. This is the case in Fig. 5.28, where the seismic section includes a small salt diapir (D) that has risen from the source salt layer (B–T), and has developed a cap-rock sequence associated with a spreading overhang at the crest; this has resulted in a very strong, convex-upward event at (C). There could also be another such event beneath, with its crest at the level of the lower of the two marked unconformities (u and U). The matter of spreading overhangs, and their relationship to unconformities, is discussed later. The strong stand-out of the cap-rock event on what is quite a noisy seismic section is noteworthy.

The seismic section in Fig. 5.29 shows two similar examples of cap-rock events related to spreading overhangs at C–C, capping two small diapirs that have arisen from the source salt interval B–T. Note that in the right-hand example, part of the left side of the cap-rock event seems to be missing; this is not uncommon where the subsidence in a secondary rim syncline has been greater on one side of a diapir than on the other, causing collapse and disintegration of part of the diapiric overhang, together with the cap rock.

The zonation in a cap-rock sequence has been studied by a number of workers, particularly in the Gulf Coast area of the USA. The upper zone of calcite is believed to be the result of reactions involving carbon sources, such as methane and a heavier carbon

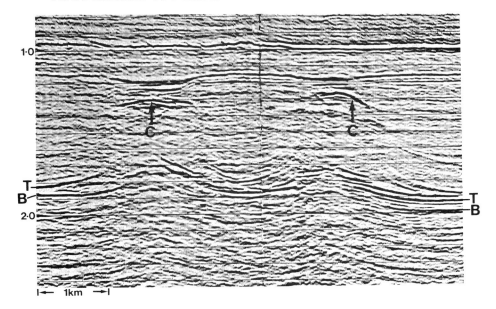

Fig. 5.29. Two more examples of cap rocks (C) associated with spreading overhangs at the crests of salt diapirs. The right-hand example seems to have most of its left side absent; this asymmetrical appearance is often due to excessive subsidence into the secondary rim syncline on one side of the diapir causing disintegration of part of the overhang. The salt diapirs are not well resolved in this example, but the cap-rock sequences stand out strongly owing to their marked acoustic impedance contrast with adjacent rocks of different lithology. B, T = base and top of mobile salt interval.
Reproduced by permission of Seismograph Service (England) Ltd

source (catagenic CO_2 or marine carbonate dissolved by CO_2) or by CO_2 and CH_4 associated with maturing hydrocarbons (Posey *et al.*, 1987). Contact of the resulting fluids with anhydrite in a high-salinity environment causes anhydrite dissolution (anhydrite, under normal circumstances, being relatively insoluble in water), and the carbon in the fluids promotes sulphate reduction by bacteria that use carbon as an energy source.

Sassen (1987) concludes that the isotopically light carbonate minerals, elemental sulphur, and sulphides characterizing cap-rock sequences of shallow Gulf Coast salt domes result from 'microbial oxidation of petroleum and reduction of sulphate'.

In a study of the cap rock of Boling Dome salt body (at the western margin of the Houston diapir province, Texas), Seni (1987) concludes that terrigenous clastics, including Tertiary sediments incorporated in the cap-rock material are present due to: (i) shearing of deeply buried Tertiary strata during growth of the salt body; (ii) aeolian processes active during deposition of the salt; and (iii) salt dissolution at the diapir crest, and underplating of insoluble materials (e.g. anhydrite) and sediments to the base of the cap rock as this rises through the overburden. Caution is urged in interpreting the provenances of materials from mineralogical information, owing to the complexities introduced by burial diagenesis over a wide depth range, and 'mineralogical simplification' of formerly deeply buried materials.

Thus the processes of development of the cap-rock sequences stem from the build-up of insolubles—particularly anhydrite—at a diapir crest as it rises through overburden formations with fluid-filled porosity (often circulating). In a number of instances—for example, in the Gulf Coast area of the USA—the dynamic process of uplift, and the environment of circulating formation fluids has provided ideal conditions for the leaching of the cap rock with its frequently irregular developments of different mineral species. This in turn has led to patchy volumes of secondary porosity and enhanced permeability favourable to the formation of hydrocarbon reservoirs and traps within the cap rock, and indeed oilfields have been located in cap-rock sequences.

Since the development of cap rocks by build-up of insolubles and subsequent diagenetic reactions is a progressive process, this should have a noticeable effect on the distribution of these features, and Halbouty (1967) points out that the occurrence of cap-rock developments in the Gulf Coast area varies greatly according to present depths of diapir crests. Cap rocks are effectively absent (as shown by drilling) from diapirs with crests currently at great depths (e.g. greater than 10,000 feet), but become progressively more common amongst diapirs with crests at increasingly shallow depths. The thickest cap-rock developments (over 1000 feet) occur with diapirs that have risen furthest through the overburden.

It has been suggested (Jenyon, 1987c) that if a cap-rock sequence of insolubles derived from the salt can develop during the *rise* of a salt diapir through the overburden, then a similar development could be expected to occur during the *retreat* of a diapir due to dissolution and removal of the salt. Such a *regressive cap-rock sequence* is only hypothetical at present, however, Fig. 5.30 may well show the seismic response to such a regressive cap-rock sequence.

The figure shows a cross-section of a salt-wall diapir (D) the long axis of which is normal to the section. The diapir has been strongly affected in this location by dissolution and removal of the upper part of the salt body, with the concomitant development of a large collapse graben in the overburden where the salt has been removed from the section. The graben is marked by faulting in a series from early (E) to late (L) times. The top of the salt source interval, a major unconformity (CU), the base of the Tertiary (A), and a shallow unconformity (U) are marked.

A strong, irregular dipping event (T) is indicated, and it is believed probable that this marks the top of a 'regressive' cap-rock sequence (the even more irregular strong event 50–100 ms below T may mark the base of the cap-rock sequence and the top of the salt rock proper in its present position). The strength of the reflection events could be a measure of the acoustic impedance contrasts involved in a salt/anhydrite–carbonate/clastic sediment 'sandwich'.

If secondary porosity can develop in some cap-rock sequences, then in a dissolution–collapse situation as here, if a 'regressive' cap rock has developed and during the collapse process has come into contact with a migratory route through the overburden at a propitious time, a hydrocarbon trap scenario may have developed. It could be worthwhile locating and testing some examples of these buried 'regressive' cap rocks.

5.6.3.1 *Overhangs*

An overhang occurs in conjunction with a salt diapir when the horizontal cross-section of the upper part of the salt body, projected downwards, falls at least partly outside the

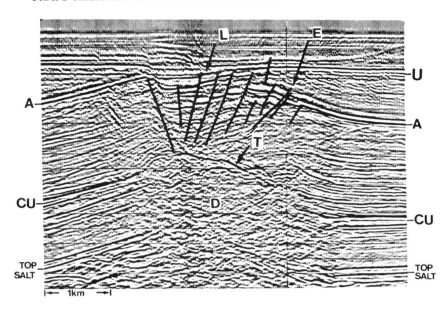

Fig. 5.30. Seismic example of a salt-wall diapir (D) with long axis normal to the section, which has undergone extensive dissolution with resultant formation of a collapse graben by early (E) to late (L) faulting. Top salt is indicated, as is CU, the Late Cimmerian Unconformity, the Base Tertiary at A, and an intra-Tertiary unconformity at U. It is suggested that the irregular event T is the top of a 'regressive' cap-rock sequence, formed by the build-up of insolubles from the salt as the latter is dissolved away and retreats *downwards*, rather than in the usual way at the crest of a *rising* salt diapir. Reproduced by permission of Seismograph Service (England) Ltd

cross-section of the root of the diapir. This may occur because of a change of direction of the rising salt piercement through the overburden, or because of an increase in the cross-sectional area of the upper part of the salt structure—i.e. the salt spreads laterally in the shallower part of the section.

Overhang features are important since they can form the basis for a trap situation where strata, truncated and upturned by the rise of the salt, are sealed by the salt beneath the overhang (where usually they cannot be traced by normal seismic methods, although 'proximity surveys' using downhole geophysical methods can help to clarify the situation—see, for example, Manzur (1985)).

One type of feature is the superficial spreading overhang, which appears as a thin 'mushroom-cap' shape at the crest of some diapirs. A well-known example associated with the Wienhausen–Eicklingen salt diapir in Germany, investigated by drilling, is illustrated by a line-drawn section in Fig. 5.31. Oil accumulations have been found in Jurassic and Cretaceous units upturned and truncated against the underside of the thin spreading overhang. The diapir is a funnel-shaped body, increasing in diameter upwards, and has a cap rock covering the crest and overhang as well as, it is believed, sheathing the diapir flanks as shown. Some of the oil wells drilled in the Wienhausen and Eicklingen areas are shown; although a relatively old discovery, it is believed that there is still some production from the latter area.

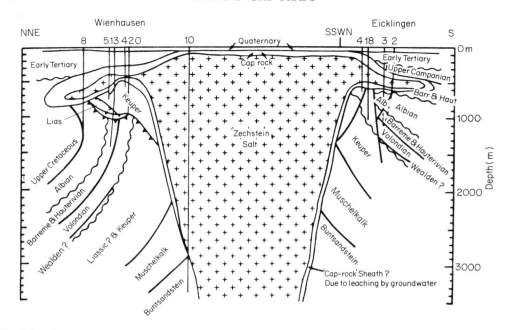

Fig. 5.31. Geological section across the Weinhausen–Eicklingen salt diapir, showing the superficial spreading overhang present at the crest. Reproduced by permission of Academic Press (from Jenyon, 1987c, after various authors)

The appearance of some examples of overhang features is shown in Figs 5.28 and 5.29 (cf. width of event C with diapirs).

Several attempts have been made to explain the mode of origin of these superficial spreading overhangs. The most favoured is, perhaps, the idea that they represent 'fossil' surface extrusions of salt analogous to the present-day emergent salt glaciers of Hormuz in Iran, and Trusheim (1960) seems to favour this view. However, the writer does not subscribe to it, since it is difficult to imagine an outflow of salt, whether terrestrial or submarine, being sufficiently permanent in the face of dissolution or deflation to be preserved by a compacted sedimentary cover. An alternative view (Jenyon, 1987c) has been put forward that involves the idea of rising diapiric salt nearing the surface, being injected laterally into a subsurface zone of weakness (e.g. a bedding plane or an unconformity surface) beneath a relatively thin overburden of at least partially compacted sediments. The latter would include impermeable beds of argillaceous material that would protect the salt injection from dissolution or deflation.

No novelty is claimed for this idea, which has been mooted previously by several workers. The novelty lies in the evidence (somewhat rare in a field rich with hypotheses) adduced from seismic data in the paper cited. This evidence is in the form of seismic examples of unusual stacked (double) collapse graben, interpreted as the results of later dissolution of such salt overhangs, which leaves clear signs in the patterns of stratal events present. Some possibilities of trap scenarios in this rather unusual situation are discussed in the paper.

5.6.3.2 Diapiric shale

It is well known that where rapid sedimentation and burial of shale and clay rock-forming materials occurs, top-down compaction of these units almost invariably leads to overpressuring. Abnormally pressured shales at depth are potentially highly mobile, and under the application of differential stress can become diapiric. Shale diapirs occur in their own right in many areas across the world, as in parts of the offshore Niger Delta. Their structural effects on the overburden can be quite similar to those produced by mobile salt, and it is sometimes difficult to determine the nature of the diapiric material from seismic data alone. Some criteria have been discussed by Lohmann (1980) and Jenyon (1986d).

Two shale (or mud) diapirs are shown marked D in Fig. 5.32, from the paper by McWhae (1986). These are related to the tectonics along the Kaltag Fault, a major element that passes through Alaska and skirts the Canadian Arctic Islands, and is postulated to have been the transform boundary between the North American and Eurasian plates since late Cretaceous times. On the section, MEU indicates a Middle Eocene unconformity. The featurelessness of the diapiric masses should be noted, in addition to the lack of any strong bounding seismic events either above or below, or any signs of velocity anomalies (positive or negative) associated with these bodies. These and other criteria can often be used to make an educated guess from the seismic data whether salt or mobile argillaceous sediment is involved.

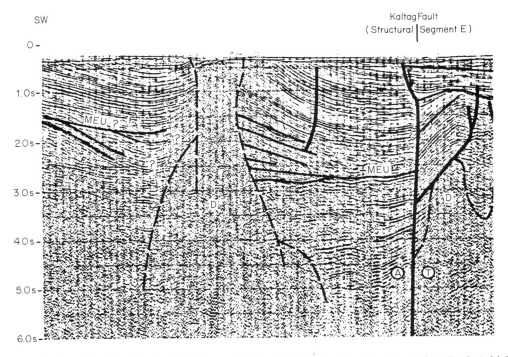

Fig. 5.32. Two diapiric shale bodies (D) associated with the tectonics along the Kaltag Fault (which is probably a transform boundary fault) that runs through part of Alaska and skirts the Canadian Arctic Islands. MEU = Middle Eocene Unconformity. From McWhae, 1986, reprinted by permission of American Association of Petroleum Geologists

It can be seen in the illustration that shale-type diapirs produce overburden deformation effects similar or identical to those produced by salt diapirs, and the range of potential traps in the overburden is effectively similar to those discussed for pre-piercement and piercement salt structures earlier, with some differences. The latter include, in general, a lack of features dependent on dissolution and collapse—effects that are normally negligible related to shale diapirs—such as major collapse graben faulting over the crest of the structure as seen associated with many salt features. Also, although mobile argillaceous material can form a seal, it is never as good a seal as that provided by the best salt seal. Further, there are considerations of seismic velocity. Salt rock has a relatively high seismic interval velocity (typically, pure halite has a compressional wave velocity around 4500 m/s, some 50% greater than typical shale velocity), while overpressured shale has an abnormally low seismic velocity when it is *in situ*, and can produce negative 'pull-down' velocity anomalies in seismic data. However, after it has become mobile, has moved and then lost mobility as in a diapiric mass, it seems likely that the low seismic velocity phenomenon may have disappeared (although no studies of this point are known to the writer), and the seismic interval velocity of the material may have reverted to that for normal bedded shale or mudstone—showing little or no acoustic contrast with the clastics in which it is embedded.

Apart from shale diapirs, mobile shale is also in some areas a 'companion' of mobile salt in diapirism. The same differential stresses that cause potentially mobile salt to move will also produce movement (at lower stress levels, usually) in plastic shales. This can result in overpressured bedded shales (through which the salt diapir passes in its rise to the surface) moving diapirically as a layer or sheath on the flanks—and sometimes partly or wholly over the crest—of the salt diapir.

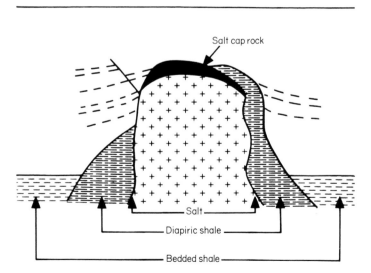

Fig. 5.33. Schematic showing typical relationship of a diapiric shale 'sheath' to a salt diapir, in a section containing normal bedded shale. Shale, clay, or mud that becomes diapiric is usually abnormally pressured, and therefore highly mobile. It will accompany a salt diapir as it pierces the overburden units

This is shown schematically in Fig. 5.33, a line-drawn section through a structure that is loosely based on a number of such features in the Gulf Coast area of the USA. In seismic data, provided a shale sheath is sufficiently thick in relation to the dominant seismic wavelength at the relevant level in the cross-section, it can usually be resolved clearly; acoustic impedance contrasts with the salt, and adjacent clastics, are normally high.

An older paper on the subject of the geophysical investigation of shale diapirism that has become something of a classic, and is still worth reading, is that by Musgrave and Hicks (1966).

PART III
STRATIGRAPHIC TRAPS

CHAPTER 6

Unconformities and Buried Topography

6.1 Introduction

Stratigraphic, or subtle, hydrocarbon traps have been increasingly engaging the attention of explorationists over the last two or three decades as a decreasing number of large structural traps are discovered (and remain to be discovered).

Most hydrocarbon traps of stratigraphic type can be classified into two groups—those occurring within seismic sequences (encompassing such features as pinchouts, porosity traps, and lateral facies changes, for example), and those occurring *at* sequence boundaries. The latter type include unconformities and buried topography, and will be discussed in the present chapter.

6.2 Unconformities

Used in its widest sense, the term unconformity refers to a surface of non-deposition and/or erosion separating an older geological sequence below from a younger sequence above (provided they have not been inverted by tectonism). Other terms may be used to describe particular types of unconformity—for example, angular unconformity, disconformity, non-conformity, non-sequence—which generally refer to the dip configuration differences, or other relationships, between lower and upper sequences.

The missing stratigraphic section related to unconformities is of great importance; it can represent the passage of a larger total geological time period than that of the stratigraphic section that *is* present. This is of particular significance in seismic data, if, as is often the case, there are many disconformities and non-sequences present in the section. It is seldom possible to detect the presence of such hiatuses where upper and lower sequences are parallel, with no angular difference to provide a clue. In such circumstances, there is no substitute for access to detailed biostratigraphic information. Van Hinte (1982) in an interesting paper has pointed out that such detailed studies may indicate intervals with continuous but extremely low rates of deposition, that on a seismic section will show as an apparent unconformity. This is a warning that great caution must be exercised in the interpretation of features of this type, and whenever possible, relevant geological input must be included.

It is most important that the limitations of resolution of the seismic method are clearly realized, with this realization requiring regular updating (as progress in the seismic technology is rapid).

Some unconformities divide sequences that are not only different geologically, but also from the seismic viewpoint. In many parts of Europe, for instance, the Base Tertiary/Top Cretaceous stratigraphic level is an unconformity separating the overlying, relatively less compacted/indurated Tertiary clastics from more compacted/indurated Chalk beneath. This lithological difference normally also has the effect of producing a demarcation between relatively low seismic velocities in the Tertiary and substantially higher velocities in the Upper Cretaceous. Similar sudden changes occur at depth in the section also—the most obvious example being the case of a clastic sequence unconformably overlying a basement complex.

Unconformities can also separate two distinct stress regimes. It is frequently the case that faulting in a sequence underlying an unconformity terminates upwards at the unconformity surface, and is not seen in the overlying sequence. On the other hand, faulting in an underlying sequence may be reactivated, and cause Riedel shears in a clastic sequence overlying the unconformity. These are some of the factors that may have an important bearing on the formation of hydrocarbon traps above or below an unconformity.

Although it has been suggested that identifying an unconformity/disconformity where the upper and lower sequences are parallel is virtually impossible, it must be allowed that it can be done under certain circumstances. Although there may be no dip change above and below the unconformable surface, the depositional environment may nevertheless have changed completely (e.g. from a shelf carbonate environment to paralic conditions). With contemporary high-resolution seismic data, there will certainly be a marked change in character (frequency content, amplitude, etc.) of the data above and below the hiatus, which may be enough to give a strong clue as to the situation. There may also be a basal reflection group associated with the upper sequence that has a character all of its own (perhaps a basal conglomerate) and accentuates the demarcation between the two sequences.

6.3 Unconformity Surfaces and Buried Topography

In Fig. 6.1(a), a mature, peneplaned surface is seen in the sketch of an angular unconformity, while in Fig. 6.1(b), an immature erosional surface is present with scarp, or *cuesta* features projecting upwards into the upper sequence. An irregular surface of this nature is termed 'buried topography' or a palaeogeomorphological surface (PGM surface for short). From the viewpoint of hydrocarbon trap potential, it can be seen that in Fig. 6.1(a) if the basal shale of the upper sequence has a sufficiently high capillary entry pressure to form an impermeable seal for a hydrocarbon column in one of the lower sequence units, and provided any of the latter are of reservoir quality, and also sealed within the sequence with suitable three-dimensional geometry, then a trap situation can arise. The same considerations hold good in Fig. 6.1(b), there being in this instance no basic difference between the two situations. A little later other instances of a similar type will be considered where the detailed geometry and lithology of units in both sequences across the unconformity will be seen to have a greater influence on the trap potential.

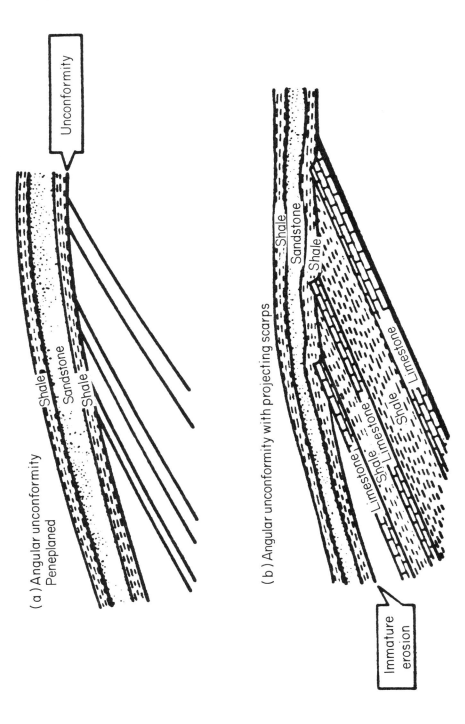

Fig. 6.1. Schematics showing angular unconformities where the surface of unconformity is (a) a mature peneplain, and (b) an immature erosion surface. From *Reflection Seismology*, K. H. Waters, ©1978, reproduced by permission of John Wiley & Sons Ltd

6.3.1 The Unconformity as a Reflection Event

In seismic data, an unconformity surface may occur as a strong continuous event, or as an intermittent event of varying strength. On the other hand, there may be *no* reflection that can definitely be assigned to the surface; all that can be seen are the stratal reflections of the underlying and overlying series.

The reason for this variability lies in the nature of a reflection event itself, and in the variety of situations possible in the juxtaposition of lithological units of various acoustic impedance values along the unconformity (in the seismic cross-section). It will be recalled that one of the important determining factors for the presence or absence of a reflection at an interface is the contrast in acoustic impedances of the units bounding the interface.

In Fig. 6.2 there are several angular unconformities and non-conformities present in the seismic section. Examining the unconformity surface (U), it can be seen that its location is well defined—to within a few milliseconds—across most of the section. Although there is no easily determined one-to-one relationship between any part of the pulse shape of a reflection and the interface (or interfaces) that give rise to it, the gently top-lapping events beneath the unconformity in the right-hand half of the section indicate the position of the surface quite precisely. However, there is an apparent hiatus at this level, just left of centre (arrowed) that is typical of the effect of an adverse juxtaposition of acoustic impedances across the unconformity surface, causing a gap in the reflection event marking the surface.

Normally, a surface of unconformity is a depositional surface during marine transgression, and as such is often overlain by a deposit typical for an environment that

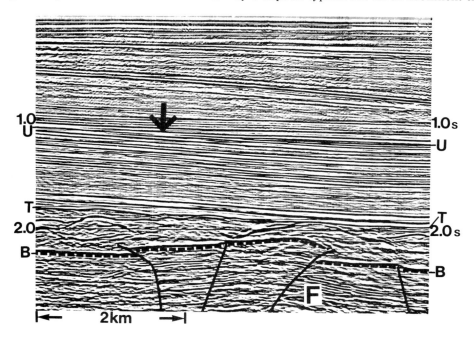

Fig. 6.2. Section showing lateral variations (arrowed) in the seismic response to an unconformity surface (U). B, T = base and top of mobile salt interval; F = fault. Reproduced by permission of Seismograph Service (England) Ltd

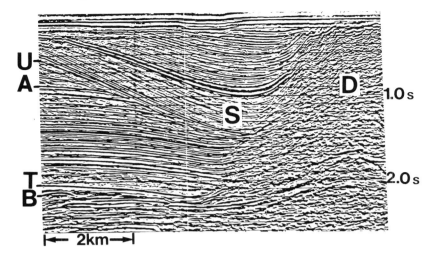

Fig. 6.3. Section with an unconformity reflection event (U) that is strong at the left-hand end, but weakens progressively to the right, where it eventually truncates against the salt diapir at D. B,T = base and top of mobile salt interval. Reproduced by permission of Seismograph Service (England) Ltd

may eventually produce a strong seismic signal. A basal conglomerate or gravel, often present either as a transgressive or regressive unit, is one instance. In Fig. 6.2, the strong band of reflections running from 0.55 s on the left to 0.8 s on the right is a typical seismic response to a basal bed sequence of this kind, although here no conglomerate is involved. The exact lithology present is not very important. The factor that produces a strong reflection is that a lithology is present which is very different from both the underlying sequence (below the unconformity) and the normal marine sequence into which it passes above.

The example in Fig. 6.3 (which is a detail from an illustration already seen in the previous chapter) also shows how an unconformity reflection (U) begins as a strong event at the left-hand end of the section, but becomes progressively weaker and intermittently absent towards the right, where it eventually truncates against a salt diapir D. The salt interval—very much reduced in thickness due to salt flow into the diapir—is marked at B–T; there are primary rim syncline deposits between A and U, and secondary rim syncline deposits at S.

Note the sequence below U, truncated in an angular unconformity; also the dip relationship between the unconformity surface and these truncated beds, which indicates that the beds were dipping steeply away from the location currently occupied by the diapir D, during part of the erosional phase of U. This dip was the result of the pillow-stage uplift of the salt, before the piercement took place.

Note also the reflections above U, gently downlapping on to the surface of the unconformity. This indicates that deposition, from a sediment source off-section to the left, was taking place during the subsidence of the secondary rim syncline stage (S). There was marked and progressive deepening of the basin of deposition in the immediate vicinity of the diapir as the salt piercement developed.

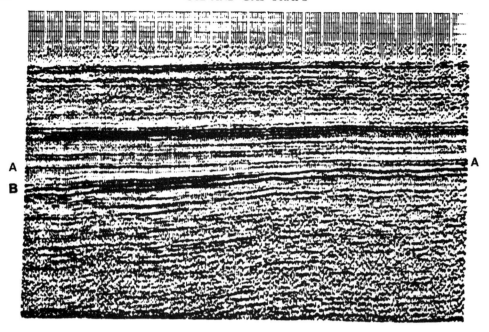

Fig. 6.4. Seismic section from Oklahoma, showing the Hunton Formation (B) truncating at an unconformity surface overlain by the post-unconformity Mississippian sequence at A. Truncations such as this frequently produce oil traps, provided that the immediate post-unconformity beds are adequate as a seal formation. From *Reflection Seismology*, K. H. Waters, ©1978, reproduced by permission of John Wiley & Sons Ltd

6.3.2 Planar and Non-planar Unconformities

Where an unconformity is a mature surface of erosion in the form of a peneplain, the most frequently observed hydrocarbon trap is in a truncated unit or sequence of the sub-unconformity rocks. A typical example of this class of structure is seen in Fig. 6.4 from Waters (1978). In Oklahoma, the Hunton Formation (B in the illustration) which is dipping gently, is truncated by the post-unconformity Mississippian sequence at A, producing good trap situations in the Hunton in this area. Again it is pointed out that the geometry normal to the section must be favourable for trap development. Waters mentions two matters requiring consideration by the interpreter in a case of this kind: (i) are the updip terminations of the truncated beds producing diffractions that have not been removed in processing, and can these apparent 'turn-overs' be distinguished clearly from the true dip?; and (ii) a reminder that line ties on a series like the Hunton should be made sufficiently far down dip for there to be a full thickness of the sequence developed. Clearly also, the post-unconformity unit immediately above the pre-unconformity sequence must be a sealing unit, undisturbed by major faults or fractures over a sufficiently large area to form a commercial trap.

More flexibility as regards trap configuration is possible where there is some relief on an essentially planar unconformity. This situation is found in the Marlin Field, situated in the Gippsland Basin, Australia. A schematic section through the field is shown in Fig. 6.5(a), where an oil accumulation with a gas cap is trapped beneath an unconformity separating a major delta complex (developed from Upper Cretaceous to mid-/late Eocene times) from overlying Oligocene units. A simplified lithostratigraphic chart of the Upper Cretaceous to Miocene sediments of the Gippsland Basin is shown in Fig. 6.5(b).

There is local relief on the unconformity due to cut-down into the delta complex of numerous channels that developed during the erosional hiatus, and the margin of one of these is seen on the right of the section in Fig. 6.5(a). Fortunately the channel-fill is impermeable calcareous mudstone (also believed to be the source of the underlying gas and oil) of the Oligocene Lakes Entrance Formation that has sealed the trap. The reservoirs are in Lower Tertiary sands of the delta complex, with average porosities of 25% and permeabilities from 400 mD to several Darcies.

The study of reflection configurations above and below an unconformity not only defines the latter as a seismic sequence boundary, but is also vital for the understanding of the tectonic and depositional history of an area. Even in a prospect where there has been no drilling, modern seismic techniques of sequence and facies analysis frequently enable educated guesses to be made as to lithologies of sequences present in the data.

It is frequently the case in seismic data that poor resolution—owing to limitations of the method or complex geology—prevents the identification of an unconformity surface, but dip changes can be observed that provide evidence of the presence of an angular unconformity in the section. A good example from a paper by Johnson and Eyssautier (1987) is seen in Fig. 6.6(b), which is from the area of the Alwyn South oilfield (East Shetlands Basin). In the central fault block beneath Well 3/14a-1 can be seen dip indications (dashed lines) that provide evidence of a major unconformity between the two lower dashed lines, although the exact location of the unconformity surface is not determinable from this section. (Note also the shallower angular unconformity at about 3.0 s at the left-hand end of the section.)

The Fig. 6.6(a) seismic example from the same paper depicts parts of the Alwyn North Field, and shows an unconformity delineated by the Base Cretaceous seismic marker. Where indicated by the arrow (and in the interpretation shown below) there is good seismic evidence of the formation of a scarp resulting from the differential erosion of variable lithological units in the Brent sequence below the unconformity. This is the type of process that results in cuesta features on a near-horizontal surface of unconformity; cuestas and scarps are examples of buried topographic features that occur in the stratigraphic section in many areas, and which will be discussed later.

6.3.3 Reflection Configuration above an Unconformity

Reflections above an unconformity surface (which may be an important element of the evidence defining a sequence boundary on a seismic section) can show a range of differing configurations as a result of variations in the environments and mechanisms of post-unconformity deposition. These variations include such factors as whether terrigenous or marine deposition is involved, and if marine, the position of the deposits in the basinal

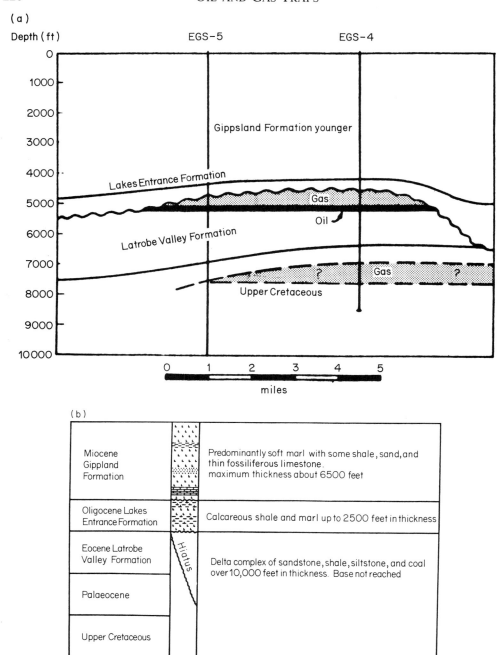

Fig. 6.5. (a) Schematic north-east section across the Marlin Field, Gippsland Basin, Australia, where an unconformity trap has resulted in an oil accumulation with a gas cap. There is local relief on the unconformity surface due to cut-down of channels—the margin of one of these being

(continued opposite)

setting; the environmental energy of deposition; the amount and type of sediment available, rate of sediment supply, and direction of transport; presence and direction of any currents, turbidity flows, or mass movement phenomena; and the tectonic stability or otherwise of the area. As a corollary, from the appearance on the seismic section of post-unconformity reflections, much can often be deduced regarding the environment of deposition present at the time, and the direction and supply rate of the sediments involved. Informed guesses can also be made on occasions as to the general lithological types present. Conclusions of a sedimentological and structural nature drawn in this way will clearly help in deciding if viable hydrocarbon traps are present associated with the unconformity and its defining sequences above and below.

In the shallow part of the seismic example of Fig. 6.7, an unconformity surface (at about 0.5 s on the left) has become a synform depositional feature, with onlap fill deposits (at B) above the unconformity. A depositional infill of this type indicates stable and uniform depositional conditions in moderate-to-deep water, probably well out into the basin towards the centre, with mainly sedimentary fines deposited in a low-energy environment. (The irregular events at S mark the location of slump structures where underlying salt movement has caused the steepening dip of the flank at the left end of the section to exceed the angle of rest of undercompacted Tertiary sediments.)

On the other hand, Fig. 6.8 shows a similar synform depositional trough (at approximately the same vertical and horizontal scales to the Fig. 6.7 example) where the infill (above the shallower unconformity U) is not uniform, but is prograding strongly from left to right into the trough. We can deduce from this that sediment transport direction is from left to right, and the sediment source is off-section to the left. From the very rapid thinning of the infill, it seems likely that the basin margin/sediment source is nearby, and that there will be some gradation of the deposits from coarser to finer in the direction of transport. Note the wedge of sediments between the two unconformity surfaces; these sediments are showing both top- and base-lap, with thinning from right to left—indications of a completely different sediment source and transport direction, with relatively low sediment supply (or sediment starvation), compared with the situation above the shallower unconformity. In all the situations mentioned, given favourable three-dimensional geometry, and the presence of adequate reservoir and seal formations, hydrocarbon traps could be found. This is especially the case where updip faulting is also present.

Reflection configuration at the lower boundary of a depositional sequence (i.e. above an unconformity surface) is referred to as *baselap*, of which two principal types are normally identified. Where effectively horizontal depositional strata terminate against a dipping surface (as at a basin margin), or dipping strata terminate against a more steeply dipping surface, the relationship is known as *onlap*. (This word has largely replaced the older term 'overlap'). An example of onlapping reflections is seen in Fig. 6.9, between the two open arrowheads at the right-hand end of the section. Cretaceous stratal events can clearly be

---

shown here. Reproduced by permission of Elsevier Applied Science Publishers Ltd (from Wallis, 1967). (b) Lithostratigraphic column from offshore wells in the Gippsland Basin, showing the geology of the Marlin Field trap. The deltaic reservoir sequence (Latrobe Valley Formation) is cut into by channels infilled with impermeable calcareous mudstone of the (Oligocene) Lakes Entrance Formation which acts as a seal and may also be the source rock for the oil and gas in Marlin. Reproduced by permission of Elsevier Applied Science Publishers Ltd (from Wallis, 1967)

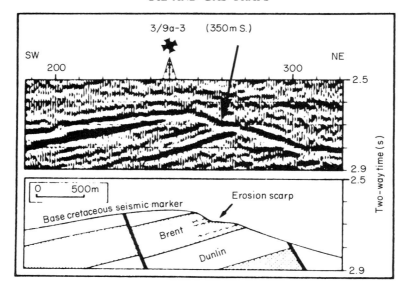
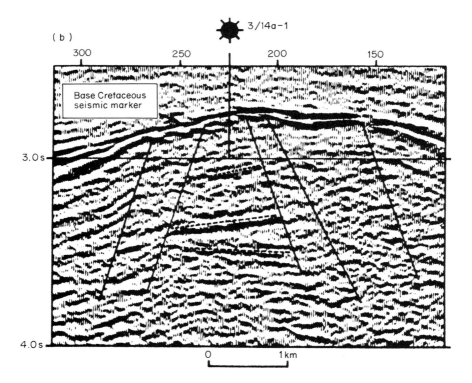

Fig. 6.6. (a) A seismic line across part of the Alwyn North oilfield in the East Shetlands Basin. (b) Section across Alwyn South oilfield. Reproduced by permission of Graham & Trotman Ltd (from Johnson and Eyssautier, 1987)

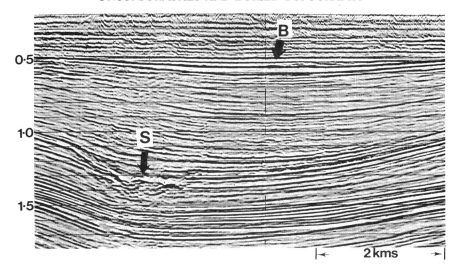

Fig. 6.7. Section showing onlap infill (B) above a synform depression, the depositional surface for the infill being an unconformity (U). The synform results from underlying mobile salt flow. S indicates an irregular feature due to slumping. Reproduced by permission of Seismograph Service (England) Ltd

seen truncating progressively up the Top Triassic unconformity (marked by the lower of the arrowheads). Where the reflections meet the unconformity, they do so while maintaining the same dip angle as they have further down-flank.

In the other type of baselap (known as *downlap*), an originally dipping stratum terminates downdip against an originally horizontal or dipping surface. Often, in the seismic section, the stratal event 'droops' in a subtle way as it approaches the terminating surface, and this is seen well in the example of downlapping events in Fig. 6.10. The solid arrows indicate two of the terminating stratal events as they dip slightly 'into' the unconformity surface. It should be noted that discussion of all these reflection configurations assumes that the examples seen are in the true dip direction—if not, then a minimum of two sections intersecting orthogonally is required to establish the true relationship.

The last seismic example is of particular interest, as it suggests how a subtle trapping configuration could come about just above an unconformity surface within strata showing the downlap effects exhibited by Fig. 6.10, provided that the three-dimensional geometry was favourable, either because of contour closure, or faulting.

A possibly applicable case is that of the Moonie oilfield in Queensland, Australia, which was the first commercial oil accumulation discovered in Australia, in the Surat Basin, 175 miles west of Brisbane. Production is from a basal Jurassic unit, the Precipice Sandstone, which unconformably overlies a truncated Permo-Carboniferous section, as described in the paper by Moran and Gussow (1963). A schematic section across the field is shown in Fig. 6.11. The Precipice Sandstone interval is sealed above by the Evergreen Formation, which consists principally of siltstones and kaolinitic shales. The oil source was not known

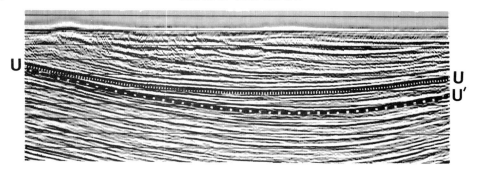

Fig. 6.8. In this example, the unconformity, U, again takes the shape of a synform depression, but with a non-uniform infill prograding (and thinning) from left to right. The sediment in the wedge between the two unconformities U and U' indicates a completely different transport direction. Reproduced by permission of Seismograph Service (England) Ltd

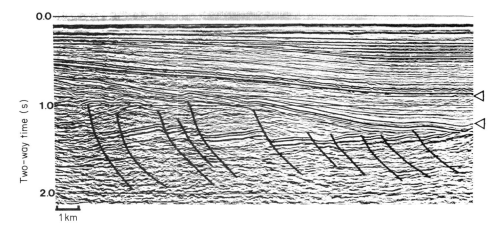

Fig. 6.9. Onlap: the stratal events in the interval between the arrowheads at the right-hand end of the section are progressively onlapping the unconformity at the lower boundary of the interval, which is a sequence boundary. The stratigraphic situation is typical of many basin margins. Reproduced by permission of Seismograph Service (England) Ltd

at the time of writing of the cited paper, but was thought to be probably Permian, possibly early Jurassic, with the accumulation occurring in mid-Jurassic times.

A seismic section across the field is shown in Fig. 6.12. Moran and Gussow say little about the structural interpretation of the trap, except in the discussion following the paper, where it is suggested that draped structure over basement topography, with closure, probably existed from the time of deposition of the Precipice sands onwards. This may well be the case. However, close inspection of the seismic line in Fig. 6.12 leads to the additional suggestion that there is at least some element of downlap of the Precipice interval immediately above the unconformity, within the area of Moonie Field. The closure on the (approximate) top of the basal Jurassic (Precipice) sands is seen as a NE–SW-trending,

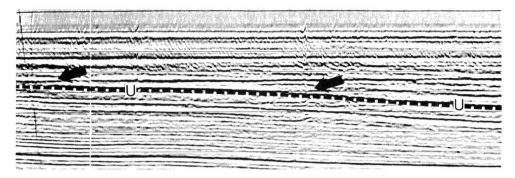

Fig. 6.10. Seismic events (arrowed) downlapping on the surface of an unconformity (U). Note the tendency of the events to curve gently downwards on to the unconformity, and compare this with the onlap terminations in Fig. 6.9 to see the difference in the two reflection configurations. Reproduced by permission of Seismograph Service (England) Ltd

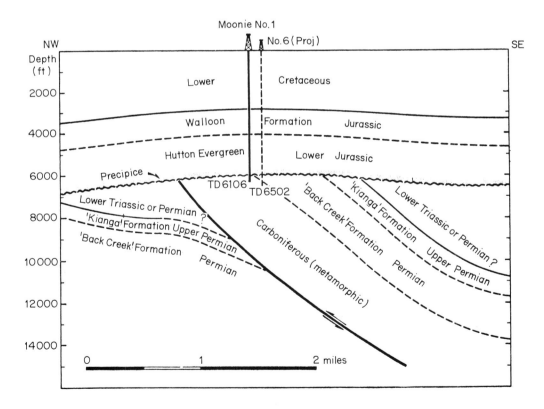

Fig. 6.11. Schematic section across the Moonie oilfield, Queensland, Australia. TD = total depth (ft). Reproduced by permission of Elsevier Applied Science Publishers Ltd (from Moran & Gussow, 1963)

elongate feature in the depth-structure map of Fig. 6.13. The fault indicated in the deeper section of Fig. 6.12 (possibly a conduit for Permian hydrocarbons?) is not seen in the map, as it is supposed to terminate upward at the unconformity.

Thus, subtle depositional features in sediments immediately overlying an unconformity can result in excellent trapping situations.

Other unconformity related reflection configurations suggest good structural and stratigraphic trap situations. In Fig. 6.14, shingled clinoform events are seen in the seismic data within the arrowed interval that is bounded above and below by disconformities (paraconformities in other terminology). These clinoform events (for fuller discussion of such events, see the later chapter dealing with deltaic and fan deposition) indicate a thin, prograding pattern with very gently dipping parallel oblique clinoforms, suggesting deposition prograding into bodies of shallow water. The individual clinoform events show little or no overlap, with variable downlap (near the left-hand end) and tangential patterns at base.

In the Fig. 6.15 example, some very clear parallel oblique clinoforms are seen. These often occur in minor, time-restricted intervals and indicate a high-energy, high-sediment-supply environment; a sea-level stillstand is usually involved, with little or no basinal subsidence. The removal of any topsets (see part on deltas in later chapter) by scour or bypass, and the high-angled downlap at the base, are typical. Sometimes small channel-fills or rapid outbuilding into shallow water are indicated by this facies. Again, the structural and stratigraphic trapping possibilities of these features are clear. The last two examples are from Jenyon and Fitch (1985). For further discussion of seismic sequence and facies analysis, see e.g. Mitchum *et al.* (1977a,b).

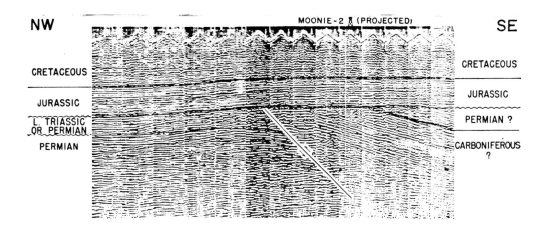

Fig. 6.12. Seismic section across the Moonie Field, showing the general drape of the Jurassic section over relief on the Permo-Carboniferous basement rock events. Reproduced by permission of Elsevier Applied Science Publishers Ltd (from Moran and Gussow, 1963)

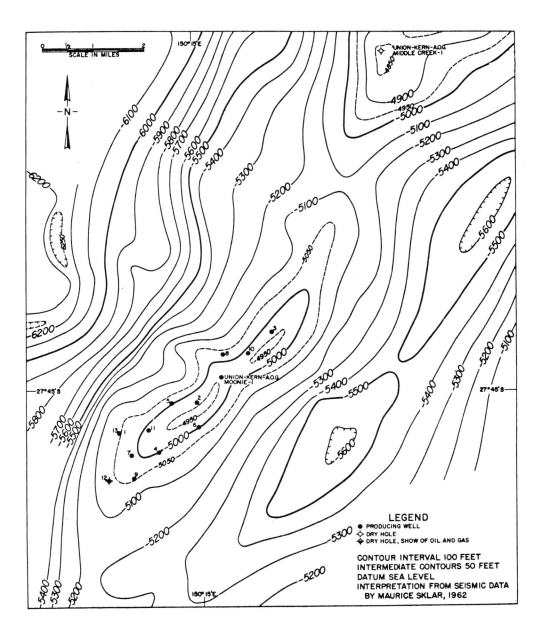

Fig. 6.13. Depth-structure contour map of the Moonie Field at the reservoir level, showing the structure to be a closed high that trends approximately NNE–SSW. From the seismic data, it seems likely that downlapping of Precipice Formation events on the unconformity surface is an important element in the trap. Reproduced by permission of Elsevier Applied Science Publishers Ltd (from Moran and Gussow, 1963)

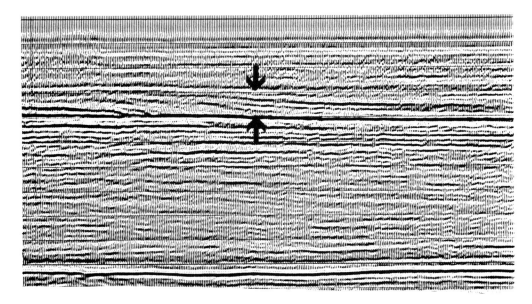

Fig. 6.14. Reflection relationships at boundaries: this seismic example shows shingled clinoform events in the interval between the two arrows (bounded by unconformities). Differing reflection configurations give indications of the depositional environment within which the sediments were laid down. (From Jenyon and Fitch, 1985, Gebrueder Borntraeger-Verlag.)

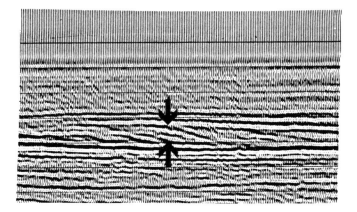

Fig. 6.15. Reflection relationships at boundaries: an example showing parallel oblique clinoforms in a discrete, thin interval (arrowed) bounded by unconformities. (From Jenyon and Fitch, 1985, Gebrueder Borntraeger-Verlag.)

6.4 Buried Topography

Buried topographic, or 'PGM' features, have been regarded with increasing interest by explorationists in recent years. Together with certain other features produced by porosity variations and pinchouts, they have come to be known as 'subtle traps'. They usually—but not invariably—involve an unconformity, and are either positive or negative in the sense that they are antiform and project upwards into the post-unconformity sequence (fault scarps, buried hills, cuestas), or synform, projecting downwards into the pre-unconformity rocks (buried channels, graben, localized depressions in karst or volcanic topography). The basic factor is that at the time of their formation, they were part of the contemporary Earth surface (Jenyon, 1987a). 'False' features that mimic buried topography can arise in a wide range of geological situations, where the features were buried under various depths of sediments when formed.

6.4.1 Positive Features

A structurally simple example of a positive feature is shown in Fig. 6.16, where a buried Triassic hill is present in the seismic section (unmarked version in (a) and marked version in (b)). The marked line in (b) coincides with the Top Triassic unconformity, overlain by

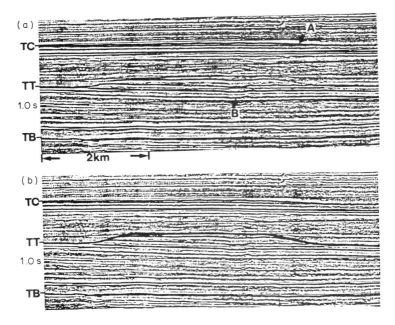

Fig. 6.16. Buried topography. In the unmarked (a) and interpreted (b) versions of this example, a buried Triassic hill can be seen (clearer if viewed obliquely from the side). TC = Top Chalk; TT = Top Trias; TB = Top Bacton. Reproduced by permission of the Journal of Petroleum Geology (from Jenyon, 1987a)

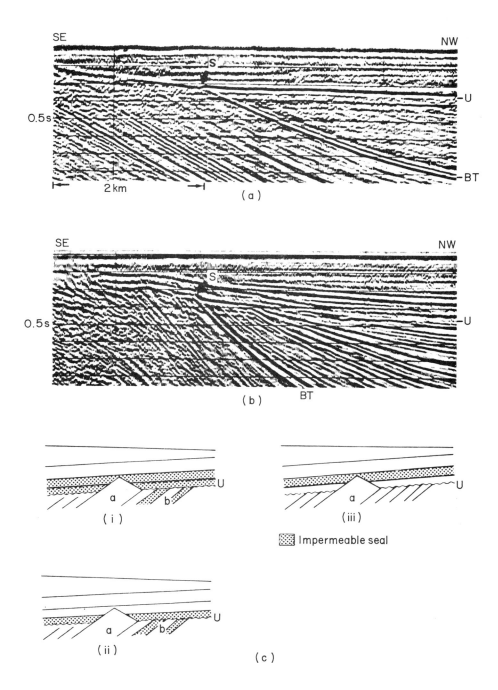

Cretaceous strata. The 'hill' is formed of nearly horizontal Triassic stratal units, and the definition of the top of the hill—see (a)—is poor because the stratal dip (and probably also the acoustic impedance of the uppermost unit) is similar to that in the overlying Cretaceous. However, the subtle upcurve of the Cretaceous stratal events on the flanks as they approach the side of the hill suggests some depositional drape as well as truncation, and is a good indication of the presence of the Triassic feature.

The reason for the existence of the Triassic hill cannot be determined with certainty, but an educated guess can be made. In this area, several thin evaporite bands occur in the Triassic, and in places these have been leached by formation water circulating through fractures. This has led to minor localized dissolution and collapse, as at B in Fig. 6.16(a). It seems likely that this process was quite widespread, resulting in a highly dissected Earth surface at the time, and leaving erosional remnants of Triassic like this 500-ft- (150 m) high example standing on a peneplain surface.

At A in Fig. 6.16(a) another PGM surface can be seen at Top Chalk level. Note the slight rugosity of the horizon between A and the left-hand end of the section, and the shallow truncation at A.

If a positive feature, such as this buried hill, contains good reservoir quality units, then it is important for a trap situation that at least the early stages of post-erosional transgressive deposition should involve suitable sealing sediments, such as argillaceous fines, or better still, evaporites and especially salt rock.

Cuesta features are produced at an angular unconformity surface that is erosionally immature, leaving erosion-resistant dipping units of the lower series standing proud above the unconformity surface. The upper illustrations (a and b) in Fig. 6.17 have good examples of this type of feature. The Base Tertiary reflection group (BT) is seen forming cuesta scarps (S) at the unconformity (U). This reflection group is produced by an erosion-resistant basal sequence of the Tertiary in this area. It is of interest to note that the direction in which the cuesta scarp faces depends on the dip of the bed at the unconformity. In (a) the scarp, which is produced by the truncated bed-ends of the basal sequence, faces south-east, while in (b), where the unit is dipping more steeply, the scarp is being produced by a bedding surface, and faces north-west.

The trap potential of a cuesta feature can often be complex, and is dependent on the detailed lithostratigraphy of the post-unconformity sequence. Some possible situations are shown by the schematics in Fig. 6.17(c). Impermeable sealing units are shaded. Diagram (i) shows a situation where the cuesta feature 'a' is totally sealed above the unconformity; however, another unit below the unconformity, 'b', is also sealed, and there is no great advantage in the cuesta type feature as a trap.

In diagram (ii) the cuesta may actually provide a breach through an impermeable seal, allowing the escape of hydrocarbons into the upper section, while a unit such as 'b' may provide an efficient truncation trap. Diagram (iii) shows a situation in which

---

Fig. 6.17. *(opposite)* (a) and (b) Two examples showing the effect of dip on erosion-resistant cuesta scarps. (a) Less steep dip of the Base Tertiary (BT) event in the lower series leads to the scarp (S) at the unconformity surface (U) facing to the south-east. (b) Where the dip is steeper, the 'scarp' is formed by a bedding plane reflection, and faces north-west. Reproduced by permission of Seismograph Service (England) Ltd. (c) Varying hydrocarbon trap potential of a cuesta feature. Reproduced by permission of the *Journal of Petroleum Geology* (from Jenyon, 1987a)

a cuesta comes into its own as a trap; the unit projects upwards into an impermeable bed that is separated from the unconformity by a permeable interval. Thus, although the other sub-unconformity units are not sealed, the upper part of the cuesta can form a trap.

In situations of this type, especially when complicated by faulting, migratory routes of hydrocarbons may be extremely involved, with fault planes, unconformity surfaces, and fracture zones acting as 'guides' to the moving fluids.

The seismic example of Fig. 6.18 raises an interesting question. The data clearly indicate a slightly angular unconformity (U) over the greater part of the section, but towards the right-hand end the dip of the sub-unconformity series steepens in an upward flexure. At the surface of the unconformity a positive feature is seen (C) having the appearance of a scarp produced by an erosion-resistant sequence (fortuitously (?) occurring just at this location in the upward flexure) that seems then to persist to the right-hand end of the section.

Although it seems likely that this is the whole explanation for the feature C, another possibility cannot be completely ruled out. Is there such a factor as what might be called 'erosional anisotropy' in some sedimentary rocks? Many rocks have quite strong directional properties imparted by the depositional environment and the shapes of the sedimentary materials from which they were formed. The most obvious example is the water-laid sediment with an azimuthal alignment of elongate mineral particles, pebbles, etc., in the original current direction.

Thus, many sediments can be considered as having a 'grain' analogous to that produced by the fibres in wood. Is it possible that, as in wood (it being easier to plane or file the material parallel to the grain rather than across it) some rocks may be eroded more easily

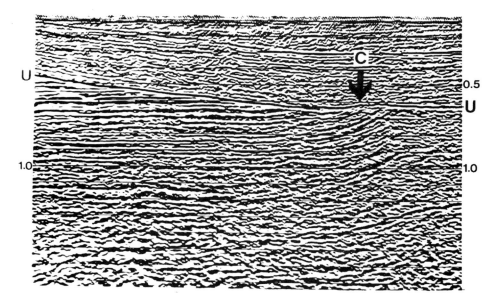

Fig. 6.18. Seismic section showing an erosion-resistant bed sequence forming a positive feature on the surface of unconformity U. This example raises an interesting question: is there such a factor as 'erosional anisotropy'? Reproduced by permission of Seismograph Service (England) Ltd

in planes subparallel to the bedding than in planes at larger angles, or normal to, the bedding?

Although this example does not give any particular support to such a concept, it does raise an interesting possibility. The axial plane of the flexure present is inclined upwards and to the left, and if there is anything in the idea, presumably the scarp feature has migrated from a location above and to the left of its present position as erosion progressed.

All other things being equal (reservoir thickness, hydrocarbon column height, porosity, permeability, etc.) it is also of interest to note the varying trap potential beneath the unconformity U in this example. There would obviously be a major difference in the volume of 'oil (or gas) in place' if there were a truncation trap beneath C, as compared with a similar trap in the left half of the section, where the truncated beds meet U at a much shallower angle.

A region in which buried topographic traps are amongst the most important types to occur is the eastern basins area of China. Oil and gas are found in mainly Mesozoic–Cenozoic reservoirs in a structural setting dominated by buried tilted blocks and horst–graben features. (In contrast, traps in the Chinese western basins commonly occur in anticlinal structures with multiple reservoirs, and in lensoid sandstones at or near anticlinal crests, serving as combination structural–stratigraphic traps—see, for instance, Fig. 3.7.)

Fig. 6.19 from Quanmao and Dickinson (1986), is a schematic composite, showing trap types characteristic of the eastern basins of China, and amongst these are several positive PGM-type traps of buried hill type, and unconformity traps.

These include: trap (1), a massive buried hill Palaeozoic or Sinian (uppermost Precambrian) reservoir trap; trap (2), a layered (stratigraphically controlled) buried hill Palaeozoic reservoir trap; trap (7), an unconformity truncated Palaeogene reservoir trap;

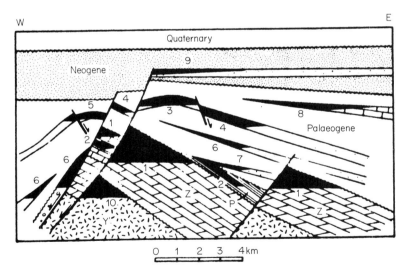

Fig. 6.19. A schematic composite showing some of the buried topographic, stratigraphic, and unconformity traps found in the eastern hydrocarbon-bearing basins in the People's Republic of China. From Quanmao and Dickinson, 1986, reprinted by permission of American Association of Petroleum Geologists

trap (9), a secondary Neogene reservoir, unconformity and fault controlled; and trap (10), a weathered granite–gneiss reservoir of buried hill type. The remaining numbered trap types are other types of stratigraphic trap, together with structural and combination traps. The key to the lettering is Y, Proterozoic basement; Z, uppermost Precambrian (Sinian); P, Palaeozoic. An unusual feature of some hydrocarbon accumulations in this area is that Tertiary sourced oil is found in Palaeozoic—or even Precambrian—reservoirs, having followed complex fault-plane-controlled and unconformity controlled routes into the traps.

6.4.2 Negative Features

A typical 'negative PGM' feature that frequently forms hydrocarbon traps is the buried channel, usually cut down through an unconformity surface and infilled with sedimentary material different from that forming the channel bed and banks. This difference usually produces a useful contrast in seismic character between infill and surrounding material on a section.

Channel sands of good reservoir quality, with a means of ingress for hydrocarbons, and an efficient seal to prevent their escape, make up an important class of traps.

As regards the nature of the infill material, there are two basic situations possible. Figure 6.20 (i) is a sketch showing the case in which the channel infill is of the same material as the post-unconformity basal transgressive unit 'a', with the unconformity surface coinciding with the channel bed beneath the infill deposits. In Fig. 6.20 (ii), however, the fill consists of some material deposited later than that into which the channel is cut, but earlier than that of the transgressive unit 'a'. It may be supposed in this case that any greater thickness of this intermediate-age sediment that existed was removed during the erosional phase of U; because of this, the unconformity surface U overlies the infill material that rests on the channel bed, which represents an *older* unconformity surface.

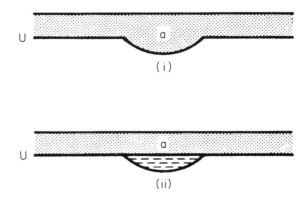

Fig. 6.20. Schematics showing two basic types of 'negative' buried topographic features. In (i) the channel infill is of the same type as the post-unconformity basal transgressive unit 'a'. In (ii), the channel-fill consists of material deposited later than that into which the channel is cut, but earlier than the transgressive unit 'a'. This infill may originally have been more extensively deposited, but removed during an erosional hiatus represented by the unconformity (U). M = multiple

The stratal relationships between the infill and the channel (e.g. whether the infill is conformable with the channel profile, or onlapping/truncating against the channel surface) may be of importance in relation to hydrocarbon trapping potential.

In the seismic section it is often possible to make some deductions about channel infill from the character of the seismic events. This can be illustrated by the following examples. In Fig. 6.21, a well-defined channel is present in the shallow part of the section. Although the seismic events in the upper part of the channel are spurious 'multiples', the lowest events from the infill are primaries, and show an onlapping relationship against the channel sides. Even from these one or two lowest events it is clear that their seismic character (basically the frequency content—dominant frequency/spacing of the reflections, and the amplitude) is different from that of the data adjacent to the channel, confirming the nature of the feature. Because of the obscuring multiple reflections (especially at M and above) it is not possible to be certain as to the relationship between the infill and the overlying material; however, the feature has the general aspect usually seen in the example shown in Fig. 6.20 (i).

In Fig. 6.22 (a) the situation is clearer. The buried channel (C) is located on the north flank of the London–Brabant Ridge, and the apparent dip of the flank can be seen across the section.

The channel C has cut down into the older rocks; note that the true base of the channel is arrowed, and the synform shapes below C at and above the level of M are multiple reflections. The unconformity clearly passes over the channel infill, which makes a subtle positive feature at the unconformity surface, indicating that it is more erosion-resistant than the pre-unconformity rocks. Stratal reflections within the infill are conformable with the channel sides. Plainly, this is a case similar to that shown in Fig. 6.20 (ii); the infill was deposited during the time period represented by U, and pre-dates the post-unconformity sediments.

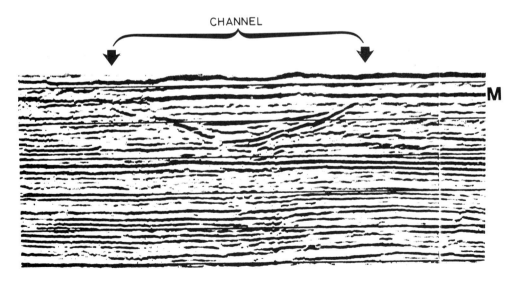

Fig. 6.21. Seismic example showing the appearance of a typical channel feature in the shallow section, obscured partly by horizontal shallow multiple reflections (M). Reproduced by permission of Seismograph Service (England) Ltd

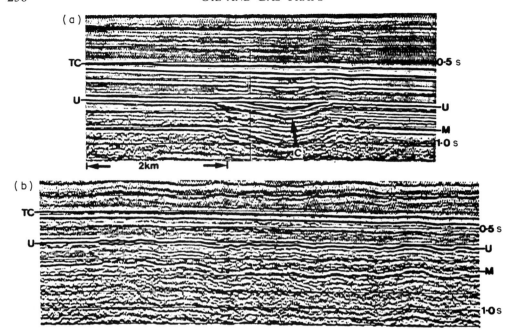

Fig. 6.22. (a) A buried channel (C) on the north flank of the London–Brabant Ridge. Unconformity U separates the base of the Cretaceous from probable Palaeozoic rocks below. The Top Chalk is indicated at TC, while M marks a strong multiple reflection. (b) The same unconformity (U) as in (a), in a nearby location (within 10 km). The considerable rugosity of the surface of U can be appreciated by comparing the character of the event with the smoothness of the reflections above 0.5 s, and also with the U event in (a). Reproduced by permission of the *Journal of Petroleum Geology* (from Jenyon, 1987a)

This channel is up-flank from the gasfields area of the UK southern basin, and could be providing traps in sand bodies within the channel; no drilling has yet taken place on this feature. The channel probably formed part of a subaerial or submarine drainage system during the Mesozoic.

Also of interest is Fig. 6.22 (b), which shows the same unconformity U in a nearby location to that in Fig. 6.22(a). Note, by comparing the U reflection with the strong reflections above 0.5 s, and the U reflections in Fig. 6.22(a), the considerable rugosity of the unconformity reflection event is seen in this immediate area. This is not apparent in Fig. 6.22(a), and seems to indicate some erosional immaturity in the area. The multiple reflection M is seen to mimic, and exaggerate, this roughness in detail.

The seismic example in Fig. 6.23 shows a very clear buried channel cut into the Pliocene–Pleistocene unconformity near Singkel, Western Sumatra Forearc Basin. This large channel was eroded into Pliocene shelf strata and filled with Pleistocene sand, gravel, and conglomerate. The contrast in the seismic data between the rather chaotic reflection events from the channel-fill, and the adjacent strong, horizontal Pliocene events, is very marked. The channel apparently disappears landward, but probably formed as a Pleistocene erosional feature cut by the Singkel River during a sea-level lowstand (Beaudry and Moore, 1985).

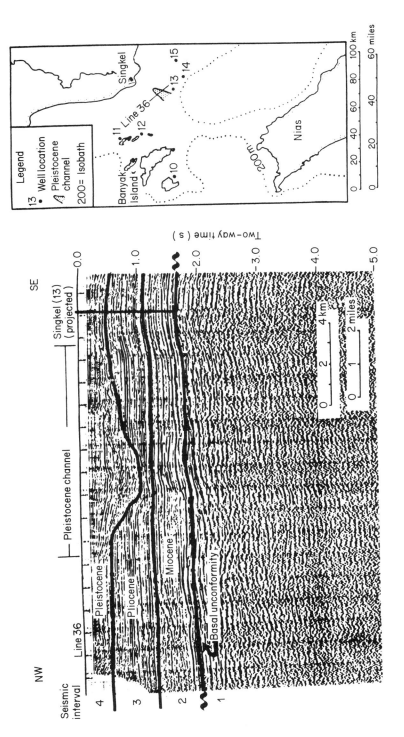

Fig. 6.23. Seismic example showing a very clear buried channel cut into the Pliocene–Pleistocene unconformity near Singkel, Western Sumatra Forearc Basin. From Beaudry and Moore, 1985, reprinted by permission of American Association of Petroleum Geologists

Whether buried channels contain hydrocarbon traps depends on many factors—the sedimentology of sandbodies of various types within the channel, the presence of adequate seals, diagenetic effects, timing of hydrocarbon migration *vis-a-vis* the formation of the channel, etc. However, situations in which certain lithologic types are encased in other lithologies must always be of potential importance to the explorationist, and will merit close investigation by both geological and geophysical methods.

This brief treatment of trap developments related to unconformities and buried topography is not the end of the subject. These elements will also be present in the consideration of some types of combination trap in Chapter 10.

CHAPTER 7

Porosity and Pinchout Traps

7.1 Introduction

Porosity traps occur where the fabric of a reservoir rock changes laterally such that the porosity diminishes until the capillary entry pressure becomes too great to permit further migration of the hydrocarbon column. The porosity may be primary or secondary, and be of inter- or intragranular, fracture, or vugular type.

Pinchout (wedge-out, feather-edge) traps are present where progressive diminution of thickness of the reservoir interval occurs laterally, until the interval, sufficiently sealed by impermeable formations, terminates updip.

Needless to say, in both types of situation, the requisite three-dimensional geometry for trap formation must be present in reservoir and seal units.

7.2 Porosity Traps

Porosity variations in a reservoir interval may be of primary or secondary origin. Lateral variations in the depositional environment, leading to changes in lithofacies or fabric, can lead to the lateral porosity changes necessary (either *in situ* or after tectonism) to develop a trap. Porosity changes cannot easily be illustrated by standard seismic data, although today there are reprocessing techniques and colour display methods that can be used to effect such illustrations.

A relatively common situation leading to porosity traps occurs where there is subaerial exposure and dissolution at an unconformity surface; this leads to leaching in various ways, including the development of vugular porosity, solution channelling, etc. Subsequent deposition of impermeable beds on the unconformity surface, followed at a later stage by migration of hydrocarbons into the leached zone (perhaps through faulting) can produce commercially viable oil/gas accumulations.

An example of this type of trap is provided in the Anadarko Basin, USA, by leaching of the Frisco Formation (a mid-Lower Devonian limestone within the Hunton Group). The overlying Woodford Shale (Upper Devonian–Lower Mississippian) is a source formation, as well as being the seal for the Frisco accumulation. In seismic data, where

Woodford Shale overlies high-velocity Frisco Formation limestone that has not been leached, a strong seismic peak of high amplitude is developed on a relative-amplitude reprocessed seismic section. Elsewhere, where the Frisco Formation has developed good secondary porosity due to leaching, its seismic velocity is reduced, the reflection coefficient between Woodford and Frisco Formations is lowered, and the strong peak is lost. This gives a basis for identifying the zones of good secondary porosity in the Frisco Formation, and hence the likely zones of trap development. Recommended is the paper by Morgan *et al.* (1982), which includes photomicrographs of solution effects in Frisco Formation limestones, together with seismic examples demonstrating the relative amplitude effects in porous and non-porous limestone zones.

The book in which the latter paper appears (Halbouty, 1982) contains many other good examples of these types of stratigraphic trap.

Other 'classic' porosity traps occur where a sandy interval 'fines' or 'shales-out' updip, and also where a sand body is completely encased in shale/clay matrix. In the latter case, in order for a hydrocarbon trap to be present, time of migration of any hydrocarbons in the area, related to time of formation of the trap, is important. Clearly, the oil/gas needs to have got inside the trap before it became completely sealed, or perhaps later owing to faulting that did not completely destroy the seal.

Diagenetic processes involving the reduction of sandstone porosity by quartz overgrowth may also be of significance in porosity trap formation. Leder and Park (1986) have studied this process using a model of fluid flow through initially porous, unlithified sands in which the fluid phase is saturated with $Si(OH)_4$ and is always in equilibrium with quartz. As the continuously circulating fluid migrates updip towards the basin margins, cooling and quartz precipitation into the pore spaces take place, with resultant loss of porosity by the sand; under favourable circumstances, a porosity trap may result.

The rate of porosity reduction is stated to depend on the following variables in decreasing order of significance: burial rate, age, initial porosity, basin size (dip angle), fluid dynamics, initial permeability, and geothermal gradient. As noted earlier in the book, there could be some argument with a process that puts the geothermal gradient in such a low position in order of importance.

7.3 Pinchout Traps

Traps due to the pinching- or wedging-out of beds updip are the cause of very many commercial hydrocarbon accumulations. The seismic method is particularly well-suited to the detection of such traps, since even quite small differences in dip between adjacent events on the seismic section are easily visible by simple inspection. Also, the effects in seismic data of bed thinning are now quite well understood, even when quite subtle.

7.3.1 Bed Thinning in Seismic Data

Two typical situations in which bed thinning and pinching-out occurs are shown in Figs 7.1 and 7.2.

In Fig. 7.1, a small half-graben structure is seen, the example in this case being a seismic section from the Manx–Furness Basin, in UK waters east of Eire. In the section, the approximate true position of the main fault zone is indicated by the steeply dipping, heavy

POROSITY AND PINCHOUT TRAPS 241

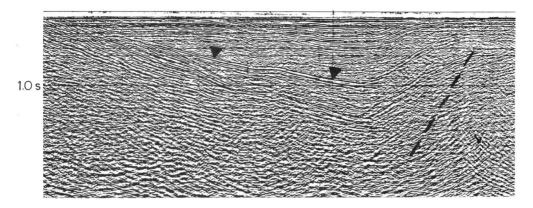

Fig. 7.1. Seismic section showing a half-graben structure with (arrowed) an interval thinning updip. This is typical in the structural setting, which is one of those frequently leading to updip pinchout traps. The thinning of intervals is the result of syndepositional subsidence of the half-graben downthrow side. Reproduced by permission of Seismograph Service (England) Ltd

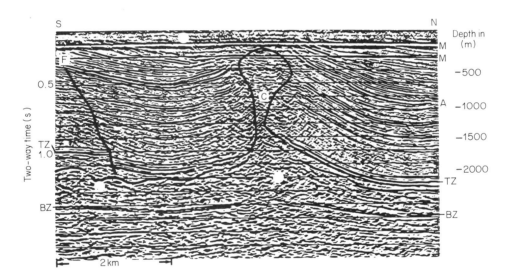

Fig. 7.2. Seismic section through a salt diapir (C). As explained in the text, owing to the salt movement, thinning intervals with updip pinchout features are common in the 'primary rim syncline' sediments if these are present. BZ = Base Zechstein; TZ = Top Zechstein; F = fault; M = multiple reflection. Reproduced by permission of Seismograph Service (England) Ltd

dashed line in the right-hand half of the section. One prominent interval thinning updip to the left is arrowed; it is obscured in its upper left part by near-horizontal shallow multiple events at the level where it must be pinching-out.

An interval that thins in this way is generally indicative of synsedimentary movement (i.e. in this case the graben was subsiding while deposition of the interval was taking place).

Whether this particular interval constitutes a pinchout trap situation depends on whether it terminates updip in a feather-edge and is sealed. (Alternatively, whether thinning or not, the interval could involve a porosity trap if the porosity 'tightens' (decreases) sufficiently in the updip direction.) Under usual circumstances, sedimentary 'fines' could be expected to occur more towards the basinal/graben centre (other things being equal) which would mean that porosities would tend to *increase* from the centre towards the margins (i.e. updip) with coarsening sediments. This would not favour a porosity trap situation. However, in a high-energy depositional environment, and depending on the scale of the basin involved, alternative depositional scenarios could be envisaged that would result in updip porosity decrease.

As regards the possible presence of a pinchout trap, it was noted that synsedimentary movement is required. This point is important in the second situation modelled, as shown in Fig. 7.2. Here, an active salt basin situation is seen, where a mobile salt layer within the marked BZ–TZ interval has intruded a diapir (C) into overlying sediments. In this situation, there is a specific level in the overburden (lettered A at the right-hand end of the section) below which (down to TZ) the overburden sediments will show no tendency to pinch-out towards the rising diapir at C. This is because they were present as deposits of uniform thickness before the start of the salt movement (i.e. they represent the pre-salt movement overburden). Above A there may be (unless removed by erosion) primary rim syncline sediments, recognizable because they *thin* towards the salt diapir, that represent deposition during early salt movement into the pillow that was precursor of the diapiric piercement structure C (see Chapter 5 for a more detailed account of salt-movement-related traps). These primary rim syncline sediments will show thinning towards the diapir (i.e. updip) and therefore the potential will be present for pinchout trapping.

Because of the uniform thickness of the overburden below 'A', the beds here will not pinch out towards the diapir (although they could include porosity traps, or truncation traps against the salt). In some cases, the zone of 'uniform overburden' above a salt pillow does include some pinchouts due to early, isolated salt movements.

Beneath an unconformity, clear distinction should be made between a truncation trap and a pinchout. The two types are modelled in Fig. 7.3, in which unmarked (a) and marked (b) versions of the same section are shown. In the shallow part of the section there is an unconformity U. Beneath this, a pinchout can be seen at A, while at B there is one of several truncation traps that are present in this example.

7.3.1.1 *Seismic resolution (horizontal and vertical)*

The observation and determination of thinning beds raises the question of resolution in seismic data. What is the smallest geological feature that can be observed clearly in the seismic section? Put another way, what is the thinnest bed for which we can expect to resolve (separate) the top and bottom of the bed visually in the seismic data?

Such questions have been addressed—and answered—by several workers in the past. For example, Widess (1973) studies the effect in seismic data of bed thinning in the paper 'How thin is a thin bed?', while general seismic resolution, with a particular study of the Fresnel zone approach to horizontal resolution in the seismic section is discussed by Sheriff (1985) in 'Aspects of seismic resolution', as Widess deals with vertical resolution.

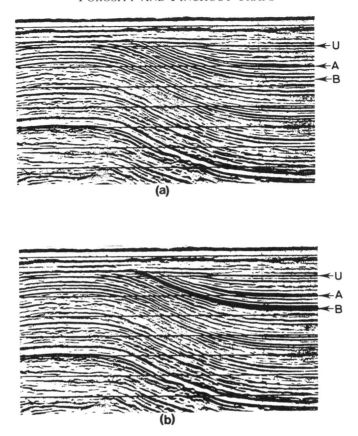

Fig. 7.3. Uninterpreted (a) and interpreted (b) seismic examples showing a thinning interval A that could provide a pinchout trap, and a truncation (B) against an unconformity U. The basic difference between these two trap types is dealt with in the text. Reproduced by permission of Seismograph Service (England) Ltd

For present purposes, it is not proposed to repeat work already done, but it may be of some value to quantize the question of thinning beds to some extent. This will be done by consideration of the sketch in Fig. 7.4, which indicates the relationship of dominant seismic wavelength to resolution of bed thickness.

A 25-Hz Ricker wavelet is used in this model of a bed that is thinning from 80 feet (lowest trace) to 10 feet (top trace). The wavelet responses to the intermediate bed thicknesses are shown on the traces in between. As the bed thins, the peak-to-trough time of the associated seismic pulse train (i.e. the time in milliseconds between the centre of the large trough, and the centre of the large (solid black) peak that form the central pair of features on each trace) decreases until the bed thickness is one-eighth of the dominant seismic wavelength (see lower five traces in the sketch). Thereafter, as the bed thins further, the peak-to-trough time remains constant (see the upper four traces in the sketch) but the amplitude decreases— the thickness is now said to be 'amplitude-encoded'.

244   OIL AND GAS TRAPS

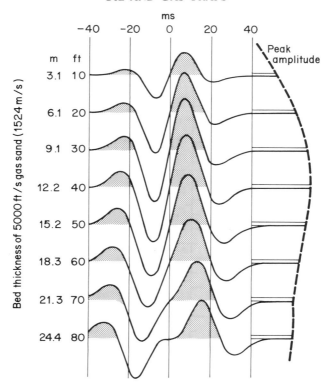

Fig. 7.4. Diagram explaining the relationship between dominant seismic wavelength and bed thickness in seismic data. A basic 25-Hz Ricker wavelet is used in this model. (From Jenyon and Fitch, 1985, Gebrueder Borntraeger–Verlag, after original work by Widess, 1972.)

In this example, the critical bed thickness—known as the 'tuning thickness', and occurring where the amplitude peaks, is between 30 feet and 40 feet. The effects seen here are easy to demonstrate in a model, or in wavelet-reprocessed seismic data using a short, simple, theoretical wavelet such as the Ricker. However, they are often difficult, or impossible, to see in field data produced by an extended and complex 'field' wavelet. Sometimes, with experience and good low-noise data, the 'thinning bed' phenomena can be recognized on the standard seismic section.

7.3.2 Seismic Example

A typical example of pinchout traps is presented by the Minnelusa Sands of the Powder River Basin, north-east Wyoming. These sands, of Upper Pennsylvanian/Permian age (Raffalovich and Daw, 1984) have provided various hydrocarbon traps that were somewhat elusive until their pinchout nature was realized and a rational exploration method devised. This involves deriving synthetic seismograms from sonic logs of boreholes drilled in the area, and arranging these in cross-section form.

Normally, this means that a seismic line is executed in the dip direction through the surface locations of several boreholes, and the synthetics derived from the sonic logs, to the same scale as the seismic section, are superposed on the latter in their correct positions. By this means, it becomes relatively easy to observe where the top Minnelusa reservoir sand disappears in a pinchout.

Figure 7.5 shows three such synthetic seismograms A, B, and C (indicated by the open arrows along the top of the section) superposed on a Vibroseis® seismic line at the appropriate locations. The character match between the real and synthetic seismic data is very good, allowing formation tops to be picked on the seismic line.

As seen here, the seismic (black) peak that correlates with the reservoir (upper Minnelusa sand) disappears updip (where indicated by solid black arrow) between the borehole locations B and C, as confirmed by the fact that the corresponding strong black peak seen on the synthetics at A and B is absent at the updip C location.

In view of previous discussion on thinning beds, it is of interest to note the marked decrease in amplitude of the black peak as it approaches the pinchout or feather-edge location between boreholes B and C. Also, downdip to the south-west, the black peak disappears again near to the left-hand end of the section, suggesting that this sand body is a discontinuous sheet or flat lens. This supports the interpretation of the Upper Minnelusa as having been deposited in a sabkha/aeolian environment and consisting of sands, dolomites, and anhydrites.

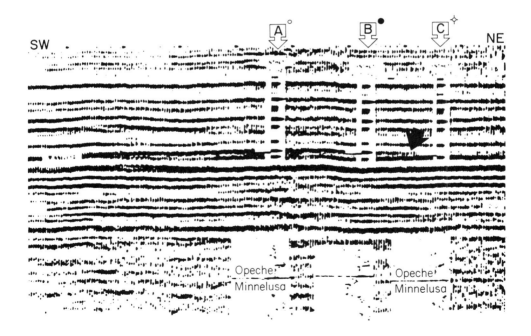

Fig. 7.5. Seismic example demonstrating the pinchout of the Minnelusa Sand, Powder River Basin, north-east Wyoming (solid black arrow), as confirmed by the synthetic seismograms at A, B, and C. (From Raffalovich and Daw, 1944, *Geophysics*.)

7.3.3 'Erosional Pinchouts'

Some trap situations, which in the terminology used in this book would be classed as 'truncation traps' beneath an unconformity, are sometimes referred to as pinchouts, or 'erosional pinchouts' elsewhere.

A situation of this type is seen in the diagram section of Fig. 7.6, which is a schematic of the situation at Guanarito, Venezuela. This is categorized as an 'erosional pinchout' by Kiser (1967), who has described the structure. In Fig. 7.6, B indicates basement; C, Cretaceous; E, Eocene; T(M–O) is a later Tertiary (Miocene–Oligocene) sequence overlying the Eocene unconformably; E_p marks the Eocene 'pinchout' updip; and C_p the Cretaceous 'pinchout'. Three major unconformities (B/C, C/E, and E/T) converge in the updip direction on to the flank of the Al Baul Arch, as expressed by the basement rocks there. The overall structure is described as a homocline with a maximum dip at the top of the basement of 2°. The thickness of potential productive reservoir sequences in the Cretaceous, Eocene, and Oligocene strata varies from a few feet to 900 feet. Light porosity oil and gas have been encountered in drill tests. The principal source beds are considered to be Upper Cretaceous dark euxinic marine shales and limestones.

Note that the situation here is effectively the same as that seen at level B in Fig. 7.3, although the stratigraphy is more complex.

Mackenzie (1972), in a paper on primary stratigraphic traps in sandstones, described primary traps as involving 'lateral termination of the reservoir as a direct or indirect result of factors related to the depositional environment'. Thus, lateral variations of porosity, 'shaling-out' and other facies changes, as well as pinchouts *sensu stricto* would all fall within Mackenzie's definition of a primary statigraphic trap. Erosional 'pinchout' at an unconformity would be excluded, however, since, as specifically stated by Mackenzie, 'the main trapping mechanism is unrelated to deposition of the reservoir sands'. This is a useful distinction, and a serviceable definition to bear in mind when considering porosity, pinchout, and truncation traps. Although not recent, Mackenzie's paper is still recommended reading, and has a useful reference list.

Another, similar trap is the wedge-out/truncation of the Sarir Sandstone (reservoir) in the giant Messla oilfield, Libya. The field lies on the east-dipping flank of a gentle Precambrian basement high, the Sarir reservoir wedging-out westward on the basement, and is truncated in an 'erosional pinchout' by an overlying basin-wide unconformity at the base of the capping Upper Cretaceous marine shales (the Sarir being a Lower Cretaceous

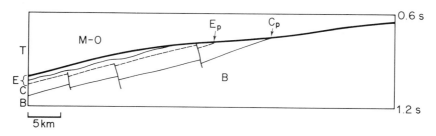

Fig. 7.6. Diagram section showing the 'erosional pinchouts' at Guanarito, Venezuela. Text discusses the difference between these, and pinchout traps *sensu stricto*, and explains the lettering. Reproduced by permission of Elsevier Applied Science Publishers Ltd (from Kiser, 1967)

fluvial sandstone). The field is described by Clifford *et al.* (1980), with an excellent seismic illustration of the feature.

7.3.4 Stratigraphic Traps in Aeolian Rocks

A special case of stratigraphic traps is that for aeolian sediments. Fryberger (1986), in an interesting paper, distinguishes between geomorphic, diagenetic, and system-bound traps in wind-deposited rocks. A system-bound trap involves an updip edge-of-permeability, in, for example, a dune sand system. A geomorphic trap involves geomorphic relief (e.g. dune tops) that has been preserved, perhaps with some reworking; oil/gas is trapped beneath closures on the tops of the dunes or dune complex. In a diagenetic trap, the trapping element is due to, for example, facies-selective cementation of sabkha deposits.

Of the three classes of trap, system-bound traps 'contain the largest known stratigraphically-derived oil pools in (aeolian) deposits', according to Fryberger, and he mentions the giant accumulations at Rangely Field and the updip edge of the White Rim Sandstone, Elaterite Basin, and Tar Sand Triangle of Utah, USA (see Heylmun, 1964; Baars and Seager, 1970).

CHAPTER 8

Carbonates, Evaporites, and Reefs

8.1 Introduction

The features to be discussed in this chapter can be found separately or as an 'assemblage' in more or less close association with each other. Such an assemblage may be found, for example, where a carbonate shelf or ramp, with reefal features at the shelf margin and in back-reef lagoonal zones, projects outwards from the margins of an evaporite basin towards the basin centre. Depositional environments of this kind, and many variations thereof, have occurred frequently throughout geological time, and are found in the stratigraphy in many areas world-wide.

As regards the relationship between these features and hydrocarbon accumulations, many different factors are involved.

Carbonates may develop as hydrocarbon source rocks, as reservoirs, and sometimes as seal formations. Where they act as reservoirs, their quality as such often depends much on post-depositional history (diagenesis and/or tectonic stress applications), which can also be said of carbonates acting as seal formations. There is good evidence that points to some carbonate formations being 'self-sourcing' (i.e. acting as both source and reservoir), which would go some way towards resolving certain problems (mentioned briefly in Chapter 1) regarding difficulties of explaining the primary expulsion of hydrocarbons from some carbonate rocks. Where they are source rocks, the details of the depositional environment are determining factors in their quality—for example, type and amount of sediment, type and amount of organic carbon content, conditions and depth of burial, and geothermal gradient.

Evaporites, where they are present, can play a variety of important roles in hydrocarbon accumulation. The plastic deformation of salt rocks and bitterns (potash salt deposits) and/or dissolution of these salts with concomitant collapse of the overburden, can produce both structural and stratigraphic traps in the overlying section. If these pre-date hydrocarbon migration through the section, they can form excellent trapping situations. Salt rock, with its ability to flow plastically, forms perhaps the most efficient seal formation found in the stratigraphic column in any area. Other evaporite types, such as anhydrite (provided it has not been subjected to tectonic stress leading to fracturing) can also act as seals.

Facies-bounded carbonate reservoirs, such as biohermal reefs, depend for their viability as hydrocarbon reservoirs on their post-depositional history. The various types of structure normally referred to as bioherms—coral/bryozoan/algal reefs, debris banks, etc., have high primary porosity (which may be over 40% in some cases), but this is invariably destroyed on burial. Any secondary porosity that develops (and this sometimes reaches 20–30% or more) is the result of diagenetic processes, such as selective leaching. The secondary porosity volumes that result are often complex in pattern and amount through the reef mass, with variations frequently controlled by original lithological zoning, expressing differences between the reef core, fore-reef debris slope, back-reef zone, etc. Some bioherms have been extremely persistent over long periods of geological time; shelf-edge algal reefs are known that extend over depth ranges of many hundreds or even thousands of metres.

Each of the types of features mentioned here produces a range of characteristic responses in seismic data, and some of the more frequently encountered of these will be discussed in this chapter, together with actual examples of hydrocarbon traps.

8.2 Carbonates

Typical ranges of seismic compressional wave (P-wave) velocities (in m/s) in some different rock types are:

Coal	1900 3500
Shale	1600 4100
Sandstone	1750 4800
Salt rock	4500 (constant)
Limestone	3900 6000
Anhydrite	5750 6250
Dolomite	4900 7100

Even from this short list it is seen that although there is considerable overlap in the velocity ranges of the clastic rocks (coal, shale, sandstone), and also overlap in the chemically precipitated rocks, there is minimal overlap in the ranges between clastic and chemical types.

The chemical rocks have, in general, higher P-wave velocities than those of the clastic rocks; this is to a great extent due to the denser and more compact nature of the chemical rocks, which have less intergranular porosity, and higher bulk density. In rocks, higher density usually means higher seismic velocity. The term 'acoustic impedance' is used to refer to the product of bulk density and P-wave velocity ($r.V_P$) in a rock, which gives a measure of the transmissibility of elastic (seismic) energy.

The contrast in acoustic impedances between two rock units separated by an interface (e.g. a bedding plane) will result in a seismic reflection being produced at the interface when seismic energy impinges on the interface, provided that: (i) the contrast is sufficiently high; and that (ii) the thicknesses of the two rock units bounding the interface are sufficiently large in relation to the dominant seismic wavelength at the level of the interface in the subsurface. This means in practice that each rock unit must have a minimum thickness not less than one-quarter wavelength of the dominant seismic wavelength. If these conditions hold good, then the strength of the seismic reflection produced will depend directly on the size of the acoustic impedance contrast across the interface. The *sign* (positive or negative) of the contrast will determine the polarity of the reflection that results.

From this, and the figures shown for the P-wave velocities, it will be apparent that with higher velocities and densities than the clastic rocks, chemical rocks, such as some limestones, the dolomites, and the evaporite rock anhydrite, should provide strong reflection groups when encased in clastic lithologies or some other types, such as the evaporite salt rock. This is found to be true, and is characteristic of many carbonate intervals in seismic data. An example is shown in Fig. 8.1, where a rather involved stratigraphic situation is seen in which carbonate and evaporite rocks play a dominant role. The example is from the Mid North Sea High (MNSH) area, and shows the situation across the south-west margin of a re-entrant channel of Zechstein Z2 salt occupying a topographic low in the southern flank of the MNSH. The complexities of the Permian geology, as observed in seismic data, have been described by Jenyon (1988e), and Jenyon and Taylor (1987). For present purposes it is sufficient to point out the strong, well-resolved seismic reflection events marking carbonate (limestones, dolomites) formations associated with evaporite (salt rock, anhydrite) lithologies, as indicated by the schematic section of Fig. 8.2. In the latter illustration, the part of the schematic between 'a' and 'b' represents the full length of the section shown in Fig. 8.1. The base of the Zechstein is indicated by BZ in both illustrations, and the Top Zechstein by TZ. Level TA is the marginal-facies equivalent of the top of the Plattendolomit, P, in basinal facies, with P being a competent dolomite/anhydrite band marking the base of the Zechstein Z3 evaporite cycle.

Above the Top Zechstein in Fig. 8.1, the horizon marked CU is the Late Cimmerian Unconformity, and above this, BT indicates the Base Tertiary/Top Chalk unconformity. Apart from the CU event, the poorly developed, discontinuous seismic reflections in the TZ

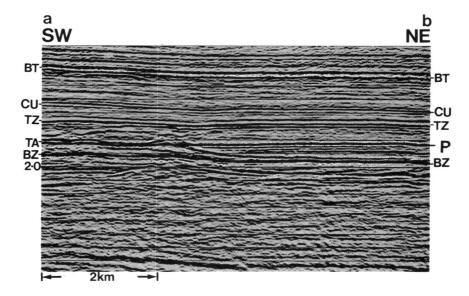

Fig. 8.1. A seismic example of a sequence that includes carbonate and evaporite rock types in close association. See the schematic in Fig. 8.2. BT = Base Tertiary; CU = Late Cimmerian Unconformity; BZ = Base Zechstein; TZ = Top Zechstein. Reproduced by permission of Seismograph Service (England) Ltd

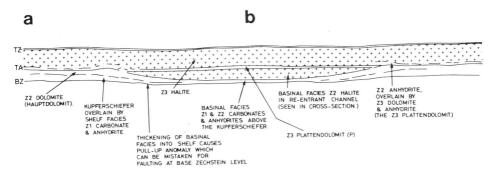

Fig. 8.2. A schematic section which, between 'a' and 'b' is the equivalent of the complete section a–b in Fig. 8.1. BZ = Base Zechstein; TZ = Top Zechstein. Reproduced by permission of the publishers, Butterworth & Co. (Publishers) Ltd © (from M. K. Jenyon, 1988e, *Mar. Pet. Geol.* 5, 352–358)

to BT interval greatly contrast with the strong, continuous seismic events within the Zechstein BZ–TZ interval.

Another characteristic sometimes distinguishing carbonate rocks is their susceptibility to dissolution ('karstification') by waters with even quite small quantities of dissolved acids, such as the carbonic acid in meteoric waters. Such a process often occurs at the Earth's surface, leading to various effects from the occurrence of limestone 'pavements' of clints and grikes, to major karst topography, including potholes, canyons, disappearing streams, underground rivers and caverns, with major collapse features. Sometimes a karst surface may develop where a limestone formation is buried beneath a permeable or fractured overburden. In seismic data the effects are sometimes identifiable, and an example is shown in Fig. 8.3.

The polarity on this seismic section is such that the Top Chalk marker consists of a multicycle seismic event containing one prominent peak (the strong black event A), with a weaker, irregular peak B beneath it. In spite of B being weak, it persists over the whole area, and in many places is defined by a series of minute, incompletely collapsed diffraction hyperbolas, and there are frequent irregularities that strongly suggest localized collapse features affecting phases A and B of the marker (e.g. D in Fig. 8.3). The effects seen can be ascribed to karstification of the Top Chalk surface, with dissolution of sink-hole and cavern type, many of the collapse spaces being filled with post-Chalk material (Jenyon, 1984a). Event A is interpreted as being (approximately) the seismic response to the Chalk surface, and B the response to the effective base of the collapse zone. The strong event marked C, which shows no sign of deformation, corresponds with a sudden change in lithology of the Chalk, there being a highly glauconitic zone present about 300 feet below the Chalk top.

Outside the area of supposed karstification, the marker event is of a relatively undisturbed character. Figure 8.4 shows an example located only a few kilometres from that in Fig. 8.3, and clearly, phases A and B show no sign of the irregularities affecting the latter section. The interpretation of karstification is supported by acoustic and lithological data from wells in the area, which show a relatively uninterrupted sequence of Tertiary shales right down to the Chalk surface (no unusual lithologies, such as basal conglomerate, being present). In one typical well in Quadrant 53, the acoustic log shows the strong shift in

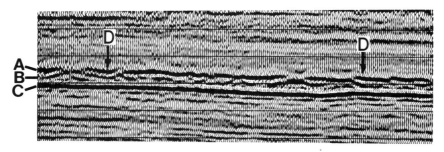

Fig. 8.3. Seismic section showing karstification of the Top Chalk surface (reflection events A and B), with collapse features, two of which are marked at D. Reproduced by permission of the publishers, Butterworth & Co. (Publishers) Ltd © (from M. K. Jenyon, 1984a, *Mar. Pet. Geol.* **1**, 27–36)

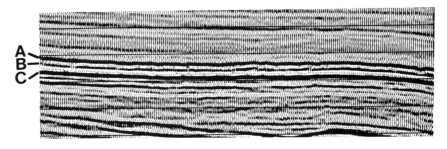

Fig. 8.4. Seismic section showing absence of karstification effects in a location near to that of the example in Fig. 8.3. Reproduced by permission of the publishers, Butterworth & Co. (Publishers) Ltd © (from M. K. Jenyon, 1984a, *Mar. Pet. Geol.*, **1**, 27–36)

transit time (representing velocity) to be expected when passing from the Tertiary shales into the Chalk, but for a depth of about 130 feet (39.6 m) below the top of the Chalk, the log trace varies widely about the mean Chalk velocity before settling down suddenly into the much narrower limits of variation to be expected. Over the same depth interval, the caliper log shows hole-diameter data consistent with that in the overlying Tertiary shales; then at the base of this zone, 130 feet below the Chalk top, it swings over sharply to hole diameters typical of the Chalk. Thus there is strong evidence of a zone with a mixture of collapsed Tertiary shale and Chalk rubble about 130 feet thick forming the top of the Chalk in this immediate area.

The reason for this fairly localized area of karstification is not difficult to find. It lies along a southerly extension of an inversion axis (Sole Pit, southern North Sea), and the movements involved would have stressed the Chalk, making it more susceptible to attack by agents of dissolution. In this general area, buried topographic features have been observed at the Chalk top, suggesting very shallow water or sub-aerial conditions.

8.2.1 Non-reefal Carbonates

Except in the case of facies-bounded carbonates (e.g. reefs), as dealt with later in this chapter, the morphological aspects of traps involving carbonates are not usually related to lithology. Carbonates can be source, reservoir, or seal rocks, but the trapping mechanisms involved normally fall into the other structural, stratigraphic, and combination categories discussed in the book. Some have already been seen in previous chapters (e.g. the Chalk fields included in the treatment of fold-related traps) and others will occur in later chapters.

Illing *et al.* (1967) have addressed the question of non-reef stratigraphic traps in carbonate rocks, pointing out the difficulty (or impossibility) of identifying these in geophysical data. They suggest, however, some considerations of facies associations that can be used to seek potential reservoir rocks. In areas of undisturbed, continuous miogeosynclinal carbonate sedimentation, uncemented primary intergranular porosity often occurs, as in the Lower Cretaceous Minagish Oolite of southern Kuwait. Where there are frequent interruptions of the sedimentary sequence, and in deformed areas, lime sediments tend to be tightly cemented by secondary calcite; in such areas, secondary (diagenetic) porosity is more important, with dolomite being the principal carbonate reservoir rock.

Microdolomites (formed probably in coastal *sabkhas*) are poorly permeable, but often highly porous—a texture that leads to their ability to form good reservoir rocks when fractured; often, they are interbedded with anhydrites (another product of the coastal sabkha), and the combination can form an impermeable cap rock.

Chalks also can, when fractured (particularly the coarser varieties), form good reservoirs; when not fractured, and of the porous crypto-crystalline variety, high capillary entry pressure allows them to act as good impermeable cap rocks.

8.2.2 Non-reefal Carbonates as Source Rocks

In Chapter 1, there was a brief discussion of carbonate rocks as possible sources of hydrocarbons, and the doubts that have been raised by some geologists as to the credibility of this type of rock as a source, particularly as regards: (i) TOC content, which is thought to be too low; and (ii) an expulsion mechanism. These doubts are supported by the ideas that: (a) depositional environments of carbonates currently being laid down are often not anoxic, and there is bioturbation on a major scale, which would militate against the survival of all but very small amounts of organic matter before it could be preserved by deep/rapid burial; and (b) ideas of Hunt (1967) and Tissot and Welte (1978) include the suggestion that early cementation of carbonate sediments would hinder mechanical (and, in some cases, chemical) compaction, thus preventing compaction processes being involved in any expulsion mechanism.

The subject is reviewed in an interesting recent paper by Ferguson (1988), who concludes from the evidence that: (i) carbonate source rocks have, indeed, made a significant contribution to world oil and gas resources; (ii) published data on the nature, quantity, and quality of organic matter in the rocks show that predominantly it is of marine algal origin, of high quality as measured by atomic H/C and O/C ratios, and present in amounts that compare well with other source rocks; and (iii) the presence of clay is not a requisite for source-rock potential.

Ferguson goes on to identify two principal environments of deposition for carbonate source rocks—deep-shelf anoxic, and shallow-water mesohaline, associated with evaporite deposition. The presence of anoxic conditions is important from the aspect of preservation of organic matter, and in the 'biochemical fence' concept resulting in minimal bioturbation, again favouring preservation of organic matter. The problem of the expulsion mechanism in carbonate source rocks may be modified by the operation of 'self-sourcing', where the source and reservoir are in the same unit. Other situations possible are where there has been limited migration within a carbonate complex, and where migration into a separate reservoir outside the carbonate has occurred. Self-sourcing leads to other problems arising, as noted in Chapter 1, and the question of primary expulsion in carbonate source rocks, which is certain to be a complex process, must for now remain open. Ferguson's concise review of the subject is recommended.

8.2.3 Special Aspects of Carbonate Reservoirs

8.2.3.1 *Chalk*

Some special features of chalk as a reservoir have been mentioned earlier, particularly its variable properties, which can result in the rock acting either as a reservoir or as a seal formation. Chalk, being made up to a greater or lesser extent of the calcite shells of microorganisms (foraminiferids, planktonic algae, coccoliths, etc.), is very susceptible to the crushing effect of burial on these minute structures. The shells collapse under pressure, which has an irreversible effect on the rock structure; chalk does not show the 'hysteresis' effect of many granular rocks, which on re-emergence after deep burial recover at least some of their original porosity (and also revert to a lower seismic velocity—the 'velocity hysteresis' effect). It is, therefore, the case that the details of the diagenetic and tectonic history of chalk are fundamental in the study of both quality and location of reservoir rock.

There is production from Danian Chalk in the Ekofisk area of Norwegian waters. In general, this chalk has high porosity (in the range 30–40%), but very low (less than 1 mD) permeability, which would normally cause it to be categorized as non-productive. However, in some locations, underlying salt has become mobile and uplifted the chalk in several field locations, causing intense fracturing, and hence improving the permeability to an average of about 10 mD. A schematic section of one of the fields in this area is seen in Fig. 8.5, and shows the reservoir of fractured Danian Chalk (solid black), with a sealing cap rock of Palaeocene mudstones.

The depositional and diagenetic factors that play a part in the development of chalk reservoirs in the North Sea have been studied recently by Taylor and Lapré (1987). They came to the conclusion that three major phases of chalk diagenesis, related to burial depth, are recognizable. These are:

(i) Early diagenetic cementation within the bioturbated top 10 m of the sediment pile, with the carbonate cement derived from marine waters. Such early cements are only a minor part of the chalk matrix, except in hardgrounds or reworked clasts within debris flows.
(ii) Sediment mass movement resulting in destruction and redeposition of early diagenetic cements as grains within a poorly packed matrix. Only the tops of the mass-flow deposits were bioturbated, cemented, and dewatered, so that prior to burial, the bulk of the allochthonous bodies had higher porosity than autochthonous bodies (see Fig. 8.6, which sketches the environment and features of chalk depositional processes in the Tor and Ekofisk areas (Norwegian waters)).

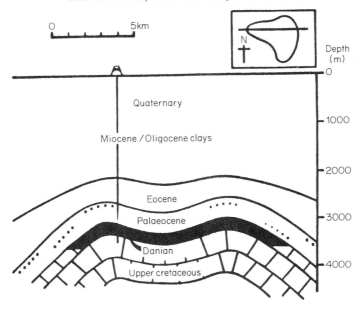

Fig. 8.5. Schematic section of a typical Danian Chalk field in the North Sea, with fracture porosity developed in the chalk owing to movement of underlying mobile salt. Reproduced by permission of Elsevier Applied Science Publishers Ltd (from Walmsley, 1975b)

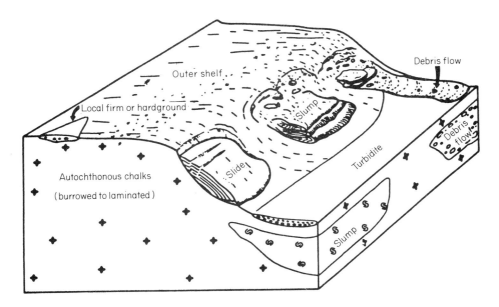

Fig. 8.6. Sketch of the environment and features of chalk deposition in the Tor and Ekofisk areas, Norwegian waters. Reproduced by permission of Graham & Trotman Ltd (from Taylor and Lapré, 1987)

(iii) Physical compaction became a major porosity destroying process, and produced chalks with a rigid framework at a porosity level of about 40%. During this phase in the Ekofisk area, warm, strontium-rich fluids migrated from the underlying Zechstein evaporites through a shear-fracture system developed in the chalks, into the chalk matrix. In the Greater Ekofisk area, there is much evidence (stylolites, flasers) of chemical compaction processes. Deep-burial cementation promoted by pressure dissolution of the chalk matrix began at a depth of about 1 km, and was the major porosity/permeability destroying process (cf. discussion in Chapter 2). The degree of burial cementation was controlled by hydrocarbon migration. As noted previously, hydrocarbon migration inhibits all diagenetic processes, and if it begins at a relatively shallow depth, either before or concurrently with the onset of the main pressure dissolution phase, will prevent the latter in large measure from destroying the poroperm characteristics of the rock. See Taylor and Lapré (1987) for a more detailed account.

8.2.3.2 Stylolites

Many carbonate (limestone) reservoirs are modified, sometimes favourably, sometimes adversely, by the development of stylolites. These are the result of post-induration compaction and pressure-dissolution processes.

Morphologically, stylolites occur in intervals or 'seams', usually located at natural horizontal surfaces in the rock (original bedding, joint, or parting surfaces), and are irregular 'intergrowths' of material across these surfaces. Projections, termed 'columns', from one side of the surface fit into hollows in the other side.

If stylolite development in a limestone progresses uninterruptedly, reservoir quality (thickness, porosity and permeability) may be adversely affected.

Dunnington (1967) has made an interesting study of the phenomenon, and concludes that if hydrocarbon accumulation originates before stylolitization, inhibition of pressure dissolution by hydrocarbons (as noted earlier) may localize stylolite distribution to original high water saturation zones, and stylolite seams may now dominate reservoir nature and behaviour.

However, if stylolitization precedes hydrocarbon migration, seams may cause, or originate, hydrocarbon migration.

Stylolite seams that cut directly or obliquely across permeable reservoir-quality limestones (reefal and non-reefal) may constitute flow barriers that are sufficiently impermeable to trap oil. This is suggested in Fig. 8.7, which is an idealized model based on real situations, where stylolite seams act as fluid-trapping barriers along the axis of a folded limestone biohermal reef complex, in which the limestone may be supposed to be of good porosity and permeability throughout, except for the stylolite seams.

There is also evidence of folding of limestones under compressional stress, not by the normally understood processes, but by the stylolitic removal of rock wedges, these being perpendicular to the stylolite seam, with the wedges thickening towards the core of the incipient fold. The closing-up of these eliminated rock-wedge spaces is a stress-relief mechanism acting under compressional stress, and this mode of folding by means of dissolution/chemical-compaction processes is an unusual and interesting phenomenon that deserves further study. Dunnington mentions the late Tertiary compressional folding of

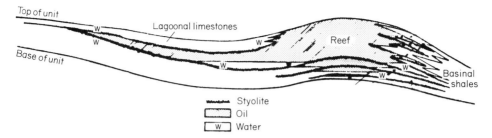

Fig. 8.7. Schematic of stylolite seams acting as flow barriers in reef limestones. Reproduced by permission of Elsevier Applied Science Publishers Ltd (from Dunnington, 1967)

a major reservoir limestone in northern Iraq as being achieved '. . at least in part . . . by stylolitic removal of wedges that increase in size and number inwards, towards the centre of the anticline'.

8.2.3.3 Limestones v. dolomites

As a generalization, it may be said that while limestones are more common than dolomites, and are likely to have better porosities and matrix permeabilities than dolomites, yet dolomite reservoirs tend to have more oil and non-associated gas in place than do limestone reservoirs. This paradox is investigated by Schmoker et al. (1985) by utilizing a large database of USA carbonate rocks, and their study supports the idea that dolomites can be regarded as the better reservoir rocks.

From information in the database, they show that dolomite reservoirs, on average, are larger and deeper, but less porous and permeable than limestone reservoirs. This can be supported by evidence from other areas in the world. A typical example is shown by Hawle et al. (1967) in Fig. 8.8, which shows curves of changes of permeability (upper) and porosity (lower) with increasing overburden pressure for a dolomite (curve 1: dolomite, Schönkirchen–Tief) and a limestone (curve 2: limestone, Baumgarten–Zwerndorf) in the Vienna Basin.

Schmoker et al. come to the conclusion that the principal reason for the paradox is that apart from any notional poroperm improvements brought about by dolomitization, effective fracture systems at reservoir depth are more likely to occur in dolomites than in limestones. As they say, '. . . all else being equal, an open fracture system connecting formation and well bore would act to increase a reservoir's effective size . . . '. It is also noted that dolomites 'are less ductile, and have greater ultimate strength, than limestones, and so are more likely to yield by fracturing than by intragranular flow. Differences in mechanical properties of the two rock types could, ultimately, explain why effective fracture systems at reservoir depths occur more frequently in dolomites than in limestones.

8.2.3.4 The carbonate 'signature' in seismic data

It has been noted previously that carbonate rocks generally provide strong acoustic impedance contrasts with other lithologies, and if the units involved are thick relative to

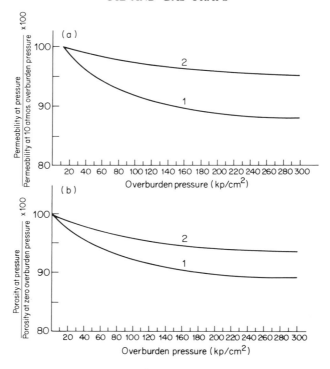

Fig. 8.8. Changes of permeability (a) and porosity (b) with increasing overburden pressure for a dolomite (curve 1) and a limestone (curve 2) in the Vienna Basin, Austria. Reproduced by permission of Elsevier Applied Science Publishers Ltd (from Hawle et al., 1967)

the dominant seismic wavelengths at any particular level in the seismic section, then strong and continuous reflections will be generated.

This is seen clearly in the example of Fig. 8.9, from the paper by Beaudry and Moore (1985) dealing with aspects of the West Sumatra Forearc Basin. In the illustration, a seismic section from the inner shelf area is seen, with depositional sequences numbered 1, 2b, 3, and 4 marking Pre-Neogene, Mid- to Upper Miocene (Lower Miocene being absent on this line due to erosion; the basal disconformity is indicated), Upper Miocene to Pliocene, and Pleistocene, respectively. The well (Bubon 3) reached TD in a Middle Miocene limestone reef, and the basal part of the 2b sequence is taken up by a high-amplitude reflection group continuous with the mounded carbonate reefal facies. This substantial limestone group is overlain by 1000 feet (300 m) of Middle Miocene marls that pass upwards into a thin-bedded limestone facies of late Miocene to early Pliocene age (the strong event marking the boundary between sequences 2b and 3). Sequence 3 consists of a deltaic depositional series. The contrast between the strong basal 2b limestone group and the seismic data immediately above (and below) is very typical, and the mounded reefal facies associated with it is additional evidence that a carbonate complex is involved.

In an important study dealing with seismic interpretation in carbonate depositional environments, Fontaine et al. (1987) discuss the seismic signature of carbonates in a range of depositional environments, from basin to supratidal zone. This covers pelagic (shales,

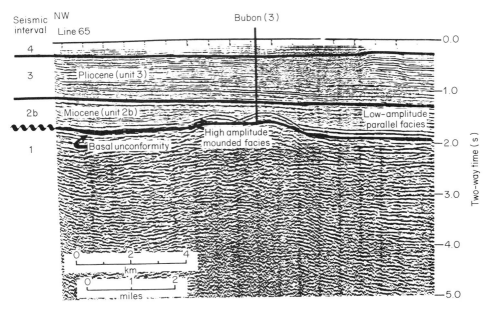

Fig. 8.9. Carbonate rock 'signature' in a seismic section. Carbonate rocks occur at the boundaries of sequences 1 and 2b (where they are continuous with the mounded, reefal limestone facies seen in the centre of the section), and also sequences 2b and 3. Note the strong and continuous reflection groups produced. From Beaudry and Moore, 1985, reprinted by permission of American Association of Petroleum Geologists

micritic limestones, chalk), talus and debris flow deposits, barrier reefs, shelf border sands, inner shelf deposits, intertidal and supratidal zones. Palaeokarst zones are also dealt with.

The relationships between seismic facies and the various carbonate environments are summarized in Fig. 8.10, taken from the cited paper; also from the paper (which will remain a standard reference for a long time) is Fig. 8.11a depicting a seismic section through a hydrocarbon field (A) producing from Barremian karst limestone offshore in the Mediterranean Basin. The karstified limestone in Field A passes laterally into compact limestone in Well C. Note again the very strong signature of the carbonate group. The appearance of the karstified top of the Chalk in seismic sections in the North Sea was referred to earlier in this chapter. The high reflectivity of chalk is also shown by the seismic example, with well data, in Fig. 8.11b. The Lower Palaeocene/Top Chalk reflection is arguably the strongest event on the section; the well location is arrowed.

8.2.4 North Sea Examples

Two well-known examples of oilfields with (non-reefal) carbonate reservoirs in the northern North Sea are the Auk and Argyll Fields. Both owe their existence to the presence of Jurassic source rocks deep in the Central Graben that have provided oil to migrate structurally updip

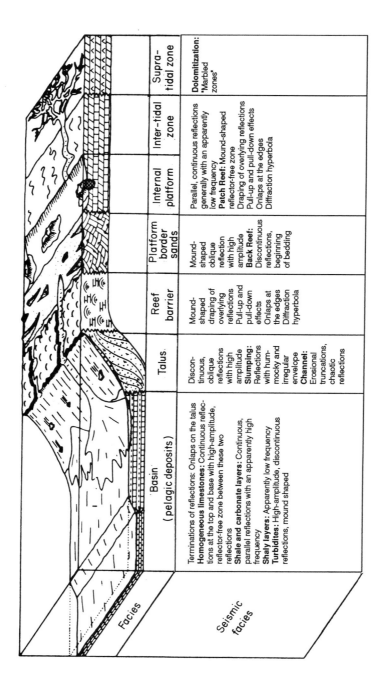

Fig. 8.10. Relationship between seismic facies and the various carbonate environments, in summary. From Fontaine et al. 1987, reprinted by permission of American Association of Petroleum Geologists

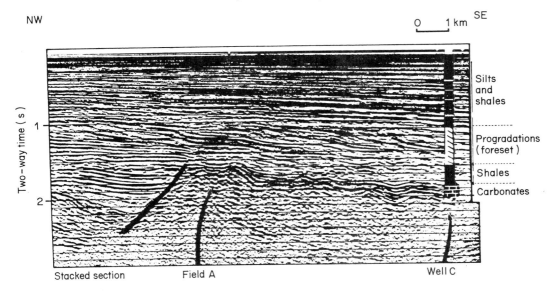

Fig. 8.11a. A hydrocarbon field (A) producing from Barremian karstified limestone in the Mediterranean Basin. The karst passes laterally into normal, compact limestone in Well C. From Fontaine *et al.*, 1987, reprinted by permission of American Association of Petroleum Geologists

and up south-west bounding faults of the graben into higher, but stratigraphically older Zechstein (Upper Permian) carbonate reservoirs.

The Auk oilfield is a fault-bounded trap located in a horst block forming part of the south-west bounding fault complex of the Central Graben just north of 56°, and about 2°E (Brennand and van Veen, 1975). Lower Zechstein dolomites were exposed to meteoric waters during Mesozoic erosional phases, producing vuggy, very porous, and permeable rocks (see Fig. 8.12(b)). At a later stage, these rocks became cavernous and mechanically unstable, leading to collapse and brecciation (with fragments of the overlying rocks included in them) of overlying dolomitic limestones (Fig. 8.12(a)).

A seismic section across the Auk Field is shown in Fig. 8.13. The seal above the Zechstein consists of thin Lower Cretaceous shales, tight Chalk, and shaly Tertiary rocks. The total Zechstein interval across the field varies in thickness from about 40 feet to 140 feet only, and so seismically the Zechstein interval is difficult to identify owing to the strong Chalk reflection group immediately above it (cf. the Argyll Field later). Note the marked flexure in the overburden reflections above the right-hand (east) bounding fault, and the weaker flexure related to the left-hand (west) bounding fault.

The Argyll oilfield is located in the crestal zone of a large tilted fault block at the south-west margin of the Central Graben, some 50 km to the south-east of the Auk field. Reservoirs in Zechstein carbonates and in Rotliegendes (Lower Permian) sandstones are present. In this case, unlike the Auk example, it was Zechstein salt dissolution and removal (see later in the chapter) that caused collapse of the dolomites, with the development of fractured and vuggy porosity.

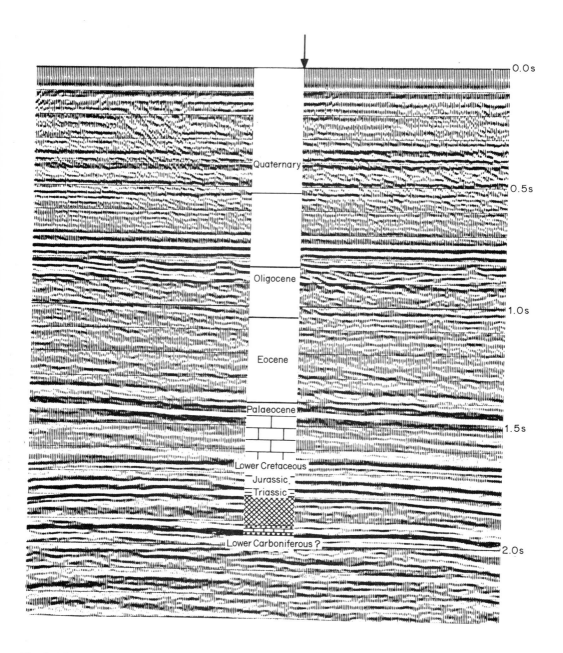

Fig. 8.11b. The high reflectivity of chalk as seen in a seismic section from the North Sea. The reflection group at about 1.5 s (Palaeocene/Top Chalk) is probably the strongest and most continuous on the section. Reproduced by permission of Seismograph Service (England) Ltd

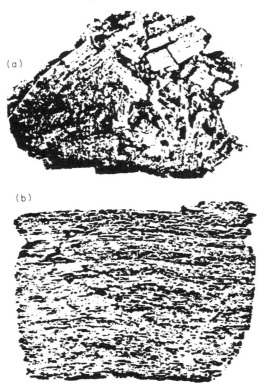

Fig. 8.12. Leached Zechstein dolomite, a very porous and permeable, vuggy rock (b); the leaching led to mechanical instability and collapse, with occurrence of brecciation of overlying dolomitic limestones (a). These rock types form the reservoirs in the Auk Field, North Sea. Reproduced by permission of Elsevier Applied Science Publishers Ltd (from Brennand and van Veen, 1975)

A seismic section across the field is shown in Fig. 8.14. Comparison with the schematic section in Fig. 8.15 (a) will show why it is not possible to make detailed identification of the Zechstein carbonates in the seismic data; the strong Top (Danian) Chalk reflector effectively obscures the details of the Zechstein section, as in the Auk example. In both fields, it is the fortuitous preservation of a thin Zechstein interval beneath the Late Cimmerian erosional surface that is responsible for the presence of the fields.

In Fig. 8.15 (b) depth-structure contours on top of the Zechstein interval are shown, the area seen covering about 5.25 by 4.75 miles (Pennington, 1975). Fault A shown on the depth-structure map is the bounding fault of the Central Graben locally, fault B (a 'scissors' fault) bounds the productive area to the north-west, and fault C bounds it to the south-east. The line X–X' is the line of section of the seismic example in Fig. 8.14.

As noted elsewhere, there is nothing specific to carbonate rocks about the trap morphology here; both fields are fault-bounded closures.

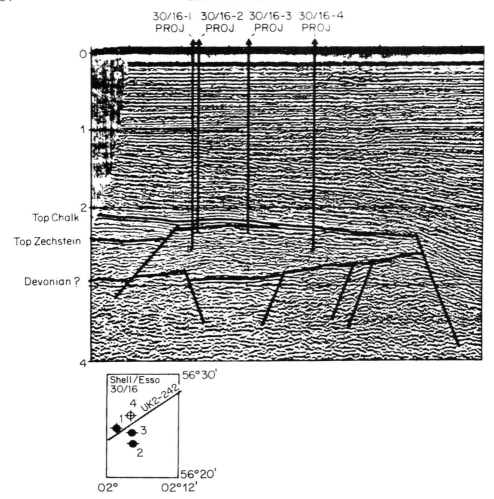

Fig. 8.13. Seismic section across Auk oilfield. Note the strong Top Chalk reflection group masking the Zechstein reservoir interval within the structural horst block forming the field. Reproduced by permission of Elsevier Applied Science Publishers Ltd (from Brennand and van Veen, 1975)

8.3 Evaporites

One important aspect of evaporites—plastic deformation in salt rock, and its effects on the overburden in relation to hydrocarbon trapping—was discussed earlier (Chapter 5). Two other matters related to evaporites that need to be discussed here are: (i) the plastic deformation behaviour of evaporites other than salt rock; and (ii) dissolution of evaporites, particularly salt rock, and the resultant effects on the overburden.

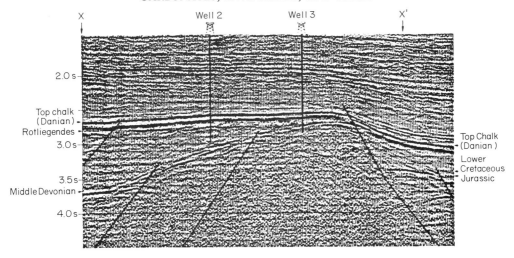

Fig. 8.14. Seismic section across Argyll oilfield. As in Auk, the Top (Danian) Chalk reflection group obscures the details of the thin Zechstein reservoir interval. In this case, dissolution removal of Zechstein salt caused collapse of the Zechstein carbonates, producing fracture/vuggy porosity, and not just leaching and collapse of the carbonates themselves, as at Auk. Reproduced by permission of Elsevier Applied Science Publishers Ltd (from Pennington, 1975)

8.3.1 The Deformational Behaviour of other Evaporites

In an evaporite basin with a progressively concentrating body of brine (i.e. a restricted, or silled basin according to current concepts), when evaporitic drawdown has progressed until only one-tenth of the original water volume remains, then salt (predominantly the mineral halite, NaCl) begins to be precipitated. As the brine concentrates further to dryness, the last 'bitterns' yield a group of minerals known as the bittern salts, or less formally, the 'potash' minerals (although they do not all contain potassium). These comprise a highly soluble mineral group, some of which are unstable at s.t.p. and readily break down into other compounds.

With progressively increasing solubility, the individual minerals are progressively less dense, and more easily deformed plastically under stress, in a series from halite onwards. The principal potash minerals in descending order of abundance are:

Chlorides
Sylvite	KCl
Carnallite	$KMgCl_3 \cdot 6H_2O$
Bischofite	$MgCl_2 \cdot 6H_2O$
Tachyhydrite	$CaMg_2Cl_6 \cdot 12H_2O$

Sulphates
Kieserite	$MgSO_4 \cdot H_2O$
Langbeinite	$K_2SO_4 \cdot MgSO_4$
Epsomite	$MgSO_4 \cdot 7H_2O$
Polyhalite	$K_2MgCa_2(SO_4)_4 \cdot 2H_2O$

Chloride–sulphate
Kainite	$4(KClMgSO_4) \cdot 11H_2O$

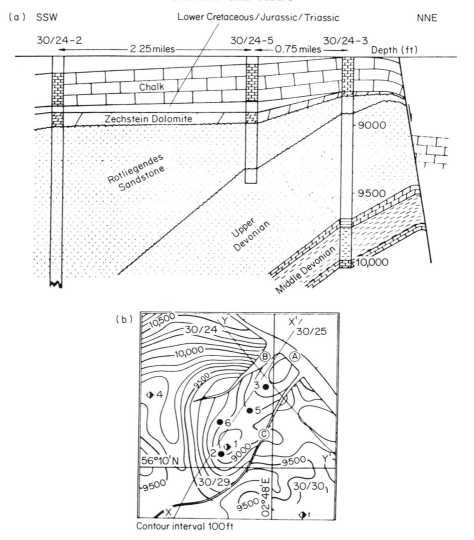

Fig. 8.15. Schematic section (a) across Argyll Field. The lower sketch (b) shows structure contours on top of the Zechstein reservoir interval, with the main controlling faults present. Reproduced by permission of Elsevier Applied Science Publishers Ltd (from Pennington, 1975)

The potash minerals include those of primary precipitation, secondary minerals that are alteration products of less stable primary minerals, and exotic mineral species resulting from the entry of solutions into the potash deposits after these have been buried. In all, over 40 different salts have been found in various deposits—those in the list above being the most abundant—but of these, relatively few are present in more than minute amounts.

The phase relationships and physical chemistry of the potash minerals are highly complex, and have a literature of their own. Standard texts are those by Borchert and Muir (1964) and Braitsch (1971). Here the concern is only with certain aspects of these minerals.

The extremely high solubility and frequent instability of the potash minerals result in their presence as discrete deposits in an evaporite basin being somewhat rare. More frequently they are absent for one of the following reasons: (i) brine in the basin never reaches the required concentration; (ii) the concentrating brine in the basin is removed by reflux before the bitterns precipitate; and (iii) the deposited bitterns are redissolved (perhaps many times), and eventually disseminated through the basin.

When they *are* present, since they are the most soluble minerals they are precipitated the latest, and therefore tend to occur in the later, upper part of an evaporite cycle, sometimes as intergrowths with each other and/or with halite or anhydrite. Since their resistance to plastic deformation is lower than halite, where they occur as intergrowths with this mineral, they are the 'weak link' in the chain resisting deformation under applied stress, and will start to flow long before halite begins to become plastic. This, together with their occurrence in the upper part of an evaporite-cycle/salt-rock interval, has interesting results.

Coelewij *et al.* (1978) referred to *differential* salt movement in North Sea Zechstein salt as taking place where plastic flow occurred in the deeper levels of a salt interval owing to the higher pressures/temperatures affecting these deeper parts, compared with conditions in the shallower part of the salt. They suggested that this would lead to thickening at certain locations in the lower part of the salt, while the upper part of the salt at these locations remained unaffected apart from being uplifted towards the surface by the thickening.

In the same paper, they recognized another type of differential movement that they called *preferential flow*, as being due to the differing deformation characteristics of different types of evaporite salts, as just discussed for the bitterns.

Strongly marked structures have recently been identified in Zechstein salt, which are interpreted as being due to preferential flow—probably of potash salts, but possibly of some other material of argillaceous type (Jenyon, 1985c). In Fig. 8.16, two orthogonally intersecting migrated seismic sections that show such a feature are illustrated. In these, letter A indicates the base of the Zechstein salt interval, B shows the seismic reflection event of the Plattendolomit competent anhydrite/dolomite band separating the Z2-cycle salt below from the Z3- and Z4-cycle salts above. Event C marks the top of the Z4 cycle (and also the top of the Zechstein interval), while D is a marker in the overlying clastic overburden that indicates the deformation of the latter locally by the underlying movement.

A broad, rather asymmetric Zechstein salt swell is seen in cross-section on line A (upper illustration), being well defined by the Plattendolomit event B, which generally follows the original shape of the swell. Just off the crest of the structure, a strong uplift of the post-Plattendolomit (post-B) intervals is seen. Referring to the lower illustration (line B) of the section along the axis of the main salt swell, it is seen that the feature defined by event T (which is intended to indicate the same stratigraphic level as C (top Zechstein) over this part of the structure) is a strongly marked hogback feature parasitic on the main swell, with the competent Plattendolomit band B showing considerable flattening, and even upward concavity, beneath—an effect which is also seen on horizon B on line A.

From the effects in the overburden, it appears that the parasitic structure developed at a late stage in the rise of the main swell, and indicates very marked thickening and flow in the upper (Z3/Z4) intervals. (The shallow overburden follows the general shape of the main swell, deformation from the parasitic feature being confined to the deeper overburden, and dying out upwards.)

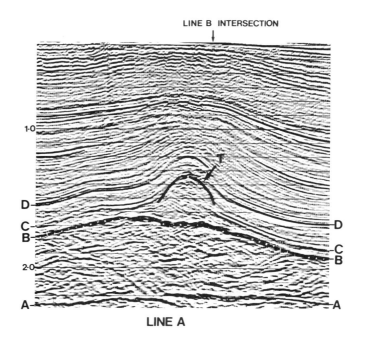

LINE A

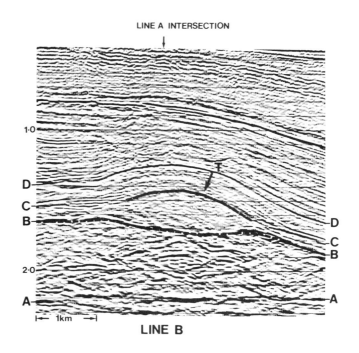

LINE B

Other, similar structures have been seen in this general area, and although none have been tested by boreholes, it seems most likely that the effects seen are due to the presence of unusual concentrations of potash salts in the upper part of the salt of the Zechstein Z3 cycle, and to a lesser extent in the Z4 cycle.

In the paper by Coelewij *et al.* (1978), figures are quoted for the yield stress of halite and two of the principal 'potash' salts (carnallite and bischofite), as determined under simulated subsurface conditions (equivalent to a depth of 2400 m, and temperature of 100°C) in the Shell research laboratory at Rijswijk, Netherlands. The yield stress quoted for halite is quoted as 40 kg/cm^2, while those for carnallite and bischofite are 4.5 kg/cm^2 and 1.0 kg/cm^2, respectively. Thus the potash salts, if present in appreciable volume, are likely to undergo much greater deformation for any given applied stress than is salt rock (halite), and could produce very marked structures.

It is suggested from the examples in Fig. 8.16 that differential overburden pressure between the flanks and the crest of the main salt swell was sufficient to initiate flow of the potash salts, from the flanks up towards the lower stress zone at the crest, leading to the marked thickening seen within the Z3/Z4 intervals, the hogback feature, and the flattening/downwarping of the Plattendolomit band B beneath the thickening.

Clearly, the mechanism proposed is of considerable interest from the viewpoint of hydrocarbon exploration, since it represents another possible process leading to structural trap formation in the overburden above an evaporite basin. Also it is of interest since it suggests locations at which considerable flow-concentrations of valuable potash and magnesium salt occur in the subsurface, perhaps amenable to some form of solution mining. Some caution must be used in interpreting such phenomena, however, since similar effects could occur if the upper part of an evaporite cycle included potentially mobile argillaceous material—a shale or clay—as is not uncommon. Nearby well data and detailed knowledge of the seismic stratigraphy in the area are obviously required for a more secure assessment to be made.

8.3.2 Dissolution and Collapse

For a long time it has been realized that because of its high solubility, salt rock (together with other evaporites of lesser importance volumetrically) buried in the subsurface may be subjected to dissolution and removal by water undersaturated with respect to the relevant salt compound—NaCl in the case of salt rock—and that such removal would normally lead to subsidence collapse of the overburden. Before the advent of high-resolution seismic data, a number of structures of this type, some involving trapped hydrocarbons, had been discovered by drilling, especially in areas of the USA and Canada, like the margins of the Williston Basin. Papers by workers in this field, such as Parker (1967), Smith and Pullen (1967), and Langstroth (1971), built on earlier work to discuss various North American features of this type.

Fig. 8.16. *(opposite)* Two orthogonally-intersecting seismic sections illustrating a small parasitic structure (T) formed by preferential flow of material (usually either potash 'bittern' salts, or mobile argillaceous material) near the crest of a gentle salt pillow (the crest being at the culmination of, for example, level B on line A). Reproduced by permission of the Oil & Gas Journal (from Jenyon, 1985c)

In the early part of the 1970s, Lohmann (1972) drew attention to the presence of dissolution collapse features as interpreted in seismic data in the southern UK North Sea, related to both Triassic and Permian salt intervals. This was skilled observation at the time, since seismic data quality was not up to present-day standards, and indeed, Lohmann categorized the data as too poor to be used as illustrative material in his paper.

With the coming of much higher resolution seismic data over the following decade, it became easier to identify these features, leading to other work, such as Jenyon (1984a, 1986d, 1988d). In one of the works cited (Jenyon, 1986d), a complete chapter is devoted to this subject, and so as to avoid much unnecessary repetition, it will be treated only briefly here; the object will be to show some of the mechanisms and situations involved that may be of importance in the development of hydrocarbon traps. In some areas, such as the Mid North Sea High, it has been shown (Jenyon and Taylor, 1987) that in the Zechstein interval, dissolution–collapse processes can result in the dominant structural style.

In order that salt dissolution may take place, certain basic conditions must be fulfilled. Means of both ingress and egress to the salt formation of water undersaturated with respect to the salt is required. This water is usually formation brine, but other sources of solvent are: (i) meteoric water entering an aquifer at a basin margin; (ii) oceanic brine entering the aquifers at the basin margin, or through fractures, or through a permeable unconformity surface (sea water has an average salinity of 35,000 ppm, of which nearly 80% consists of NaCl—however, hypersaline brines, such as Dead Sea water, can reach salinities as high as 250,000 ppm); (iii) 'bound' water released in the gypsum–anhydrite conversion, and some clay mineral reactions; and (iv) entry of juvenile and/or connate waters into the system. Fracture/fault systems may allow the circulation of water under artesian pressure and brine under gravity control (brine density flow—see Anderson and Kirkland (1980)). An impermeable layer at or close to the ocean floor may or may not reach to the basin margin, which can be another control over access of undersaturated brine to the aquifers.

In order that salt dissolution may take place, a circulatory system must develop, so that the undersaturated water has access to the salt layer, and can remove it in solution. Standing water in contact with a salt interval will not remove the salt: once the water becomes saturated with respect to the NaCl, no further net removal of salt takes place, but only ion exchange between salt and water.

The schematics of Fig. 8.17 show various stages of development of a 'sombrero' structure, so-called because of the morphological resemblance to a hat. In sketch (a), the movements of a deep fault at f(?) allow access of fresh water from below—perhaps under artesian pressure—to a salt layer (cross symbol). At (b) a localized dissolution feature of sink-hole type results, with subsidence of a clastic overburden plug (dot shading). At (c), during subsidence, further deposition takes place, with compensatory thickening (represented by the cross-hatched lenses) over the low. At some later time (d) after dissolution of all the surrounding salt, general overburden subsidence leaves the original low as a positive feature. This is the explanation given by Smith and Pullen (1967) for the development of the Hummingbird structure in Saskatchewan, and an example seismic section across the structure is shown in Fig. 8.18, with the two pay zones of the oilfield marked with dashed lines. The subtle nature of the trap is clear. A depth structure map at the lower of the two pays (the Upper Devonian Bakken–Birdbear Formation) is shown in Fig. 8.19, which underscores the highly localized nature of the structure—a not uncommon feature of this type of structure in the area—and which is very puzzling. After all, access of water owing

CARBONATES, EVAPORITES, AND REEFS 271

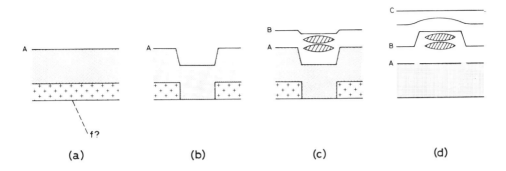

Fig. 8.17. Schematics showing stages of development of a 'sombrero' structure resulting from multistage, localized, salt dissolution, overburden subsidence, and sediment deposition. Reproduced by permission of Elsevier Applied Science Publishers Ltd (from Jenyon, 1986d)

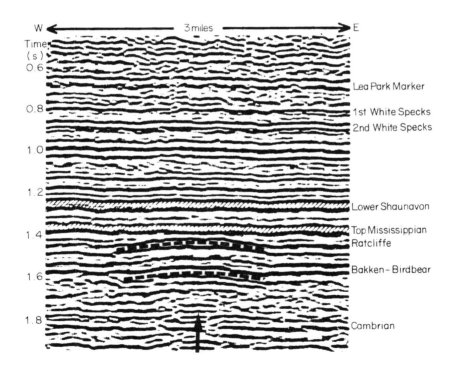

Fig. 8.18. A seismic section across the Hummingbird oilfield—a 'sombrero'-type structure in Saskatchewan, Canada. The two pays in the oilfield are marked with dashed lines, and the subtle nature of the trap is clear. Reproduced by permission of the Canadian Society of Petroleum Geologists (from Smith and Pullen, 1967)

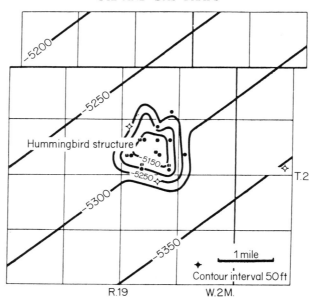

Fig. 8.19. A depth-structure contour map at the level of the lower pay in Fig. 8.18, which is the Upper Devonian Bakken–Birdbear Formation. The contours emphasize the highly localized nature of the structure. Reproduced by permission of the Canadian Society of Petroleum Geologists (from Smith and Pullen, 1967)

to fault movement implies an elongate feature, since faults are linear in plan. Smith and Pullen advance the idea that such structures are located at intersections of two different fault systems, where additional and more extensive fracturing may provide localized conduits of higher permeability for circulating water. Multistage dissolution features (sometimes combined with several depositional and compaction phases of a highly complex nature) arise, it has been suggested, by intermittent movements of deep-seated faults 'turning the tap on and off' with respect to circulating water.

The Prairie Evaporite salt of the Hummingbird structure is also involved in the example of Fig. 8.20, which shows a schematic section across the Outlook Field, Montana, as described by Parker (1967). Two oil reservoirs were located, the lower being in a Silurian interval (OWC at C2), and the upper in a Devonian interval (OWC at C1). Because of the stratal conformation overlying the Silurian–Devonian (S–D) unconformity surface, the C1 oil pool originally lay above the C2 accumulation. Subsequent dissolution of the Prairie Evaporite salt (cross symbol) led to collapse of the overburden and migration of the C1 oil to its present position. This interpretation is supported by the results of a drill-stem test in Well 2, which recovered minor oil and gas shows with sulphurous brine in the same stratigraphic interval at A as that in which the C1 oil accumulation is found in Well 3. This is a case where salt dissolution and overburden subsidence have not created a trap for hydrocarbons to migrate into, but have changed the location of an already existing oil accumulation—a thought to conjure with in other salt areas perhaps.

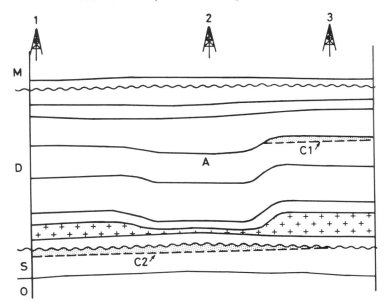

Fig. 8.20. A schematic section across the Outlook Field, Montana. a lower oil reservoir (OWC at C2) in a Silurian interval (S), and an upper reservoir (OWC at C1) in a Devonian interval (D) are present. Oil at C1 was originally above C2, but migrated owing to salt dissolution and overburden subsidence. Prairie Evaporite salt shown with cross symbol; O = Ordovician; A = subsidence feature; M = Mesozoic. Reproduced by permission of Elsevier Applied Science Publishers Ltd (from Jenyon, 1986d); simplified after Parker (1967), reprinted by permission of American Association of Petroleum Geologists

8.3.3 Salt-edge and Areal Dissolution

The schematic in Fig. 8.21(a) shows three different possible modes of dissolution of a salt layer (cross symbol) feather-edging at a basin margin, and immediately overlain by a permeable clastic interval (dot shading).

The first mode, indicated by the arrow on the right, involves water moving intraformationally down the basinal dip above the salt. This will attack the salt edge, and also probably the top surface of the salt layer near to the salt edge. Assuming that there is a circulatory system, and the resulting brine is removed (e.g. through fractures under artesian pressure), salt will be removed, and after some time, the situation in (b) will develop. The salt has retreated down the basinal dip, and now terminates at a dissolution salt slope (S), leaving behind a zone of collapsed overburden, which is small where the original thin feather-edge of the salt has been removed, but increases in amplitude downdip to a maximum at the salt-slope location, where the removed salt layer was considerably thicker.

The second possible mode of salt dissolution/removal is by areal dissolution, indicated by the arrows above the salt interval. There may be overlap of effects from the previously mentioned mode—that is, the water producing the effect may be from the same source as that effecting the edge dissolution. It may also, however, be percolating downwards

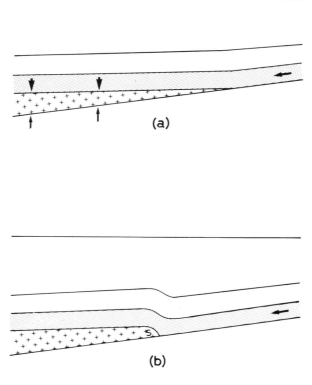

Fig. 8.21. (a) Schematic diagram showing possible modes of dissolution of salt layer (cross symbol) feather-edging at a basin margin, and immediately overlain by a permeable clastic interval. Diagram (b) indicates the situation after edge and/or areal dissolution of the salt has taken place. The salt edge has retreated down the basinal dip, and now terminates in a salt-edge dissolution slope at S, with progressive collapse of the overburden. Assuming favourable geometry normal to the section, the overburden structure above S may form a trap just where the salt edge would allow hydrocarbon migration into the trap location from beneath the salt. Reproduced by permission of Elsevier Applied Science Publishers Ltd (from Jenyon, 1986d)

through a 'leaky' unconformity surface, and through permeable clastic sediments, to remove the salt over a wide area. The top surface of the salt would then be a dissolution surface (there being a specific German geological term for this—a 'Salzspiegel', translating as 'salt mirror'). Although no dramatic overburden subsidence effects may be observable in seismic data attendant upon an areal dissolution process of this type, the slow settlement of the overburden in a large area can lead to major 'blanket' brecciation in the formations immediately overlying the disappearing salt, with the effect dying out upwards. In some marginal areas around the Williston Basin of USA and Canada, where the Prairie Evaporite salt has undergone such processes, there is a zone of such brecciation around and beyond the present edge of the salt, indicating that the latter was originally more widespread than it is now.

The third mode of dissolution, indicated in Fig. 8.21(a) by the arrows beneath the salt interval, is what might be termed an inverted *Salzspiegel*, produced by water moving in an aquifer *below*, and in contact with, the salt layer. It is believed that, in general, the effects of this would be similar to those discussed in the previous section.

In Fig. 8.22a, an actual example of salt-edge dissolution is seen, with the original thick development of salt rock represented seismically in the interval between B and T at the south-western end of the section. The basinal dip towards the south-west is obvious at the level of the base salt, 'B'. At the north-east end of the section, the whole salt interval has disappeared, and the two markers B and T have come together as a single reflection group marked B/T, which represents remnant basal carbonates and anhydrites, with some anhydrite rubble from the original top of the salt layer in places.

The salt, which formerly extended to the north-east almost to the end of this section, has retreated downdip and now terminates at a salt slope. Between S and almost to the north-east end of the section, clear indications of overburden collapse can be seen in the interval between B/T and the Late Cimmerian Unconformity, CU. To the south-west of S there are further signs of some very restricted subsidence in the same interval, which points to limited areal dissolution terminating before the end of the section is reached, in addition to the edge dissolution that has taken place.

The seismic data thus strongly support the interpretation that undersaturated water moved down the basinal dip in an aquifer in the immediate post-salt clastic formations (or possibly along the actual overburden–salt contact), dissolving and removing salt from the depositional edge, and also partly in a fairly narrow areal zone on top of the salt interval around the salt basin margin.

Fig. 8.22a. Seismic example of salt-edge dissolution. The original salt thickness is seen between B and T at the south-west end of the section, while at the north-east end, all the salt has been removed by dissolution and only residual 'marginal facies' lithologies make up the B/T interval. Basinal dip is to the south-west, as seen clearly at the level of the Base Salt, B, and the salt slope is seen at S with clear evidence of overburden collapse in the interval between the B/T level in the north-east half of the section and the overlying Late Cimmerian Unconformity (CU). A small remnant feature seen at R has had a considerable structural effect on the subsided overburden above it. This major effect, which the writer terms *subsidence drape*, will be discussed further. Reproduced by permission of Seismograph Service (England) Ltd

Another very interesting observation that is clear in this example is the effect of the small remnant feature R, originally at the basal salt level, on the subsided overburden deposits between B/T and CU. This small feature has had a structural effect in the overburden out of all proportion to its apparent size (further, the physical object producing the seismic feature is probably much smaller even than the indications on the section). The possible nature of remnant features like R will be discussed later, but for the moment, the major drape effects it is causing in the subsided overburden should be noted, as should the fact that this is *principally* the effect of what can be termed *subsidence drape* rather than the more commonly encountered depositional/compaction drape.

Returning to the matter of the main overburden subsidence related to salt slope S, note that at S this produces a counter-basinal dip slope in the B/T–CU interval, and if the original basinal dip (to the south-west) is still maintained, then potentially an excellent trap situation exists in the overburden above S, provided that the geometry normal to the section is favourable. Not only does there exist, potentially, a linear high (perhaps with local culminations) running parallel to the basin margin and normal to the section above and to the left of S, but as the salt has been completely removed by dissolution to the north-east of S, then the salt seal no longer exists. Any hydrocarbons that are migrating up the basinal dip from the south-west beneath the salt seal, will inevitably find their way into these overburden trap(s) in the B/T–CU interval around the basin margin.

Once again, the Williston Basin of North America is mentioned as a salt basin around the margins of which oilfields are found associated with overburden drape structures over the salt edge (see e.g. Langstroth, 1971).

Although in many instances the dip reversal and the basinal dip combination discussed will tend to form traps, in other instances where later tectonism may have reversed, or at least removed the original basinal dip, trap situations may no longer exist, as will be seen by tilting the Fig. 8.21(b) diagram or the Fig. 8.22a example, to the right. This is shown in a somewhat exaggerated way by the schematics in Fig. 8.22b, where in sketch (a) the favourable situation exists whereby the original basinal dip has persisted, or even increased, leading to a trap development at A in the subsidence-drape fold; whereas in (b), tectonism has reversed the basinal dip and the trap situation no longer exists. In the latter situation, depending on the maturation and migration history of any hydrocarbons in the area, it is possible that there was an accumulation in the erstwhile trap position, but that it has now migrated updip.

In Fig. 8.23 there is an example shown where salt-edge dissolution has occurred near the margin of a salt basin, but where no distinct salt-dissolution slope has resulted—perhaps because the rate of thickening of the salt into the basin was rather low. The basinal dip from left to right at the level of BZ is clear (in the direction of the arrow). The dip reversal from X to Y above the level of TZ is due to edge dissolution and removal of salt from within the BZ–TZ interval, and can be seen to have affected the whole of the shallower section. (The graben faulting (F) is not directly related to the salt dissolution discussed here (Jenyon, 1984a).)

The Mid North Sea High (MNSH) provides many examples of salt-edge and areal dissolution. The MNSH is a major, E-W-trending, Palaeozoic antiform structure that separates the northern from the southern basins of the North Sea in the British sector. Zechstein salt is not present over the crest of this feature, but is present in substantial thickness further down both the northern and southern flanks. Recent studies (Jenyon and

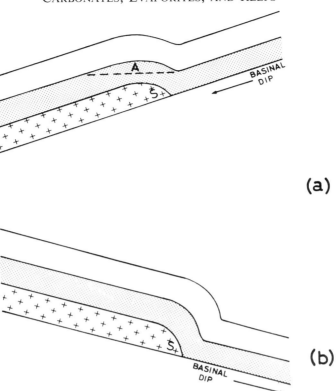

Fig. 8.22b. Salt-edge dissolution in (a) is a favourable situation, where the original basinal dip has persisted, for trap development at A in the subsidence drape overburden fold. In (b), tectonism has reversed the basinal dip, and the trap situation no longer exists. Hydrocarbons *might* at an earlier time have been trapped in the (b) situation, before the basinal dip was reversed, and may now be found in some other trap configuration further up dip (cf. the situation in Fig. 8.20). Reproduced by permission of Elsevier Applied Science Publishers Ltd (from Jenyon, 1986d)

Taylor, 1987) have indicated that the cause of the absence of the Zechstein salt over the crest is principally dissolution removal by undersaturated water, probably by entering aquifers above the salt through fracture systems in the crestal region of the MNSH and moving down both northern and southern flanks. This movement has caused the retreat of salt down both flanks (initially, there was probably only a minimal salt thickness over the crest of the feature), and salt-edge dissolution slopes are now to be found facing upwards towards the structural crest on both flanks.

This is shown schematically in Fig. 8.24, taken from the paper cited above. Sketch (a) shows a north–south section across the MNSH, before any dissolution has affected the salt, with the pre-Permian surface of the MNSH overlain by the Zechstein salt interval thickening into both northern and southern subbasins from the crest of the MNSH. The unit marked A is supposed to be an indicator band in the overburden.

It is believed that at some stage, as noted above, undersaturated brine entered Mesozoic units resting on the salt in the crestal area of the structure, and flowed intraformationally

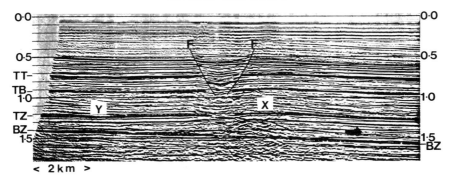

Fig. 8.23. Seismic example of salt-edge dissolution where no distinct salt dissolution slope has resulted—perhaps because of the thinness of the original salt layer in this marginal location, with a low rate of thickening into the basin. However, removal of the salt by dissolution has produced a distinct dip reversal (X to Y) in the overburden above the interval containing the salt layer (BZ–TZ). Basinal dip from left to right at the BZ level is indicated by the arrow. Counter-basinal dip reversals like the one seen here may result in important structurally-high trends paralleling a basin margin and producing traps along a salt edge—cf. the Williston Basin in North America. BZ = Base Zechstein; TZ = Top Zechstein; F = fault. Reproduced by permission of Seismograph Service (England) Ltd

(perhaps at the salt–clastic contact) down both flanks, dissolving and removing part of the salt interval as suggested by the arrows marked W. There are several possible reasons why this could have occurred:

(i) Compaction of Mesozoic across the MNSH axis may have induced fracturing along the axial zone of the structure.
(ii) Pre-Permian faults trending along the axial zone of the MNSH may have been reactivated, resulting in zones of weakness and fracturing in the compacting Mesozoic sediments.
(iii) The cut-down of the Late Cimmerian Unconformity (CU) may—perhaps only in the later stages—have allowed water to percolate down through the unconformity surface in the axial zone of the MNSH. (Some relatively minor subsidence effects are seen at CU level, but on a much smaller scale than the main collapse features.)

(NB Salt may also have been dissolved from *below*, aided by dislocations in the pre-Zechstein strata, and producing an inverted *Salzspiegel* situation.)

As the salt was removed, the Mesozoic overburden would have subsided, leading to the situation illustrated in (b), where the retreating salt terminates on both flanks in a salt-edge dissolution slope D, and the Late Cimmerian erosion has cut down into the Mesozoic overburden (CU). This is the simplest situation to be envisaged, and assumes a completely uniform Zechstein salt interval.

Diagram (c) shows what the *actual* situation seems to be, from interpretation of the seismic data. The Zechstein interval was *not* uniform; positive irregularities (R) are present at Base Zechstein level, which, after removal of the salt, produce strong effects on the seismic section, having imparted subsidence drape to the collapsed Mesozoic clastics. This drape persists up to the CU surface (and above, at some locations). The salt-edge dissolution slopes, and the collapsed Mesozoic clastics are clearly identifiable in the seismic data.

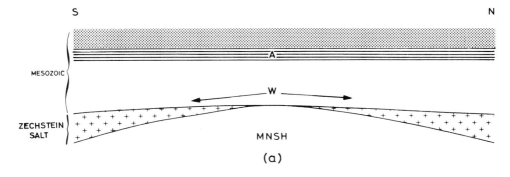

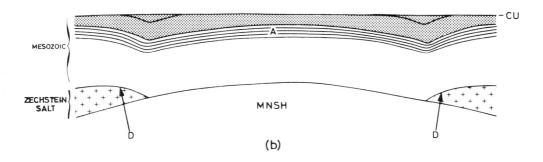

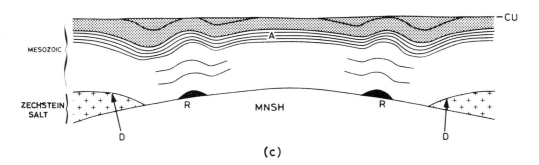

Fig. 8.24. Schematic sections across the Mid North Sea High (MNSH): (a) before any salt dissolution has taken place; (b) an imaginary situation where simple salt-edge dissolution and overburden collapse has taken place; (c) a simplified sketch showing what is believed to be the actual situation, with subsidence drape in the overburden above remnant features (R) at original base salt level. CU = Cimmerian Unconformity. Reproduced by permission of Springer-Verlag (from Jenyon and Taylor, 1987)

This is illustrated in the seismic section (a) and line-drawn interpretation in Fig. 8.25. The seismic example is at a location a short distance down the north flank of the MNSH from the crest. The salt dissolution slope S is indicated on the interpreted section, facing up towards the MNSH crest, which is just off-section to the south-west. The north-east end of the section has the base and top of the Zechstein salt interval marked at BZ and TZ, with the Late Cimmerian Unconformity at CU. Well 38/18-1 shows a thickness of nearly 1000 feet (300 m) of Zechstein Z3 salt at this location (indicated by the cross symbol), whereas another well just off-section to the south-west shows a complete absence of any salt. A seismic section to the south-west of this one shows a 'mirror image' situation on the south flank of the MNSH in the Zechstein interval, with a similar salt-edge dissolution slope facing upwards towards the crest to the north-east. At the south-west end of the Fig. 8.25 section, it will be noted that the TZ horizon seen to the north-east of slope S is no longer represented. The basal Zechstein horizon BZ represents the whole remnant

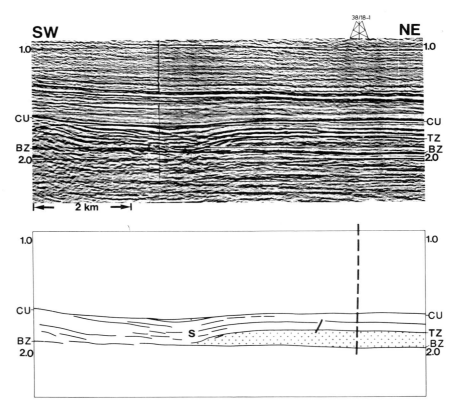

Fig. 8.25. A seismic example located a little way down the north flank of the Mid North Sea High (see Fig. 8.24 (c)), where the salt slope D on the right is represented here by the slope S in (b). Well data to the south-west of the slope S prove a complete absence of salt, while the released well 38/18-1 marked on this section, logs nearly 1000 feet (300 m) of Zechstein Salt between BZ and TZ (Base Zechstein and Top Zechstein). The collapse features to the south-west of S, and the subsidence drape over a probable reef at the south-west end are seen clearly. CU = Late Cimmerian Unconformity. Reproduced by permission of Springer-Verlag (from Jenyon and Taylor, 1987)

Zechstein interval (carbonate/anhydrite units) where the salt has been removed. The clear overburden subsidence features in the BZ–CU interval can be seen to the south-west of S, whence the salt edge has retreated to its present position.

It is noted at this point that for a full understanding of features of this type, precise knowledge is needed of the operation of the circulatory system of the water involved in the dissolution process (i.e. specifically, where it comes from, and where it goes to after effecting the salt dissolution). Unfortunately, this is a subject that has received very little attention, apart from a few admirable studies, such as that by Anderson and Kirkland (1980), of the effects of water flow in aquifers of the Delaware Mountain Group and the Capitan Reef in the tilted Delaware Basin of West Texas and south-east New Mexico, USA, on salt rocks of the Castile and Salado Formations. This work, which examines the operation of brine density flow, is strongly recommended.

8.3.4 Remnant Features and Subsidence Drape

The seismic example in Fig. 8.22 showed what was termed a 'remnant feature' at the original salt base level. This feature had become apparent due to the removal of the salt, and the subsidence of a clastic overburden that became draped over the small protuberance. It was noted that the effects in the overburden seemed grossly out of proportion to the size of this remnant feature.

In the process of areal dissolution and removal of a salt interval resulting in subsidence of the overburden to fill the incipient void left by the disappearing salt, remnant features like that mentioned are often seen, and may be of diverse origins. While the salt is still in place, these features may not be obvious, becoming revealed eventually because of their capacity to produce drape structures in the overburden that subsides upon them.

The schematic diagrams in Fig. 8.26 model the situation, and show a selection of the most common types of remnant feature that may be encountered.

The model used is a familiar one of a faulted carbonate ramp (a) before and (b) after an overlying salt interval (cross symbol) has been partially removed by areal dissolution. This has been dealt with elsewhere (Jenyon, 1986d), but will be repeated briefly here as it may be of considerable significance for hydrocarbon trapping and reef-trend mapping in some areas that include evaporite intervals.

In diagram 8.26(a), carbonate build-up features (e.g. biohermal reefs) marked R, a cuesta feature, C (due to a more erosion-resistant bed in a truncated, dipping series below the unconformity), and a fault scarp F, have all been buried by a salt interval (numbered 1). The salt layer, lying unconformably on the carbonate ramp, is overlain by clastic deposits 2–4. In the lower diagram (b), a period of erosion has caused the cut-down of unconformity U into the clastic sequence; during this process, water may have percolated downwards through the unconformity surface and the permeable clastic units (or entered aquifer 2 further updip, and flowed down under gravity) effecting areal dissolution of the salt. The clastic bed 2 now lies directly on the pre-salt unconformity over the greater part of the model section, except that some salt, which terminates in an edge-dissolution slope S, remains on the left. To the right of S the remnant features R, F, and C have resulted in drape structures in the collapsed clastic overburden, as has also the remnant feature P, which may be a pod of undissolved salt, or evaporite insolubles. Late-stage dissolution may also cause subsidence at and above the unconformity U.

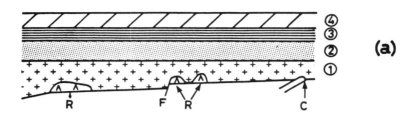

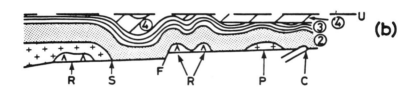

Fig. 8.26. Schematic sketches modelling the situation where a salt interval (unit 1, with cross symbol) in (a) with various remnant features buried at its base, is partly removed by dissolution (b), showing the subsidence drape in the clastic overburden units 2–4. Possible remnant features marked as R (carbonate build-ups); C (a cuesta feature resulting from an erosion-resistant unit in the pre-unconformity series); F (a fault-line scarp); P (a remnant pod of evaporite insolubles, such as very anhydritic halite). U = unconformity. Reproduced by permission of Elsevier Applied Science Publishers Ltd (from Jenyon, 1986d)

It can be seen that any remnant features that are present after removal of the salt will, owing to the drape effects in the overburden, advertise their presence clearly. Edmunds (1980) noted the fact, commenting that the process '... significantly affects the mappability of underlying reef accumulations and ... mapping of prospective trends.', quoting the Middle Devonian Keg Reef trend of Northern Alberta and the Silurian Niagara Reef trend of Michigan as examples.

Following from these considerations, an interesting and potentially very important matter arises with regard to the forms of drape structure.

It has been noted that in the type of situation just discussed, remnant features of quite limited size produce surprisingly large drape effects in the subsided overburden. This is the result of the condition of the rocks in the overburden.

Under normal circumstances, depositional drape occurs either in a growth structure, such as an anticline that is undergoing synsedimentary uplift, or in a situation where sediments are deposited over a positive irregularity (such as a reef) on an otherwise more or less plane surface. Owing to processes such as winnowing of sediment from highs with redeposition in lows interacting with compaction, both types of drape structure tend to die out upwards quite rapidly, as seen schematically in Fig. 8.27(a) related to a lenticular irregularity A.

However, the situation is quite different in the areal dissolution–collapse context. The overburden that eventually subsides on to the remnant feature is already compacted and perhaps well indurated, since the subsidence is not 'synsedimentary'—the contemporary depositional surface may be many hundreds of metres above the basal beds of the subsiding overburden. When the salt layer has been completely removed, therefore, the shape of the remnant irregularity persists upwards in the section through the overlying stratigraphic units, as shown in Fig. 8.27(b), again related to a small lenticular irregularity A resting on the original salt base level.

In this diagram, the salt layer originally interposed between the small irregularity A and the basal overburden unit B. The latter has subsided, taking up the antiform shape of A. The situation demands that units C and D take up the same shape. The geometry involved demands, however (*vide* the bedding plane traces between B, C, D, etc.), that the zone W between the inflection points (i.e. points of inflection between the horizontal bedding and the arc of the drape structure) on each side of the drape structure widens with increasing

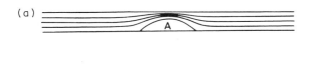

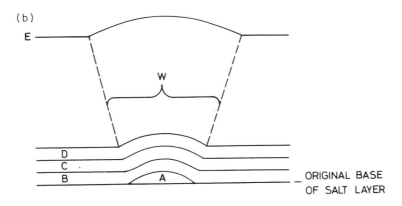

Fig. 8.27. Diagrams showing the differing effects of a lenticular feature on the overburden. In (a) the irregularity produces normal depositional/compaction drape that dies out fairly quickly upwards in the overlying deposits; in (b) the same form and size of irregularity produces a subsidence drape feature that persists upwards in an overburden that was already compacted and indurated when it subsided on the irregularity

upward distance above A. At some later time, horizon E will still be showing a curve concentric with that of the top of A, but over a much wider arc than is present at A. This could be of importance in a trap situation as it leads to increasing areal extent of a closure upwards. The effect can be seen in Fig. 8.22, and usually the remnant feature is much more irregular than the feature A in Fig. 8.27, where the situation has been simplified.

This matter may also be of importance in another sense. As noted, drape structures in a subsided interval will give an exaggerated impression of the lateral extent of the initiating irregularity; for example, a very small patch or pinnacle reef may produce such a large drape structure in the subsided overburden that an erroneous idea of the initiating irregularity may lead an interpreter of seismic data to discount the possibility of the presence of small reef features. Very small features of this type may actually be below the limit of seismic resolution at a given depth on the cross-section, only their exaggerated drape effects in the overburden being visible.

It deserves to be stressed that subsidence-drape structures related to salt interval dissolution are of great potential importance since they occur exactly in a situation (i.e. removal, or at least breaching, of a sealing salt layer) where hydrocarbons migrating upwards from beneath the salt may be able to move into the subsidence structures formed. As far as is known, little attention has been paid by explorationists to this type of feature, perhaps because the processes controlling its development have not been well understood.

8.4 V_P and V_S: a Concluding Note on Carbonates

In recent times, research into seismic stratigraphy has been paralleled by other studies into relationships between seismic data and the petrophysical/petrographical characteristics of rocks, with some significant results.

In particular, there have been some important and suggestive findings related to the ratio of seismic compressional-wave to shear-wave velocities (V_P/V_S) in a variety of rock types.

One fascinating aspect of these studies is the suggestion that measurable variations in this ratio occur before the onset of an earthquake, and therefore with further study it may be possible to use this as an earthquake prediction method.

In the present context, it has been found that laboratory studies of relationships between certain acoustic properties and the petrographic character of brine- and air-saturated carbonate rocks indicate that porosity is the major factor influencing both P- and S-wave acoustic impedance and velocity. Combined use of P- and S-wave velocity data can be valuable in discriminating between porosity and lithology changes, while the ratio V_P/V_S can be used to discriminate between limestones and dolomites (Rafavich et al., 1984).

Wilkens et al. (1984) stress the importance of this velocity ratio as being controlled by the pore geometry in homogeneous carbonate rocks, and describe the use of the ratio as a discriminant of composition in siliceous limestones that have varying proportions of carbonate and silica. They note that the behaviour of the ratio V_P/V_S of individual samples during increasing confining pressure is consistent with crack-closure theory.

It can confidently be expected that this particular field of study, integrating seismic velocity parameters, pore geometry, and lithological variation, will become increasingly important in the future.

8.5 Reefs

The term 'reef' encompasses a wide range of features of varied origin, morphology, lithology, and (palaeo)physiographic location. It can be (and has been) used to describe biogenic structures with a definite morphology (bioherms), or stratified structures interbedded with other rocks (biostromes), as well as other types of structure, such as shell-banks, debris mounds, etc., resulting from wave and current action.

Since biostromes have no distinctive form and usually show little or no difference (in seismic velocity and other attributes) from the rocks in which they occur, they cannot be identified by seismic means alone, but require borehole sampling. For this reason, they will not be dealt with here. The principal concern will be with bioherms and other carbonate build-up features exhibiting recognizable characteristics in seismic data.

8.5.1 Identification of Carbonate Build-ups in Seismic Data

There are two basic aspects of build-ups that can assist in their seismic identification; these are morphology and location. A typical bioherm is a mounded or lenticular structure, with a roughly convex-upwards top surface and a plane lower surface (frequently the upper surface is of irregular shape, although tending overall to be convex upwards). Some classifications of build-ups are based on morphology.

On the other hand, within a basinal setting, there are certain preferred locations for different types of reefal bodies. These locations are related directly to ranges of water depth, temperature, salinity, and light levels within which various types of reef-building organisms can survive, and for a specific type of organism, these factors tend to vary somewhat according to geographical latitude. (In passing, it should be mentioned that certain factors are inter-related; for instance, the matter of water depth is probably more to do with the levels of light, and perhaps also temperature, required by the reef-builders than it is to pressure.) A present-day coral reef assemblage in the tropics, for instance, is observed to grow vigorously only in water depths between 20 m and 100 m, while the water-depth limits for reef-building calcareous algae are different.

Palaeoreef structures are often found with overall heights of hundreds—even thousands—of metres; this (unless the survival requirements of the organisms involved have altered drastically, which can be discounted) clearly implies subsidence. This is in line with the original work of Darwin on the coral islands and atolls of the Pacific; he saw their development in terms of an initial fringing reef just seaward of the land–water boundary of a volcanic island, becoming a barrier reef offshore, and subsequently an annular atoll, due to progressive—and eventually total—subsidence of the original island.

Reefs may be classified in terms of basinal location combined with morphological considerations, as suggested by Bubb and Hatlelid (1977) who noted that the conventional classification of reefs and carbonate banks is not easily applicable to seismic data. Many types of carbonate build-up are recognizable by a combination of morphological character and basinal location. They recognized offshore barrier reefs, which are massive, elongate, rather symmetrical (in cross-section) structures some distance into the basin from the shelf margin, and asymmetrical shelf-margin (fringing) reefs of similarly elongate form around the shelf edge, but with definite fore-reef (towards the basin) and back-reef (away from the basin) developments. Between barrier and shelf margin, developments of pinnacle reefs (small bodies with substantial vertical, rather than areal, dimensions) are recognized, while behind the shelf-margin reef, a shallow-water back-reef lagoon may be found in which small patch reefs occur (the areal dimensions of which are greater than the vertical dimensions, in contrast to pinnacle reefs: the difference, perhaps, signifying contrasting water depths). Thus shape, size, symmetry, and location can all be used to identify the particular type of reef structure present in seismic data.

The illustration of carbonate environments shown in Fig. 8.10 indicates the type of setting in which reefs are frequently found, although the terminology used in the illustration is slightly different from that used here. The zone indicated in the figure as 'reef barrier' coincides with the zone of the shelf-margin reef just discussed, while into the basin, what we have referred to previously as a 'barrier reef' is absent in the figure. Behind the shelf-margin reef in Fig. 8.10, the platform-border sands, internal platform, and intertidal zones of the carbonate environments make up the area of the shallow back-reef lagoon referred to earlier, in which patch reefs may form.

8.5.2 A Shelf-margin Bioherm

To discuss a specific situation that might be expected in seismic data, a shelf-margin reef will be used as an example, since this is perhaps the type of structure most commonly thought of as a 'reef'. In Jenyon and Fitch (1985) the typical seismic cross-sectional expression of such a reef is said to include the following characteristics:

(i) A lenticular shape, with the upper face convex upwards, the lower essentially plane.
(ii) A steep fore-reef slope, sometimes steep enough to develop hyperbolas at the fore-reef side of the reef core.
(iii) A more gentle back-reef slope.
(iv) A difference between fore-reef and back-reef sedimentation, so that the seismic sequences do not correlate in detail.
(v) Draping of sediments over the reef: the convex-upward form of the upper reef surface persists, with decreasing relief, into higher beds (but see the earlier part of the chapter, dealing with *subsidence drape* over build-ups).
(vi) The detailed structure of a typical reef is a reef core, massive and unbedded, with fairly steep dips away from the core. Evidence of steep, persistent reflections can often be seen. The reef core represents the actual biogenic growth part of the feature, the remainder of the structure being talus debris or penecontemporaneous sediments interacting with the growing reef.

Often, seismic stratal events related to basin fill in front of the fore-reef slope and over the back-reef area, when taken together, assist in identification of the reef structure. Stratal events from basin fill in front of the reef frequently show clear onlap or truncation in relation to the rather steep fore-reef talus slope, whereas the sediments in the back-reef area, while showing some thinning towards the reef body, tend to be relatively conformable with the underlying strata.

In Fig. 8.28, an actual example of a shelf-margin reef is shown in which some of the characteristics mentioned can be seen. The vertical scale of the seismic section is, as usual, in two-way time, and depth indications in feet have also been marked. The coral reef

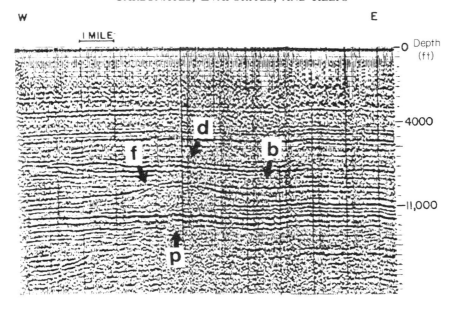

Fig. 8.28. Seismic example showing several characteristics of a shelf-margin reef: the fore-reef face (f), back-reef zone (b), subtle depositional drape above the feature (d), and a velocity pull-up anomaly (p) beneath. (after Jenyon and Fitch, 1985)

producing the effects here has its floor at about 1.8 s and its crest at 1.35 s. Note the following points:

(i) A small hyperbola is developing at 'f', marking the steep fore-reef face; the back-reef face on the east side of the reef is smoother, and with a gentler dip. Note the difference in the seismic response to fore-reef and back-reef deposits (cf. point (vi) above). The back-reef area is marked 'b'.

(ii) The 'velocity pull-up' anomaly—an apparent gentle anticline in the floor of the reef above 'p' is an expression of the higher velocity of the reef material than that of the off-reef material (i.e. it causes a shorter reflection time for seismic energy passing through the reef than for seismic energy passing through the off-reef material). This 'pull-up' is an artifact in the seismic section, and has no counterpart in the 'real Earth'. Note that a 'pull-down' anomaly below a reef can indicate the presence of gas-filled porosity.

(c) There is a subtle arching above the reef also, at the level of 'd', owing to depositional drape over the reef. It can be detected as high as 0.65 s, and is a 'real Earth' structural feature.

To the west of the reef, at about 1.2 s, there is an odd, pod-shaped mass bounded by strong reflections. This also could be a carbonate build-up; possibly a debris mound, or a biostrome. Clearly, from the lack of a pull-up anomaly beneath it, the velocity of the material in it is not higher than that of the adjacent sediments, and so it is very unlikely

288   OIL AND GAS TRAPS

to be a smaller version of the main reef feature. Small pod-like build-ups of this type are often seen close to larger reefs.

8.5.3 Lithology, Porosity, and Seismic Velocity

All of the complex discussed above—reef core, reef flanks—is likely to be, or have originally been, of limestones with a wide range of textures, biological content, and accessory minerals, within a range of high seismic velocities.

The original reef-building organisms had extensive cavities in their hard parts (all being lime-secreting animals). Also, parts of many reefs consist of relatively loose intergrowths of hard and soft colonies of corals of varying shapes, with loose calcareous sediment trapped within the coral framework, or between intergrowths of algae. Because of this type of organization, primary porosity in the original reef was very high, and this is retained in a young reef until increasing burial depth and cementation result in collapse of the original framework of the reef, destruction of porosity and permeability, and increasingly high seismic velocity—frequently higher than that in adjacent and/or encasing bedded carbonate sediments, such as lime silts and muds, marly limestones, etc.

At a later stage some reefs may undergo diagenetic changes that can be favourable to the redevelopment of porosity and permeability. Uplift may bring the structure within the range of fluids that produce selective leaching. Dolomitization, with production of secondary porosity, may occur. Tectonic stresses may be applied that improve permeability, etc. In such instances, porosity increase accompanied by decrease of bulk density may cause the seismic velocity within the feature to decrease relative to surrounding rocks, with production of a velocity 'pull-down' anomaly beneath the feature. Where such an effect exists related to a reef, it must be investigated carefully, since it may establish the commercial interest of the structure. Many major oil- and gasfields have been discovered in reef structures that have developed good secondary poroperm characteristics.

An unusual example having two structures in close proximity, one of which is showing a 'pull-up' seismic velocity anomaly and the other a 'pull-down' anomaly, is seen in Fig. 8.29. This seismic section shows an unmarked version at (a) and a marked version at (b). The features are interpreted as being due to the presence of two small patch reefs approximately at the level of the horizon 'P' indicated on the right of the section. The left-hand feature shows a positive 'pull-up' velocity anomaly beneath it, and drape of the overlying section. The right-hand feature, on the other hand, shows negative 'pull-down' effects beneath it, but normal drape above it.

It seems likely, from their close proximity and identical stratigraphic level, that these two features are reefs of the same type and approximately the same size. Clearly, however, if this is the case, then it seems likely that the right-hand feature must have developed marked secondary porosity sufficient to cause the strong pull-down effect, whereas the right-hand feature has not.

Unfortunately, no well data are available close enough to enable more detailed investigations of these features to be made. However, the right-hand structure must be of some potential interest commercially; as at the time of writing, it has not been drilled on.

In Fig. 8.30, a similar situation is seen, with a now somewhat familiar section showing where an originally thick Zechstein salt interval (TA–TZ at the south-west end of section) has been removed by dissolution and circulation, so that there is no salt to the right of

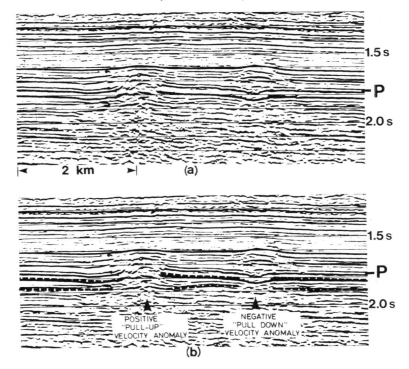

Fig. 8.29. Seismic example showing two carbonate build-up structures in close proximity; the juxtaposition is unusual in that one exhibits a 'pull-up' seismic velocity anomaly, and the other a 'pull-down' anomaly. The illustration shows the unmarked section (a) and a marked version (b). A negative 'pull-down' anomaly like the one seen here must always be of interest, since it is a possible indicator of relative density deficiency in the feature above that can be the result of the presence of gas-filled porosity. Reproduced by permission of Seismograph Service (England) Ltd

'A' as shown on the interpreted line-drawn section below the seismic example. As described in the previous part of this chapter, the collapsed Mesozoic overburden is now encasing, by subsidence drape, two features (R), interpreted as patch reefs resting on the basal Zechstein beds, probably of Z2 age. The actual reefs are very small, and may be outside the range of resolution of the seismic data here. However, their effects on the overburden can be seen, persisting up to the Late Cimmerian Unconformity 'CU'. The interesting point about these two reefs (which may be an atoll), apart from the small 'pull-up' anomalies present beneath them, is the presence of two 'flat-spot' direct hydrocarbon indicators at FS, one above each of the two features. These are interpreted as the response to gas-water contacts in the collapsed Mesozoic sediments above the reefs. They are strong, flat, and have end polarity reversals, which make it highly likely that they are DHI events. This shows that apart from the possibility of secondary porosity in the reefs themselves, there may be trap potential in drape structures above the reefs. In this instance, the removal of the salt seal by dissolution has left a migratory path open for gas from the underlying Carboniferous Coal Measures to move up into the structural traps indicated.

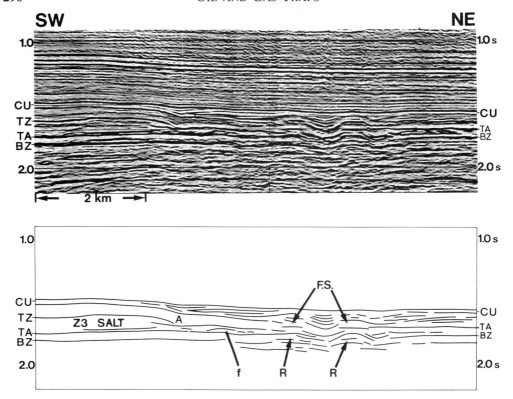

Fig. 8.30. In the seismic data (a) here, there is a salt slope, indicated at A in (b), and two features (R) interpreted as reefs (perhaps an atoll in cross-section), which are revealed clearly by the subsidence drape of overburden units after dissolution removal of the Z3 salt layer to the north-east of the salt slope A. In both cases they are showing signs of 'pull-up' anomaly, but in this case, flat-spot events (FS) are indicated (possible GWC events) in the antiform overburden subsidence features. This shows that apart from the possibility of secondary porosity in reefs themselves, there is trap potential in drape structures above reefs. BZ = Base Zechstein; TZ = Top Zechstein; CU = Late Cimmerian Unconformity; f = fault. Reproduced by permission of the Canadian Society of Petroleum Geologists (from Jenyon, 1988d)

As regards velocity anomalies below carbonate build-ups (as also in the case of salt bodies), sometimes the build-up material is *not* very different, density (porosity) wise, from adjacent sediments, in which case no anomaly will be present. This is seen in the Fig. 8.31 seismic example from the Rainbow–Zama area of Alberta, Canada. The occurrence of two small pinnacle reefs of the (Mid-Devonian) Keg River Formation is affecting the overlying Slave Point (s) and Sulphur Point (p) Formations at (d), resulting in a so-called 'eyebrow structure'—a double antiform shape. These formations have subsided on the reef features owing to the dissolution and removal of Muskeg Salt that formerly occupied the inter-reef areas (and perhaps was also deposited over the reef structures). The situation here is quite similar to that in the Zechstein over the Mid North Sea High, as discussed earlier in the chapter. Oil is found in numerous small traps of the type seen in Fig. 8.31, and

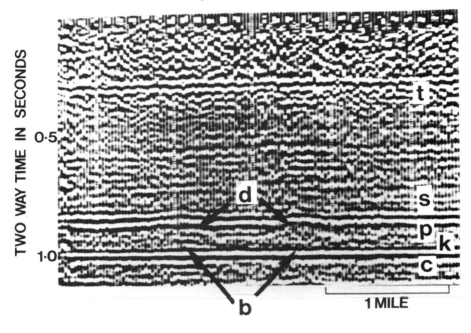

Fig. 8.31. Seismic example of an 'eyebrow structure' from the Rainbow–Zama area of Alberta, Canada, believed to result from the dissolution removal of Muskeg Salt. The double antiform shape of the feature is indicated at 'b', and subtle collapse and drape features are seen in the zone 'd'. Reproduced with permission of Elsevier Applied Science Publishers Ltd (from Jenyon, 1986d; after Bubb and Hatlelid, 1977)

recoverable reserves in the Rainbow–Zama area in the late 1970s were estimated at some 800 MMbbl of oil. The Keg River Formation rests on the Cold Lake Salt Formation (c), which is close to basement. Note that beneath the pinnacle reef locations (b) there is no sign of any velocity anomaly, either positive or negative. This may be due to a lack of differential between the seismic velocity in the reef materials, and that in the collapsed Sulphur Point sediments. On the other hand, at the scale of the section, the pinnacle reefs are very small (less than 30 ms two-way time vertically, and 200 m horizontally), and so it could be difficult or impossible to see any effects on this scale, and with this resolution. It is of interest to note that reef features occur in pairs with great frequency in some areas, and must therefore be regarded as possible atolls, the pair of features being the seismic expression of the opposite 'rims' of the atoll, across a central lagoon.

8.5.4 Larger Features

A much larger example of a reef is seen in the seismic example of Fig. 8.32, which is a section across the reef holding the Intisar 'D' Oilfield of the Sirte Basin, Libya (Brady *et al.*, 1980). The discovery well in this field tested 75,000 BPD in 1967.

This reef, which is an isolated feature roughly circular in plan, 5 km in diameter and 1262 feet (385 m) thick at its maximum, is of coral and algae with grain and mud-supported

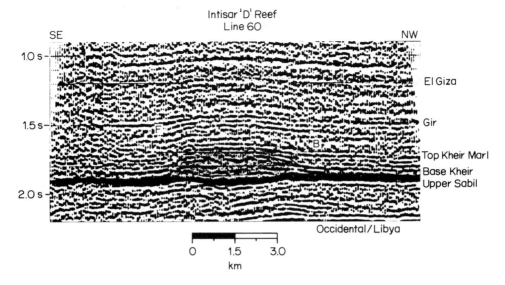

Fig. 8.32. The Intisar 'D' oilfield, Sirte Basin, Libya, in a large coral/algal reef (dark shaded) of Late Palaeocene age. Fore-reef (F) and back-reef (B) zones can be seen, with well-marked drape in the overlying deposits. The strong pull-down anomaly beneath the reef is an important feature that suggests an interesting density deficiency in the reef structure, which is, in this case, evidence of a major development of secondary porosity. From Brady et al., 1980, reprinted by permission of American Association of Petroleum Geologists

biomicrites. It developed in a late Palaeocene embayment bounded on three sides by carbonate banks. Solution and intergranular secondary porosity has developed, averaging 22%, with measured permeability averaging 87 mD, maximum about 500 mD. The main reservoir is remarkably homogeneous, without the noticeable layering typical of other reefs in the area. The OWC appears to be a simple, plane surface across the whole reef structure, again unlike many larger reefs, which are often compartmentalized into differing porosity–lithology volumes, sometimes in a very complex configuration.

The reef was full to spill point when production began, with 40° API oil in a maximum oil column of 995 feet (291 m), and cumulative oil production to 30 September 1978 totalled 777 MMbbl.

The isolated nature of the feature leads to its categorization as a pinnacle reef in Brady et al. (1980). However, such reefs tend to be smaller and rather symmetrical in development. Inspection of Fig. 8.32 suggests some asymmetry—the reef body has been shaded slightly darker than the adjacent section—and even though a roughly circular body, it appears to show distinct fore-reef (F) and back-reef (B) aspects, as discussed earlier. The configuration of the strong basal reflection event indicated at Base Kheir level implies that the supposed fore-reef face is, indeed, facing into the basin, with quite steep drape of overlying formations above the edge of the reef-face, while the much gentler drape and dips below B indicate the presence of a back-reef zone. In view of this, it may be more correct to regard the feature as an isolated part of a discontinuous shelf-margin reef.

It is interesting to note the quite marked pull-down anomaly in the events beneath the reef.

This suggests that the homogeneous reef material (perhaps as a consequence of having developed good secondary porosity throughout the feature) is now less dense, and with lower seismic velocity, than the adjacent off-reef carbonates. This is a typical case in which a pull-down velocity anomaly associated with a carbonate build-up feature is of great significance, being often a direct indication of good porosity.

Lithologies and textures in barrier and shelf-edge reefs are, in general, similar. Barrier reefs, however, tend to be more symmetrical in cross-section, since depositional environments on the basinward and landward sides of the reef are similar, with similar water depths. Shelf-edge marginal reefs, as noted, have very different depositional environments adjacent to them; on one side, a shallow shelf lagoon and on the other side, much deeper water going down into the basin. This strongly asymmetrical depositional situation results in the type of structure seen as a schematic in Fig. 8.33, which is a generalized profile across the (Upper Permian) Middle Magnesian Limestone bryozoan/algal shelf-edge reef in north-eastern England (Smith, 1981). The profile is typical for this kind of structure, with the steep fore-reef talus slope (F) facing into the basin, the reef-flat overlying the massive core that rests on a basal coquina, and the back-reef, shallow-water area located in the direction of the arrow at B. The structure is formed of a basic bryozoan/algal reef (the core), which passes upwards, with increasing algal content, into an extensive stromatolitic biostrome. The whole feature reaches heights in excess of 100 m, and widths of about 800 m in places, in an elongate complex running for at least 30 km (at surface outcrop), and probably much more is buried. The original limestones underwent almost complete dolomitization, with loss of fine detail, but locally, dedolomitization has taken place. In some areas, small patch reefs are found in the back-reef lagoonal zone associated with the main reef.

From the seismic viewpoint, such a structure cannot be regarded as sufficiently large to show any internal details on standard seismic sections, although the overall shape of the reef should be detectable, if sufficiently large acoustic impedance contrasts exist relative to surrounding rocks. The upper part of the steep fore-reef slope at F, where it meets the reef-flat, may produce a diffraction; on the other hand, depending on the dip of the talus

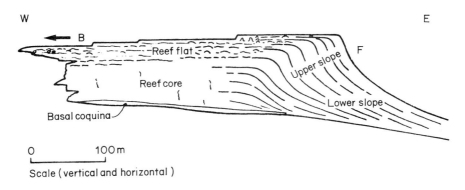

Fig. 8.33. Generalized sketch profile across the (Upper Permian) Middle Magnesian Limestone bryozoan (algal shelf-edge reef in north-eastern England, showing the steep fore-reef face (F) and the reef-flat, passing to the west into the back-reef zone (B). Reproduced by permission of Society of Economic Palaeontologists and Mineralogists (from Smith, 1981)

slope face of the fore-reef slope, and the velocities, the actual slope face may produce a reflection. In addition, overlying sediments may show depositional/compaction drape over the feature, underlying sedimentary stratal reflections may show either pull-up or pull-down anomalies, depending on the velocities/densities involved, and the back-reef fill might indicate typically gentle, continuous dips.

The seismic example in Fig. 8.34 has been produced, not by the standard techniques, but by a VSP-related procedure, using downhole data collected at the wells shown. This has resulted in excellent indications of a reef structure with a clear fore-reef slope at F, and a more gently dipping back-reef zone at B. The slopes of the upper surface of the reef are well defined by the method, owing to contact between the porous limestones of the reef and the enclosing evaporites (salt and anhydrite). This gas-bearing reef in the USSR has overlying stratal events showing gentle drape; it is difficult to decide whether there is an underlying velocity anomaly, since the basal reflection at the floor of the reef is irregular, and broken by what are possibly several faults. However, the overall morphology is sufficient for an identification of the feature as a reef (Karus et al., 1975).

Provided that physical conditions remain within the limits of tolerance of the reef-building organisms, and given the action of subsidence or sea-level rise at a favourable rate over a geologically extended time period, some reefs can persist over long time intervals, and show continuous growth through vertically large distances—in some cases, hundreds or even thousands of metres. This is often true of shelf-edge reefs in a tectonically unstable area, and the great vertical extent of such features can frequently be observed on seismic sections at basin edges.

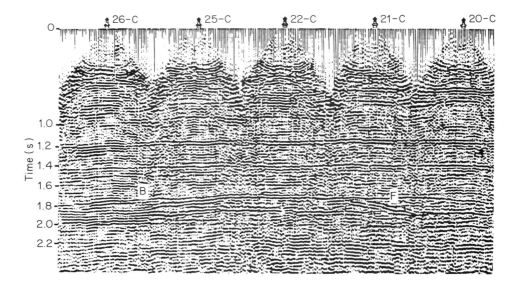

Fig. 8.34. Seismic section produced by a VSP-related technique, and showing reef features. Fore-reef (F) and back-reef (B) zones are seen, together with some overlying drape structure. This is a gas-bearing reef in the USSR. Reproduced by permission of Elsevier Applied Science Publishers Ltd (from Karus et al., 1975)

8.6 Concluding Comment

As is the case with the deltaic environment, an area that includes a typical assemblage of carbonates, evaporites, and reefs will often have everything necessary for the presence of a hydrocarbon province—source rocks in the vicinity, reservoir and seal formations, as well as a vast range of possible structural and stratigraphic traps, some of them peculiar to this kind of environment. Wherever these lithologies and facies-bound structures are found together, seldom (perhaps never) are hydrocarbon accumulations far away. The subjects of this chapter have been dealt with only very briefly; each has a very extensive literature of its own which should be consulted.

This is particularly true of the part of the chapter on reefs. Carbonate build-up features can take on a bewildering variety of different forms in seismic data, and only experience will enable identification to be made in many cases. The reader is urged to study the literature references cited here, and to follow up other citations in those books and papers referred to, in order to study as many examples of carbonate features as possible. Space availability herein has allowed only a very few examples to be shown.

CHAPTER 9

Deltas and Fans

9.1 Introduction

It is difficult to exaggerate the importance of deltaic and fan depositional environments in hydrocarbon exploration.

The deltaic environment, occurring at the interface between fluviatile and marine zones, provides some of the most prolific sand bodies that exist for hydrocarbon accumulations. The formative elements in delta development are so many, each of potentially prime importance, that their permutations and combinations in different cases ensure that no two deltas are alike.

In general, lithologies covering reservoir, 'carrier', and seal beds are present, and a wide range of trap situations of both stratigraphic and structural types occurs. There are, typically, connections between the deltaic reservoir beds, and potential source units in swamp/paralic sequences on the landward side, or black marine shales on the basinward side of the delta complex. Developmental aspects of deltas to be discussed later lead to some distinct seismic responses that can greatly aid the identification of the deltaic environment in the data.

Fan deposits are also of importance to hydrocarbon exploration. Within these features is found a range of sand bodies, principally turbidite sands, but also including slump and slide structures, grain- and inertial-flow deposits together with other mass movement features. These sands have provided many major commercial hydrocarbon fields world-wide.

Submarine fans are, as the name implies, fan-shaped, mounded bodies elongate approximately in the direction of maximum inclination of the shelf/slope where they are located. Morphologically, they are somewhat analogous to the dry 'outwash fans' seen on land in piedmont areas where rock detritus has been funnelled down through large gullies in the mountain areas, under gravity and by flash floods, to fan out on the plains below. Submarine fans consist either: (i) of sediment funnelled down individual large submarine canyons on the outer continental shelf and slope; or (ii) of sediment funnelled down a series of canyons that develop basinwards of a delta front due to the action of the major distributary channels of the delta. The latter type of fan deposits may coalesce to form a fan complex along the delta front. In some cases, the delta builds outwards over these coalesced fan deposits.

It is often the case that fan deposits are not easy to identify unequivocally from seismic data alone. However, they tend to occur in rather predictable locations within a basin (as noted above), and this fact, combined with some morphological characteristics recognizable in the seismic section, may aid in their recognition. Often, drilling information will be available to assist in the identification of a feature as a fan deposit, and this will increase confidence in other identifications in an area without further drilling results.

9.2 The Development of Deltas

A delta is a major sedimentary feature, and exists because of the ability of a river to discharge sediment-laden water into a lake or sea over an extended period of time. The discharge must be at a greater rate than will allow the removal of the sediment by such processes as wave or current action in the body of water into which the river flows.

Some rivers discharging into large bodies of water (such as oceans) develop deltas while others do not. If the body of water into which the river flows is assumed to be currentless, tideless, and wave-free, jet theory can be used to study the types of deltaic body to be expected. Bates (1953), on the basis of jet theory, recognizes three different situations depending on whether the sediment-laden inflow water is more dense (hyperpycnal inflow), equally dense (homopycnal), or less dense (hypopycnal) than the water in the 'permanent' body of water that receives the inflow. The inflow characterized as 'homopycnal', where the density of the fresh, sediment-laden water of the inflow is approximately equal to the density of the oceanic brine receiving the inflow, is that which produces the 'classical' deltaic deposits of bottomset, foreset, and topset beds.

From a depositional viewpoint, delta development can be separated into two stages— the constructive, and the destructive (Scruton, 1960). In the constructive stage, a river builds embankments seaward, while in the destructive stage, the sea encroaches on the new land so formed.

The constructive stage requires subsidence in order that an increasingly thick sediment pile may accumulate. It is incorrect to think that such subsidence is brought about solely by the weight of the sediments themselves, although this becomes an increasingly important factor with the passage of time. In the early history of the delta, before any appreciable thickness of sediment has built up, *tectonic* subsidence is required. Thus, the deltaic environment is—at least in the early stages—related to tectonic instability. The common occurrence of growth faulting associated with deltaic sediment accumulations (see the discussion in Chapter 4) also points to an unstable situation.

Deltas can also be classed as: (i) fluvial dominated; (ii) wave dominated; and (iii) tide dominated (see e.g. Galloway, 1975). The classification depends on the dominant form of energy in the depositional environment at the delta front. In general, prograding clinoform reflections are recognized as typical of the fluvial-dominated delta (forming the foreset beds), while shingled reflections (imbricate reflections with little or no overlap between reflection events—see Fig. 6.14, and accompanying text) are believed to be typical of wave-dominated deltas. Tide dominated deltas like the present-day Fly River Delta, Papua New Guinea have not been well studied so far, but are known to be difficult to identify by seismostratigraphic methods (Berg, 1982). It seems very likely that there is considerable overlap between effects deriving from the different types, and it can be assumed that the fluvial dominated delta is the type under discussion in this chapter, except where otherwise stated.

9.2.1 Deltaic Reservoirs

Hydrocarbon reservoirs in the deltaic environment are almost invariably sand bodies and sand intervals of various types, occupying a number of different physiographic positions in the contemporaneous delta. Distributary channels carrying sediment to the delta front are favoured locations for sand bodies of reservoir quality: various types of channel sands, point bar, bar finger, shoe-string and other fluviatile features can be of importance. Eventual silting-up of a distributary channel will often cause overflow of the sediment-laden water across the natural raised levees (formed of silty sands and clays) that bound the distributaries. Sometimes such overflow results locally in a 'crevasse-splay'—a fan of sediment spreading laterally from the break in the levee, or sometimes as a broader, more general overflow. In both cases, coarser sediments will be deposited close to the breached distributary (due to the reduction in stream velocity), while finer silts and clays will be carried further away to form the main interdistributary areas of the delta plain, into which the channels cut down.

At the delta front, the flow of the sediment-laden water is checked by the main body of water into which it is discharging (e.g., the ocean), and along-shore sand bodies of various types may occur depending on a number of factors, including: (i) the amount and the type(s) of the sediment being discharged; (ii) the strength and direction of any tidal ocean currents; and (iii) the influence of wave action. Upon such factors will depend the presence or absence, and the form of beaches, sand spits, barrier bars, etc. in the basin of discharge of the sediment. As the delta builds outwards, it may override and preserve at least some remnants of these frequently ephemeral, erstwhile 'offshore' sand bodies.

The sedimentology of deltas is a major subject with a large literature, and will not be dealt with at length here, since we are concerned in this context only with the 'art of the possible' as far as identifications of deltaic features in seismic data are concerned. However, a brief discussion of a modern delta—that of the Mississippi River—will help to put matters in perspective.

The Mississippi Delta is fairly typical, as regards sedimentology, of those 'fluvial' deltas having the classical tripartite form of deposits as described by, for example, Scruton (1955,1960) who divides them into topset, foreset, and bottomset beds. The two papers by Scruton are recommended for a detailed description of the sedimentology involved.

The topsets consist of marsh deposits together with delta-front silts and sands, but also including a complex array of channel, natural levee, and crevasse-splay sediments together with clays and shell beds of the interdistributary bays, lakes and tidal streams. Scruton also notes that deep scour, torrential deposition, and slow marsh deposition lead to sharp sedimentary breaks and general heterogeneity in these (topset) beds. In the seismic data of ancient deltas, the effects of scour and sedimentary bypass can often be seen in the topset stratal events, and will be seen illustrated later.

The foreset beds grade from prodelta silty clays to relatively coarse sands, silts, and clays deposited in and by the major distributaries in the system. These beds show primary depositional dips to a greater or lesser extent in most deltas, resembling cross-bedding on a large scale, due to the rapid outbuilding and 'tip-dumping' of sediment at the advancing delta front. In some instances, the usage 'foreset' and 'bottomset' is not entirely appropriate. In the case of the delta of the Fraser River in British Columbia, Mathews and Shepard (1962) point out that at the delta front, the northern muddy slope might be called 'bottomset' and the southern sandy slope 'foreset' even though both slopes have the same inclination.

The bottomset beds are offshore clays, at least partly resulting from the 'fines' deposited ahead of the delta front from the distributary/river-borne sediments. Distally, these may be turbidite silty sands that were earlier funnelled through slope-channels in front of the advancing delta, forming turbidite fans on which the later deltaic deposits (later bottomsets and foresets) were laid down. A further comment on this will be found in the later section on fan deposits.

Figure 9.1 shows the tripartite arrangement in a much simplified schematic block diagram. It should be clear from this that the depositional situation must include active subsidence of the deltaic deposits, which progressively build up a large sediment pile. For example, as noted by Scruton (1960), the Mississippi River carries an average daily sediment load of 1–1.5 *million tons*, consisting of some 50% clay, 48% silt, and only 2% sand. From these figures, one could be forgiven for thinking that to search for sand bodies under these circumstances would be like looking for the proverbial needle in the haystack. However, some of the best reservoirs occur in discrete sand bodies, the location of which is to some extent predictable in many deltas.

It will be noted that old, buried infilled distributary channels will be seen relatively easily in cross-section in the plane A of the diagram (i.e. in a section tranverse to the main direction of advance of the delta). In the plane B, longitudinal with respect to the form of the delta, the distributary channels will be less obvious, or even apparently absent altogether, because of the geometry of the situation.

In general, it is seldom possible from standard seismic data alone to identify sand bodies of some of the types mentioned; after burial and compaction, the vertical extent of even an important reservoir sand body may be only a few metres, or tens of feet, which will be below the resolving power of the normal seismic reflection method. Again, the acoustic impedance contrasts involved may not in any case be adequate to produce sufficiently strong reflection events to delineate the reservoir satisfactorily.

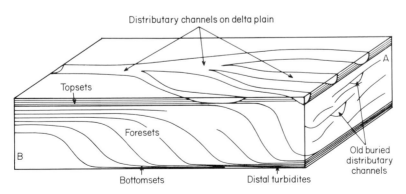

Fig. 9.1. Simplified block diagram of a fluvial-dominated delta showing the tripartite development of bottomset, foreset, and topset beds as discussed in the text

This having been said, however, good seismic data can often delineate the major elements of a delta system, and thus indicate the probable locations of many of the types of sand bodies that form reservoirs. Also, seismic data, perhaps in some cases specially processed, together with wireline log analysis, can now often be instrumental in delineating individual reservoir sands.

9.2.2 Seismic Identification of Deltaic Systems

A simple, but striking example of the power of seismic data to portray the overall deltaic situation is seen in Fig. 9.2, which shows a rather old seismic section of poor quality (although state-of-the-art at the time it was acquired). The vertical scale is marked in two-way time in seconds, and the horizontal scale is as shown. Two strong reflection groups are indicated at B and T (right-hand end of the section). Between these two groups, reflection events are seen showing strong but varying dip to the left. These events have a generally sigmoid shape, and they tend to approach the strong reflection groups at B and T tangentially. They resemble large-scale cross-bedded units. In fact, as might be guessed, the band of events marked B is from a group of bottomset beds, the band T is from a group of topset beds, and the sigmoid dipping events in between are foreset stratal events. The whole sequence is the seismic expression of a deltaic system.

The data quality in this example does not permit any detailed examination of the relationships between the foreset bedding stratal events and the overlying topsets/underlying bottomsets. However, in the next example, some typical variations in these relationships can be seen.

The Fig. 9.3 example is part of a seismic reflection profile across the Meulaboh Shelf, in the West Sumatra Forearc Basin, as described by Beaudry and Moore (1985). The section depicts several seismic sequences (1, 2a, 2b, 3, and 4), of which sequence 3 consists largely of a Tertiary deltaic interval clearly showing the tripartite divisions discussed earlier. Note that in places, particularly in the south-west half of the section, it can be seen that the

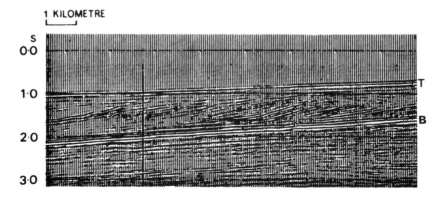

Fig. 9.2. Longitudinal seismic section through a buried delta, showing bottomset (B) and topset (T) stratal reflections, with the steeply dipping clinoform events marking the foreset bedding located in between B and T (cf. the simple diagram in Fig. 9.1). (From Jenyon and Fitch, 1985, Gebrueder Borntraeger-Verlag.)

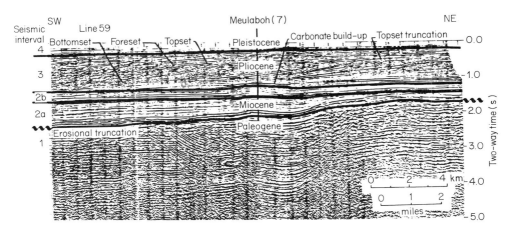

Fig. 9.3. Seismic reflection section across part of the Meulaboh Shelf, West Sumatra Forearc Basin, showing a typical deltaic situation (of Tertiary age) in the sequence numbered 3. Note the changing levels and relationships between the bottomset, foreset, and topset events from south-west to north-east across the section. From Beaudry and Moore, 1985, reprinted by permission of American Association of Petroleum Geologists

topset stratal events appear to be passing smoothly into the steeper dipping foreset events, and the latter in turn also pass smoothly into the lower dipping bottomset events. At any given level, a flat, sigmoid depositional surface is suggested as passing from the topset through the foreset to the bottomset zone, in ideal circumstances. In practice, this simple arrangement is often disturbed; some of the reasons for this have been mentioned (scour, sedimentary bypass, high-energy environments of deposition, unusually low or high rates of deposition, etc.). The clinoform seismic events seen take a number of different forms, and these can be helpful in diagnosing the environmental conditions at the time of deposition, the location within the delta system, and something of the sedimentary history subsequent to deposition. A detailed discussion of clinoform events, and variations therein, can be found in the paper by Mitchum *et al.* (1977a) on the stratigraphic interpretation of seismic reflection patterns in depositional sequences.

When the north-east half of the section in Fig. 9.3 is examined, however, although the foreset stratal events pass downwards smoothly into the bottomsets, the related topset events are seen in several locations to *truncate* the foreset beds quite sharply, while elsewhere some foreset events truncate others within the same zone. The heterogeneous—almost chaotic—nature of these relationships has been occasioned by different phases of erosional scour or reworking of the deposited sediments. According to the lithologies involved, such sedimentary features can be the cause of stratigraphic trapping of hydrocarbons within the foreset beds in some deltas.

Another rather subtle effect seen in this example is the gradual thickening into the basin of the zone of bottomset beds, and a concomitant thinning into the basin of the topset bed zone. The latter effect is often greater than the former, leading to a progressive thickening of the foreset bedding zone into the basin, as seen often in seismic data containing deltaic sequences. This thickening reaches a maximum, and then the delta sequence begins to thin again, towards the delta 'toe'. In some cases, the thinning delta grades into a distal, deep-water turbidite sequence further out into the basin.

In some deltas, the environmental energy of deposition may have been so high as to result in: (i) extensive scour at the deltaic surface and/or sedimentary bypass, with extensive removal or non-deposition of any appreciable thickness of delta plain (topset) sequences over varying time periods; and (ii) high rates of deposition of a large sediment supply at the delta front, with rapid outbuilding into the basin of discharge. These effects may be combined in some delta sequences, and their resultant expression in seismic data in an example area is seen in Fig. 9.4, where the arrowed interval shows very good evidence of foreset-bedding clinoform events, but no clear evidence of any appreciable development of topset or bottomset sequences. In the upper part of the sequence, there is a lack of flat-lying topset events passing into the foreset clinoforms, the latter being simply truncated in a fairly uniform manner across the section. The lower part of the sequence shows the parallel oblique clinoform events of the foreset bedding again terminating abruptly at the lower boundary of the interval, with no good evidence of smooth passage into bottomset beds as seen in parts of previous examples. Such a set of relationships indicates what might be called 'tip-dumping' of large amounts of sediment at the delta front (perhaps in relatively shallow water), with high deposition rates. Again, depending on the lithologies present in foresets and topsets, foreset-bedding reservoirs may result.

Note that in this example also, the overall interval as indicated by the arrows, increases in thickness to the right. The zone of foreset bedding clinoforms is thickening into the basin—by some 30–40 ms across this small portion of the section—although the basal reflector of the interval becomes less easy to follow in the right-hand half of the section.

Some interesting light is shed on deltaic morphology by the seismic example in Fig. 9.5(a), which shows the termination of a distal deltaic lobe in the northern North Sea. The delta lobe is indicated between C and D, and is affected by a small normal fault in the centre of the section. There appears to be some indication of imbricate deltaic structures here,

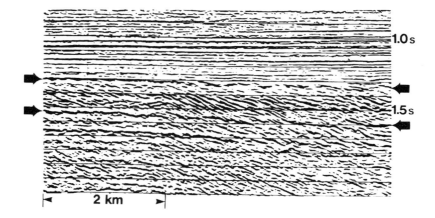

Fig. 9.4. An interval of foreset bedding events (between the arrows) seen in seismic data. These events can be classed as parallel oblique clinoforms, and probably represent a shallow water, high energy, high deposition rate type of environment. Reproduced by permission of Seismograph Service (England) Ltd

since there is some suggestion that the section between levels D and T could include another, earlier lobe of the same delta system, although the evidence is not clear.

However, careful examination of the internal structure of the lobe C–D will yield clear examples of the clinoform events typical of foreset bedding, passing fairly smoothly into both topset and bottomset stratal events; both sigmoid and tangential oblique clinoforms can be identified. Other lettering in the illustration refers to a flood basalt surface (B), a fissure vent in cross-section (V), and one of the unconformities present (U).

A slightly more detailed view of the delta lobe of the previous figure is seen in Fig. 9.5(b), in which the key lettering is different. The delta lobe is now defined by the letters L–L′, with L′ defining approximately the boundary between the foresets and the bottomsets, which occupy the interval L′–U. Above L, horizon P marks the top of the delta plain topsets, and C indicates an infilled distributary channel. At S, a sigmoid clinoform event can be seen that is truncating beneath a topset event, while at O, another clinoform is truncating beneath a topset event at a different level. Both of these clinoform events are tangential oblique reflections, which approach the bottomset events in smooth curves. The topset event against which event O truncates itself, if followed to the right, begins to pass smoothly into a sigmoid clinoform, but is interrupted by what is probably another infilled channel (arrows). Clearly, the upper zone of a deltaic sequence may be stratigraphically highly complex, and may contain many potential truncation trap situations of a subtle type. It should also be remembered that the deltaic environment is essentially unstable tectonically, and localized movements may cause sudden changes in sea level. In some areas, such as the Central Appalachians (Dennison, 1971), there were also many eustatic sea-level changes that could cause a shoreline to shift from tens to hundreds of miles in a 'geologically instantaneous' time period. Clearly, this would have fundamental influence on the stratigraphical detail in the upper zone of a delta.

As will be realized from the discussion so far, the detailed shape of the prograding clinoform events that are the seismic expression of the deltaic foreset bedding can be of considerable analytical value in determining the environment of deposition of these units. In particular, the upward and downward terminations of these events tell much about the environmental energy of deposition. Where an individual clinoform can be traced continuously from a near-horizontal topset through a foreset into a near-horizontal bottomset event, this is good evidence for a low-energy environment, and one likely to be 'shale-prone' rather than sandy.

Where clinoforms terminate upward in truncations, in various patterns (e.g. oblique, complex sigmoid oblique, oblique parallel, and shingled), several other conclusions may be drawn. Upward truncations generally suggest higher energy environments that may be sand prone. The oblique parallel pattern (cf. Fig. 9.4) and the shingled pattern referred to earlier (particularly the latter) are taken to be typical of foreset beds in wave-dominated deltas, such as that of the River Rhone in France.

No more will be said in this chapter about clinoforms, although they are of great importance. Very full and detailed discusssions can be found elsewhere (e.g. Mitchum *et al.* (1977a) and Berg (1982) for a brief summary).

The distributary channels carrying sediment to the delta front across the delta plain are cut down into the silts and clays of the latter. The material they are carrying includes coarser sandy sediment, which is the first to be shed at lesser distances from the source, and where any checks occur in the flow of channel water. Thus the infill material in the channels

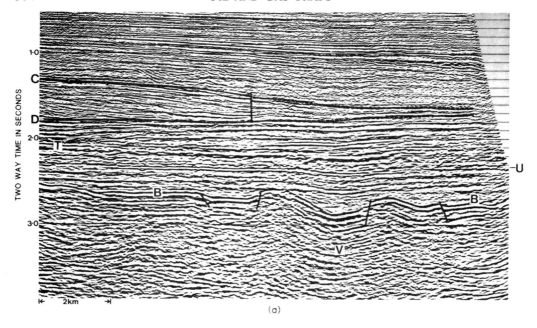

Fig. 9.5. (a) Seismic section showing the termination of a distal delta lobe (C–D) in the Central North Sea area. Reproduced by permission of Seismograph Service (England) Ltd

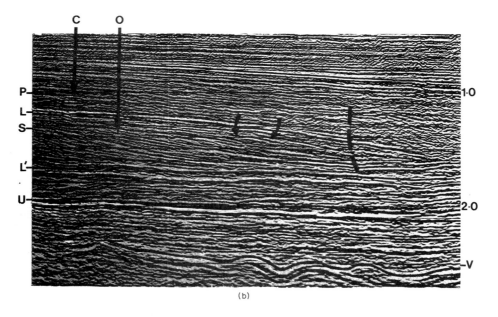

Fig. 9.5. (b) A more detailed view of the delta lobe seen in Fig. 9.5 (a). Reproduced by permission of Seismograph Service (England) Ltd

tends to be different from the adjacent delta plain materials. On the seismic section, this can produce distinctive features where the distributary channels are present, and particularly where they are seen in cross-section.

Since the major distributaries will tend to flow longitudinally along in the main direction of extension of the delta into the basin of discharge, there will be a tendency for any cross-sections of these major channels to be seen in seismic sections in directions *across* a delta lobe, rather than along it. This is illustrated by the examples of Fig. 9.6 and Fig. 9.7.

In Fig. 9.6, a seismic section longitudinally along a delta lobe is seen (cf. a section parallel to plane B in Fig. 9.1) with horizon T indicating the top of a zone of deltaic topset beds. The letter A denotes the top of a zone of foreset beds with somewhat irregular clinoform events dipping to the right over parts of the section. (NB there is a second, earlier delta lobe beneath, at TL (top of topsets) and AL (top of foresets), rather less well resolved.)

Just below the level of A, there is evidence of a major 'buried' distributary channel within the foreset sequence (arrow); from the shapes involved, it seems that the section is crossing the remnants of this major distributary at a fairly acute angle. Where it might reasonably be expected that other evidence of more recent distributaries would be present (i.e. within the sequence of delta plain topsets between T and A), there appears to be no recognizable vestiges of such features. It is, of course, possible that some fragmentary portions of

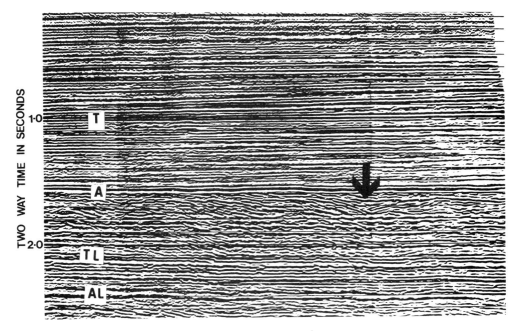

Fig. 9.6. Seismic section running longitudinally along a delta lobe. T is the top of a zone of topset bedding events, and A the top of a foreset bedding zone. What is thought to be another, thinner delta lobe is present below, between TL and AL. The arrow indicates some evidence of an infilled distributary channel of large size in the foreset zone; this channel seems to be crossing the line at a small angle. There is no good evidence of infilled channels in the delta plain topsets at and below T, however (cf. Fig. 9.7). Reproduced by permission of Seismograph Service (England) Ltd

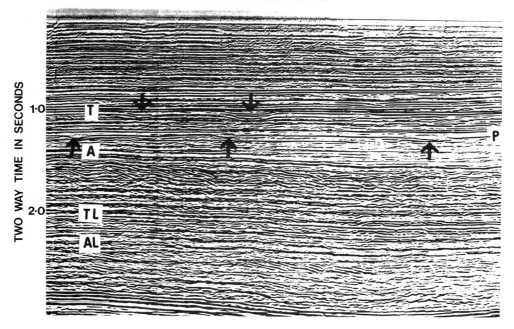

Fig. 9.7. Similar seismic section to that in Fig. 9.6, but this section is running in a transverse direction *across* the delta lobe (same notation as in Fig. 9.6). In this case, there is clear evidence of infilled distributary channels in the top of the delta plain topsets (downward pointing arrows indicate examples), with older infilled and buried channels at greater depths (upward-pointing arrows). Note the mounded, or hummocky appearance of the foreset bedding events (below A) in this section across the lobe. Reproduced by permission of Seismograph Service (England) Ltd

longitudinal sections of channel may be present which, because of the infill geometry and the direction, do not stand out against the background of flat topset events.

In Fig. 9.7, however, a different situation is seen. This is a transverse section *across* the same delta lobe as for the example in Fig. 9.6, and the lettering is the same, in general, as for the latter example. The section is, in other words, parallel to the plane A in the Fig. 9.1 schematic block diagram. Note the clear indications of the infilled distributary channels in cross-section at the upper surface T of the topsets (as indicated by the downward-pointing arrows) and also earlier, part-eroded and buried examples (upward-pointing arrows) at an earlier level (P) of the delta plain depositional surface. Note also the irregular, mounded appearance of the foreset events below A, which are now being seen in transverse section across the delta lobe.

In passing, it should be stated that it is often difficult to recognize the bottomset sequence in a delta, or where part of it is recognizable, to determine the vertical extent of the sequence. The bottomsets in more distal locations are usually formed of turbidites that represent the 'fines' carried out at an earlier stage beyond the delta front, which are later overlaid by the advancing foreset zone bedding. Sometimes slope instability in these sediments results in slump/slide structures (expressed often as an apparent ripple where the interval is a thin one), which may help in their identification.

The tripartite division of a delta system is not always obvious, and the expression of a delta in seismic data is frequently less clear than in the examples seen so far. Sometimes the only clues are the presence of a relatively rapid thickening in a certain interval on the section, a mounded appearance, and (perhaps the best indication) prograding clinoform events within the thickened interval, marking a zone of foreset bedding.

Examples of this type are seen in the next two illustrations. In Fig. 9.8, a Lower Cretaceous delta in the Porcupine Basin to the west of Ireland is indicated in the line-drawn interpretation below the seismic section, and beneath the Top Chalk marker horizon. A thickened mound of material, inside which there are the traces of some truncating, prograding clinoform events, can be seen. The normal faulting affecting the Top Jurassic and lower levels may be the result of the progressively increased loading of the deltaic material downdip into the basin. As in the next illustration, the bottomset and topset zones are indicated by the letters B and T.

Another Porcupine Basin area example is seen in Fig. 9.9, in which a longitudinal section through a Tertiary delta sequence is present (see the line-drawn interpretation below the section). There is a fairly gentle thickening together with clear developments of prograding clinoforms within the deltaic interval. In this, and in the previous example, there is first a general thickening, then a thinning, in a basinward direction. Both these examples are from the paper by Croker and Shannon (1987). These deltas owe their origins to the rapid deposition of plentiful sedimentary material from the margin of the basin. In the Porcupine area, the Lower Cretaceous and the Tertiary deltaics contain sands of reservoir quality in the foresets and topsets, and sealing cap rocks of claystones and shales are present. The full prospectivity of this area has yet to be determined.

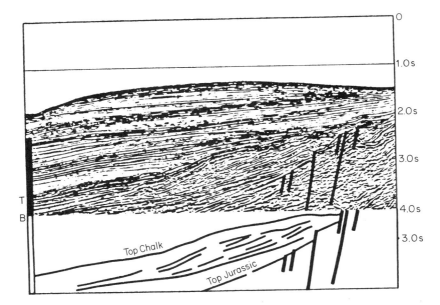

Fig. 9.8. Seismic section showing a Lower Cretaceous delta in the Porcupine Basin, with a line-drawn interpretation below. T = topset; B = bottomset. Reproduced by permission of Graham & Trotman Ltd (from Croker and Shannon, 1987)

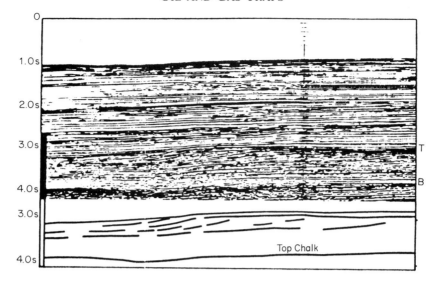

Fig. 9.9. The dips of the clinoform events in the foreset zone signify the presence here of a Tertiary deltaic sequence. From the same area as Fig. 9.8. Reproduced with permission of Graham & Trotman Ltd (from Croker and Shannon, 1987)

The reservoir units that may be associated with a delta system are many and varied. The sand bodies associated with distributary channels, offshore sand bars along the delta front (beach bars, barrier bars, stream-mouth bars, etc.), and foreset bedding have been mentioned as but a few of the potential reservoirs. As noted, a great many of these require detailed wireline log analysis in addition to, or instead of, seismic interpretation, which can normally only identify the overall context into which these details are fitted.

The two collections of papers on Modern and Ancient Deltas edited by Le Blanc (1976a,b) are strongly recommended for further reading on the geology of deltaic sequences in various parts of the world. In the context of this chapter, there has been no discussion of growth faulting, which is commonly related to delta development; the subject of growth faulting was, however, dealt with separately in Chapter 4, and should be referred to in conjunction with the material in this chapter.

9.3 Submarine Fans: Introduction

As noted earlier, submarine-fan sediments can be regarded as marine equivalents of the outwash fans present in many piedmont areas on land, where sedimentary materials being transported under gravity by wind or water are funnelled through a gully or notch in a highland area, and spread out on to adjacent lowland plains.

There are two types of environment in which submarine fans quite commonly develop. In the first of these, a canyon is cut down into a shelf zone, and traverses part of the shelf, usually crossing the shelf break to funnel sediments down the slope. Sometimes the resulting sediment fan is located partly on the shelf, partly on the slope, and sometimes wholly on the slope or partly on the floor of the basin at the bottom of the slope. The basin may be shallow and inshore, or a deep ocean basin. Frequently, canyons of this type are solitary features, and the associated fan is therefore a single, large body of sediment.

In such cases, the upper (inshore) part of the fan comprises one channel (or a small number of channels) of large size, through which, typically, turbidite sands are transported out to the middle and lower parts of the fan. These sands eventually form the channel infill that evolves as the potential reservoir buried within the more silty general matrix of the fan.

As the sediments are transported to the middle and lower fan, the channel(s) become a braided and meandering network, tending to spread the sands in flattened, lenticular pods known as 'suprafan lobes'—forming patches of reservoir quality in the lower fan, rather than the restricted channel sands found in the upper fan.

Turbidite sediments are not found solely in such morphologically distinct fan features. In oceanic basins, turbidite sheets lacking any distinct channels may develop. These often form the argillaceous fines and oozes found in this environment, and do not normally develop any good quality reservoir intervals or bodies.

Occasionally, contourite mounds found at the base of a slope in a marine basin are mistaken for fan deposits. They can usually be distinguished on seismic section by their geometry, relation to adjacent seismic events, and position in the basin.

The submarine fans associated with canyons tend to be large single features, as noted. The second common type of environment—fronting a delta system—produces a different set of features. Here, delta-front canyons of relatively small dimensions tend to develop on the slope in front of the delta, adjacent to the delta, or in a small delta-front trough. This leads to the presence of several small fan deposits that often coalesce laterally to form a fan complex that extends along the delta front just basinward of the latter. These fan complexes are found associated with a number of deltas, where sand carried to the delta front by the deltaic distributaries is funnelled down the canyons ahead of the delta. Sometimes the outbuilding delta overrides the earlier developed fan complexes, which then become the prodelta turbidites and silts that can be seen often in the seismic data beneath the foreset bedding events. Sometimes sand is also carried by longshore currents laterally to be captured by canyons that exist offset laterally from the delta proper.

A simplified sketch of the two environments just mentioned is shown in Fig. 9.10. Note that these are far from being the only type settings for fan deposits. Series of coalescing fans may form along the base of a submarine fault scarp, for instance. The sand bodies in these fans may, after burial, form reservoirs (as in the Brae Field, North Sea, where the downfaulted block formed a shallow basin in which the fan deposition took place).

9.3.1 Brae Oilfield—a Fan-complex Reservoir

The depositional environment of the Brae oilfield, in which an Upper Jurassic reservoir is developed on the western margin of the southern Viking Graben, is an interesting example of fan deposition. It has been studied by, for instance, Stow *et al.* (1982), who describe the situation as that of three or more small overlapping submarine fans forming 'a complex sediment apron along the faulted scarp margin' (at the edge of the Fladen Ground Spur). These small fans, which extend from 5 km to 10 km to the eastward of the base of the fault escarpment, were deposited mainly below wave-base in a shallow basin. Their development seems to have been controlled by tectonism (fault movements and basin

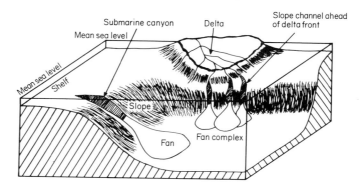

Fig. 9.10. Simple diagram showing the two most common situations in which submarine fan deposition takes place: (i) single, large fans form downslope from submarine canyons that cut down into the shelf/slope zone; (ii) smaller, multiple fans, often coalescing to form a fan complex, form downslope from a delta front, or in a delta-front trough

subsidence beginning in the Oxfordian coincided with coarse clastic sedimentation). Repetition caused six (overall) upward-fining megasequences to evolve.

Reservoirs present include breccia–conglomerate facies deposited as rock-fall avalanches at the base of the fault scarp, inner fan channel and lobe conglomerate–sandstone assemblages due to grain flow/liquefied flow/turbidity flow processes, and massive sandstones–pebbly sandstones deposited in channel mouths.

Turbidity currents in interchannel, interfan, outer fan and basinal settings (see earlier discussion) resulted in mudstone facies deposition.

Stow *et al.* (1982) summarize in a cartoon (see Fig. 9.11) the various controls that influence submarine fan development. These are represented in the illustration as: 1a, original source-rock type; 1b, climate and weathering; 1c, continental plain and shelf widths; transport methods; 1d, transitional source-sediment type; 1e, resedimenting process; 1f, slope, gradient, and width. 2a, tectonic style in original source area; 2b, tectonic style in transitional source zone; basin subsidence; 2c, margin type; 2d, receiving basin type (size, water depth, shape, etc.); 2e, sea-level variation. 3a, fan geometry, feeder system; 3b, marine conditions; 3c, marine biota. Refer to the original paper for a discussion of fan classification, and of generalized fan types resulting from variations in the controls mentioned here. It is of interest to note how many of the controls listed are also applicable in the case of a deltaic deposystem.

9.3.2 Identification in Seismic Data

The recognition of turbidite bodies and submarine fans in seismic data has been discussed by several workers, such as Sangree and Widmier (1977). The basic concepts of sediment supply through channels or topographic lows in the shelf/slope areas, together with mounded seismic facies features due to the build-up of sediment lobes, are of importance in recognition of fans and turbidites on seismic sections.

The topographic lows—canyons or notches—traversing shelf or slope zones will be seen more easily on seismic sections that traverse these features at right-angles, i.e. seismic lines

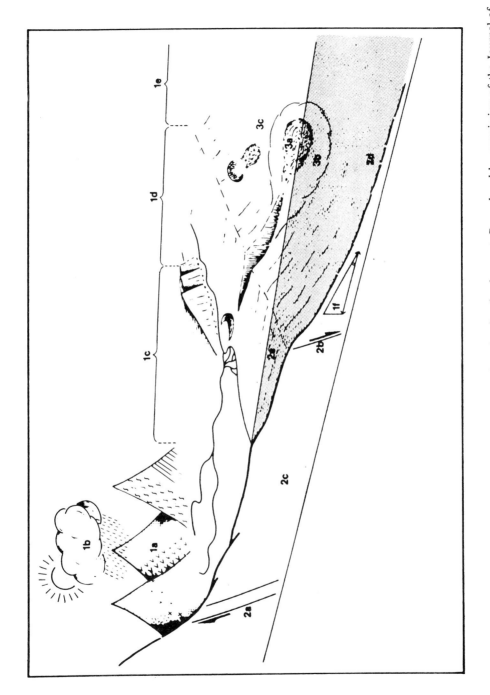

Fig. 9.11. A cartoon summarizing the controls that influence submarine fan development. Reproduced by permission of the Journal of Petroleum Geology (from Stow et al., 1982)

that are nearest in direction to the shelf or slope contours, following topographic 'strike'. In some instances, where the canyons or notches occur in the edge of a buried escarpment (a situation that exists in places on the western margin of the Central Graben of the North Sea), the canyon profiles may be seen as ghostly 'side-events' on seismic lines running parallel to, and near, the buried escarpment. These canyons, which usually end by being filled with clays and silts, tend to show onlap-fill in transverse section after the fan deposition has ceased. In longitudinal profile, they often show a prograding, shale-prone configuration.

The upper, middle, and lower fan zones may all show mounding in the seismic data, this being one characteristic of turbidites that is frequently seen. Amongst these mounds, overlying events tend to onlap or downlap on to lower events (see Mitchum, 1985).

The upper fan usually consists of one or two thick channels, which may be sand-prone, and which are bounded by levees of shale-prone material. The channels and levees will increase in number, but decrease in size, further down the fan. Channels and bounding levees in this zone tend to form concave-upward shapes in their early history. With increasing depth of burial and increasing compaction, however, they often tend to become lenticular and flatten out in cross-section, normally ending as more mound-shaped features.

Often, the lower fan is sand-prone and an overall mound-shape, with a transverse profile that is convex upward. It is here that the increasing number of sediment-transporting channels form braided and meandering patches or pod-like zones ('suprafan lobes') of sandy material that form reservoirs, contrasting with the channel-fill reservoirs that occur in the upper fan areas.

Where a major fluvial-type delta deposystem has built out into a basin for a large distance, it has usually developed outwards on top of a turbidite sequence or a series of such sequences. In the earlier part of this chapter, dealing with deltas, it was noted that the prodelta zone of turbidites could become increasingly thick in the distal parts of the system. These sheet-like turbidite sequences can be very difficult, or impossible, to identify without the benefit of well data, as seen in Fig. 9.12. This shows two turbidite intervals marked t–T in an area covered by a major delta/turbidite complex. The letter A marks the top of a delta plain sequence, and B the top of a delta front sequence; the seismic line is crossing the delta lobe here at an oblique angle. Note the indications of infilled distributary channels at level A across the section.

Any discussion involving talk of 'sand-prone' and 'shale-prone' features and the like, must always be accompanied by some cautionary words. It must be remembered, when attempting identifications in seismic data, that if the sediment source is not sandy, then any 'sandy-prone' seismic facies identified in the data will not contain much sand. Seismic interval velocities may help in some cases, but often, there is no substitute for drilling information.

9.3.2.1 *Effects of burial compaction*

On more than one occasion, seismic data has been categorized as having a hummocky, or mounded appearance (on the small, as well as the large, scale), and intervals showing mounded internal reflections can be typical of some turbidite deposits. There are many possible reasons for this type of seismic facies, both during and after deposition. During deposition, changes may occur in current speed and direction, mass movement (slumping, sliding) of materials may take place downslope, and other processes typical of high-energy

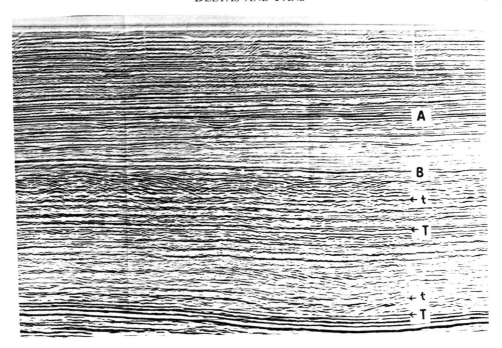

Fig. 9.12. The difficulty of identifying turbidites from seismic data alone. From information in the adjacent areas, it is noted that the two intervals marked t–T are turbidite sequences. Normally, it is difficult or impossible to identify turbidites without the benefit of well data. Sometimes, location related to shelf/slope, or to a deltaic sequence may give a clue. See Berg (1982) for some other examples in seismic data. Reproduced by permission of Seismograph Service (England) Ltd

depositional environments may lead to this type of seismic interval. After deposition, differential compaction may lead to mounded features. For instance, a buried channel, depending on the type of infill material, the burial depth, and the average overburden density, may become an inverted, mound-shaped feature, and there can be many of these in a given stratigraphic interval. In Fig. 9.13, the sketches show how, owing to excessive compression and differential compaction under large burial stresses, this type of inversion may occur. Although many channels retain (approximately) their original cross-sectional shape (Fig. 9.13 (a)) with only minimal compaction effects, some channels (infilled, perhaps, with clay or shale that later becomes plastic) may become lenticular (Fig. 9.13(b)) in certain cases, or even biconvex upwards, as in Fig. 9.13(c). In a seismic section, the seismic velocities involved may have at least some influence on the final shape of the feature.

Such shape changes are not uncommon with deep burial of 'negative' features like buried channels, and may also occur in relation to 'positive' features, such as channel margin levees. Often the deformed shapes are very difficult or impossible to recognize in standard seismic sections, due to their relatively small scale, and can be almost as difficult to interpret

314   OIL AND GAS TRAPS

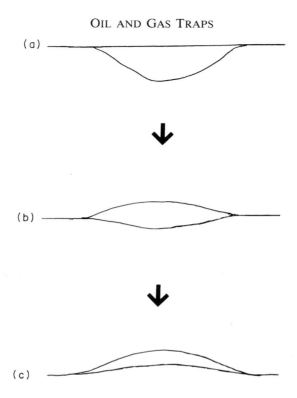

Fig. 9.13. Diagram showing how the effects of burial compaction on the infill material in some buried channels can invert the channel relief from a basically negative feature as in (a), through a lenticular shape (b), eventually to a mound shape as at (c)

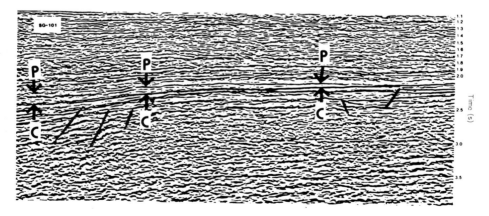

Fig. 9.14. Seismic section longitudinally through the distal part of a fan lobe that feather-edges to the right. See text for explanation. Reproduced by permission of the Journal of Petroleum Geology (from Fagerland, 1984)

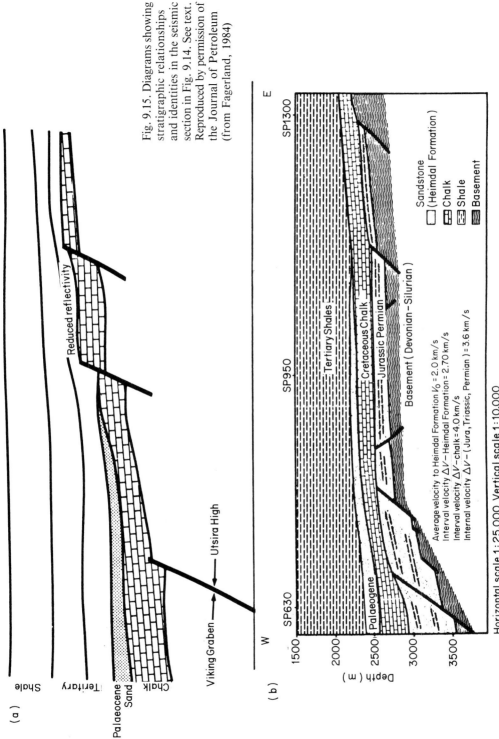

Fig. 9.15. Diagrams showing stratigraphic relationships and identities in the seismic section in Fig. 9.14. See text. Reproduced by permission of the Journal of Petroleum (from Fagerland, 1984)

from borehole information, even when the subsurface sampling is good. A very interesting case of this type is discussed by Héritier *et al.* (1981) in connection with a deep-sea fan channel in the Frigg gasfield, North Sea, where gas occurs in a lobate submarine fan of Eocene age.

It is noted here that a number of features mentioned elsewhere in the book in relation to other trap forms (such as Frigg, Forties, and Brae Fields, for example) involve reservoir intervals consisting of submarine-fan sediments; in these cases, the fact that fan deposits are involved is largely of sedimentological interest rather than from the viewpoint of fan morphology.

9.3.2.2 Longitudinal seismic section of fan

Not proved at the time of writing (Fagerland, 1984) as containing a hydrocarbon accumulation, but showing the characteristics of a good structural–stratigraphic trap, is a fan feature in Norwegian offshore block 16/4. In the southern part of the Viking Graben, the feature consists of a submarine fan sequence of Heimdal Formation sandstone (Palaeocene) capped with a Palaeocene tuff and Tertiary shales above, forming an adequate seal.

Axial zone subsidence in the Viking Graben in late Palaeocene/early Eocene times, combined with some faulting down into the Graben, led to the present configuration of the sand wedge of the fan, seen in approximately longitudinal profile in Fig. 9.14, and indicated by the P–C intervals.

The fan deposition took place from the Shetland Platform area off-section to the west (left), towards the Utsira High (towards the right in Fig. 9.14). In the seismic section, we see part of the distal lobe of the fan (the 'lower fan', probably made up largely of 'suprafan lobes' of sand, feather-edging towards the right-hand end of the section. The graben subsidence and faulting have resulted in the fan showing the 'critical' west dip seen, which in association with the pinching-out of the lobe updip, local closures on the body of the fan, and the seal formations, make for a trap situation. From evidence in adjacent blocks, there is both oil and gas potential in the area.

This example illustrates the good potential for trap formation in these wedging sands, especially where tectonism has effectively reversed the normal dip situation so that the wedging-out occurs in an updip direction in a basin, or fault traps occur.

The sketch in Fig. 9.15 (upper) shows the relationship of the Palaeocene sand wedge to the Top Chalk (which is the reflection event marked C in Fig. 9.14). The sand can be seen to terminate, effectively, against a fault displacing the Top Chalk, which is observable in the Fig. 9.14 seismic section. Other stratigraphic relationships to be seen in the seismic section of this example are indicated in Fig. 9.15 (lower). The two strong reflections seen below event 'C' in Fig. 9.14 are the Base Cretaceous and the Base Zechstein. Note that in transverse section, the fan is a wide, shallow mound shape.

PART IV
COMBINATION AND COMPLEX TRAPS

CHAPTER 10

Combination and Complex Traps

10.1 Introduction

In the terminology used in this book, combination traps are those that involve strong structural *and* stratigraphic elements in their make-up, while the term complex traps simply indicates that the traps have been formed in what are recognizable as complex geological situations, such as intense overfold and thrust belts.

A combination trap is the result of a multistage geological history. Sometimes the stratigraphic element is the first to develop, perhaps in the form of an edge-of-permeability zone in the reservoir rock (due to facies change, for instance) followed by deformation and the introduction of the structural element. This may perhaps be combined at a late stage with fluid-flow effects, such as the downdip flow of formation brine in the reservoir interval.

On the other hand, the structural element may be developed first, followed at a later stage by diagenetic effects that provide the stratigraphic aspects of the trap. Sometimes the latter may be influenced by the inhibitory effects on diagenetic processes of hydrocarbons present, or migrating into the trap.

Many of the situations discussed in earlier chapters—for example, certain trap types associated with salt movement and diapirism—are in effect combination traps, although they were placed in other classes to assist in the development of the discussions.

'Complex traps' is a somewhat equivocal classification, since there is no clearly marked division between these and some other trap types. In the context of this book, however, it is applied to traps found in thrust and fold belts, such as that found in the Rocky Mountains of Alberta, Canada and parts of the USA. The trap may be associated with an overthrust fold, or with the thrust fault plane itself, either above or below. Again, there is the possibility that some of these traps could be reclassified as 'combination traps'.

10.2 Combination Traps

10.2.1 Permeability Effects

When an 'edge-of-permeability' occurs updip in a reservoir, and this intersects with a fault, then a combination trap may result. This is modelled in the seismic example of Fig. 10.1, where there is supposed to be a permeability trap (W) against a large, right-dipping fault bounding the graben H above a salt diapir D. The fault should be imagined as crossing the seismic line at an angle (which may distinguish it from a simple fault trap).

Levorsen (1967) sketches a similar situation (Fig. 10.2(a)), which he relates to the structure of part of the Rodessa oilfield of Louisiana and north-east Texas shown in Fig. 10.3, where an oolitic limestone member of the Glen Rose Formation (Lower Cretaceous) is cut by the Rodessa Fault, forming one of the pools in the field in the angle between the fault and the edge-of-permeability. Depth-structure contours on top of the oolite member are seen in Fig. 10.3.

The diagram in Fig. 10.2(b) shows arching across an updip edge-of-permeability. A combination trap may be found where, for example, the updip edge-of-permeability is provided by the intersection of two unconformities that contain between them the permeable reservoir interval. Such an intersection is modelled in Fig. 10.4(a), where the white arrows indicate two unconformity surfaces converging updip in a Triassic series. Another example is shown in Fig. 10.4(b).

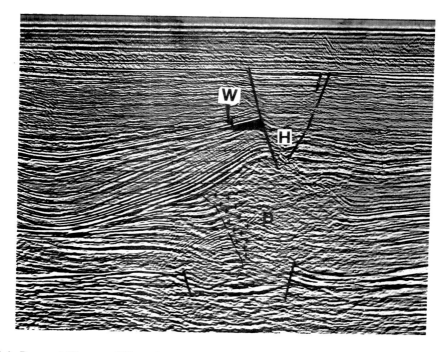

Fig. 10.1. Permeability trap (W) against a fault that is crossing the seismic section at an angle. H is a collapse graben above the salt diapir D. Reproduced by permission of Seismograph Service (England) Ltd

COMBINATION AND COMPLEX TRAPS

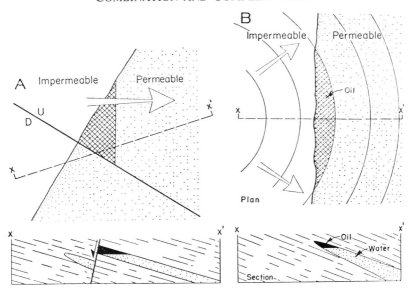

Fig. 10.2. Sketch (a) shows an edge-of-permeability trap, as modelled in the seismic line in Fig. 10.1. Note the angle at which the fault crosses the section line. Sketch (b) shows the effect of arching across an updip edge-of-permeability, which could be due, for instance, to the convergence of two unconformities, with the permeable zone between them. From *Geology of Petroleum*, 2nd edition by A. I. Levorsen. Copyright 1954 © 1967 W. H. Freeman and Company. Reprinted with permission

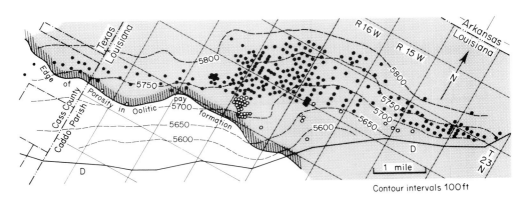

Fig. 10.3. A fault-related edge-of-permeability situation, as modelled in Fig. 10.1, and sketched in Fig. 10.2(a). This is the Rodessa oilfield of Louisiana and north-east Texas, where the Rodessa Fault (D) cuts the oolitic limestone member of the Glen Rose Formation (Lower Cretaceous) that forms the reservoir. From *Geology of Petroleum*, 2nd edition by A. I. Levorsen

An actual example of a trap of this type is illustrated in Fig. 10.5, again from Levorsen, after Minor and Hanna (1941). The depth-structure contours are on top of the (Cretaceous) Woodbine sand reservoir of the East Texas oilfield—one of the largest oilfields ever discovered. The smaller inset sketch shows the convergence of the two unconformities updip

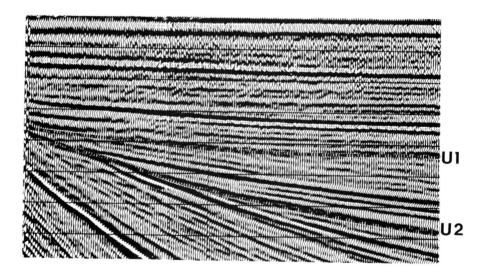

Fig. 10.4. (a) Convergence of two unconformities modelled in a seismic section (indicated by the white arrows). This is one of the situations that can lead to the presence of an updip edge-of-permeability. Reproduced by permission of Seismograph Service (England) Ltd. (b) Another example of convergent unconformities (U1 and U2). Reproduced by permission of Seismograph Services (England) Ltd

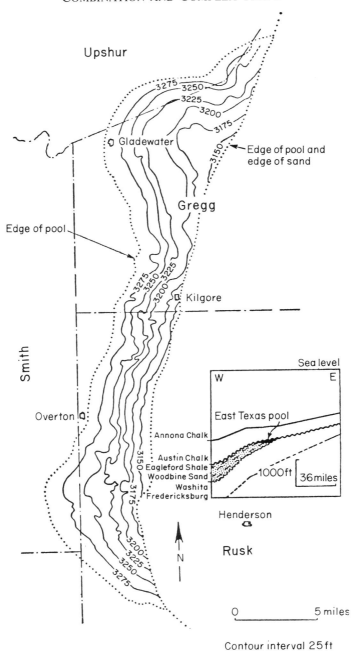

Fig. 10.5. Sketch map and section of depth-structure contours on top of the Woodbine Sand reservoir in the East Texas Field. Two converging unconformities (see section inset) enclose the reservoir, and form an edge-of-permeability trap. From *Geology of Petroleum*, 2nd edition by A. I. Levorsen. Copyright 1954 ©1967 W. H. Freeman and Company. Reprinted with permission

that contributes to the trap as the interval between them feather-edges along the eastern margin of the hydrocarbon accumulation.

The combination of an updip permeability wedge-out, and folding or faulting, is responsible for many important combination traps, such as the Poza Rica Field in Mexico, and Mene Grande, in Venezuela. Another element that may enter into combination trapping is that of hydrodynamic fluid flow, and little work has been done on this aspect. Another factor influential in combination trap formation is the presence of mobile salt in the stratigraphic column. Many of the traps discussed in the chapter devoted to plastic deformation (Chapter 5) are effectively combination traps.

10.2.2 Faults and Block Faults

A typical and important combination-trap situation occurs in eroded and tilted fault blocks. In the North Sea area these are often referred to as 'Brent-type' traps, after the eponymous Brent Field structure, which is a large example of the type.

Tilted (sometimes rotated) fault blocks tend to occur in areas of extensional teconics—zones of rifting at plate margins being typical. Such zones often form basinward of a shelf edge, and in the North Sea Jurassic include three basic kinds of trap (as noted by Selley, 1976): (i) sands thickened and banked on the downthrow side against a major fault scarp; (ii) traps above faulted, horst-block anticlines; and (iii) traps in rotated or tilted fault blocks, sometimes with the crestal part of the structure planed off by erosion at an unconformity. In such cases, both structural and stratigraphic elements thus come together to form combination traps.

In the North Sea, tilted fault-block traps occur in several fields, such as Brent, Hutton, Dunlin and Thistle. In these, the main reservoir is of (Jurassic) Bajocian–Bathonian sands, the structural crests being truncated by the 'Late Cimmerian' Unconformity (late Jurassic–early Cretaceous). The movements that produced the block faulting were also of 'late Cimmerian' age, and the blocks were only partly eroded before the Cretaceous transgression and deposition. Where erosion in post-Jurassic time was not severe, Jurassic deposits were left on crests and/or on flanks of the buried, tilted blocks. The overlying, unconformable Cretaceous sediments provide the sealing formation.

Selley (1976) notes that the reservoir thickness in these structures is not always proportional to its structural elevation; altogether there may be a reservoir section at the crest of the tilted block structure, it may be attenuated or totally absent down one flank of the faulted block (usually the steeper, faulted side in the asymmetric structure). In another paper (1975b), Selley suggests that this may be due to the major fault movement plane migrating intermittently towards the structural crest (i.e. faulting 'towards the upthrown block' typical of the western margin of the Viking Graben in the North Sea) as the structure evolved.

Other, similar structures are present in the general area, but with different reservoir sands. The lower reservoir of the Brent oilfield, for instance, comprises Rhaetic basal transgressive sands of the Statfjord Sandstone Formation, while (rather rare) Upper Jurassic Oxfordian/Kimmeridgian shoal sands some 80 m thick form the reservoir in the Piper oilfield (Williams et al., 1975), another Brent-type field, although with very complex structure and stratigraphy.

Of the three trap types mentioned previously as being found in the northern North Sea

Jurassic (i.e. (i) thick sands banked against a major fault at the margin; (ii) anticlinal traps associated with horst block faulting; and (iii) tilted, eroded block traps), the first two types may be classified in various ways. For instance, type (i) may be a complex fan-deposit trap, and type (ii) a pure, anticlinal trap. However, owing to (e.g.) sealing by an unconformable unit, complication by faulting, facies change, or diagenetic effects, both types may also be combination traps. In the present context, since types (i), (ii), and (iii) are often found in the same regional situation, they will be treated here as combination traps. Examples of types (i) and (ii) will be shown before dealing with type (iii).

An example of where Middle and Upper Jurassic sands are piled against a major fault is found in the so-called 'transitional shelf' area. This is located at the western margin of the Viking Graben in UK North Sea Quadrants 2 and 3, east of the Shetland Islands. This shelf is juxtaposed in the west with the East Shetland Platform, from which it is separated by a large fault throwing down to the north-east. Against this fault, on the downthrow side, a substantial thickness of Upper Jurassic sandstones (together with more restricted Lower and Middle Jurassic sands beneath) has been deposited. A Middle Jurassic (Callovian) sandstone—the Emerald Sand—is productive in a fault-controlled oilfield (Emerald) at the northern end of the shelf. This unit is overlain and sealed by the Heather siltstones (upper Middle Jurassic) and, in places, the Kimmeridge shales (Upper Jurassic). There are no mature source rocks on the transitional shelf. The main oil source is (from biomarker evidence) the Kimmeridge shale at depth in the Viking Graben to the east, from which the oil migrated up the bounding faults. A diagram section is shown in Fig. 10.6, which depicts the shelf, the major bounding fault to the west, and the location of the Emerald oilfield (which is purely fault-controlled, and not a combination trap). A corresponding seismic section across the shelf is seen in Fig. 10.7, and has the location of one of the Emerald Field wells (2/10a-4) marked. It will be noted that the horizontal scale on the seismic section is compressed, so that in comparison with the sketch in the

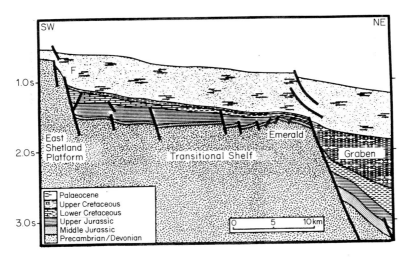

Fig. 10.6. Shelf edge with major down-to-basin fault F, showing thickened pile of Upper Jurassic sands on the downthrow side of F—a situation in which combination traps frequently occur. Reproduced by permission of Graham & Trotman Ltd (from Wheatley et al., 1987)

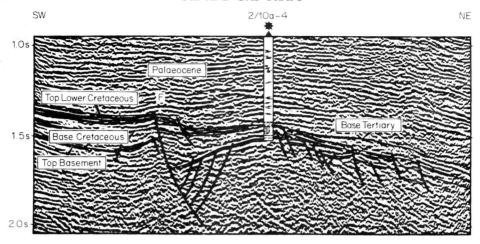

Fig. 10.7. Seismic section, equivalent of the schematic in Fig. 10.6, showing the corresponding fault F together with the thickened Jurassic interval on the downthrow side of the fault. Well 2/10a-4 is drilled on the Emerald Field structure. Reproduced by permission of Graham & Trotman Ltd (from Wheatley et al., 1987)

previous figure, the shelf in Fig. 10.7 appears laterally shortened. However, the main fault on the left, separating the shelf from the East Shetland Platform to the south-west is clear, as are the thick shelf deposits of mainly Jurassic sediments on the downthrow side of the fault, and the faulted high at the well location. This, then, is a model of the type (i) situation, where combination trapping is possible in the sands on the downthrow side of the fault and elsewhere. The area is well described by Wheatley et al. (1987).

Another, similar situation is seen in the diagram section of the South Brae oilfield in Fig. 10.8. This feature, again at the western margin of the Viking Graben, has Upper Jurassic Brae Formation rocks downfaulted against the Fladen Ground Spur to the west. The reservoir units comprise a thick pile of sand-matrix conglomerates and sandstones alternating with other thick units of mudstone and sandstone. The coarse-grained units have sharp, erosive bases, and occur in both extensive sheets and channel-fill bodies extending radially basinward. They appear to have been deposited by a range of subaqueous transport mechanisms (high-density debris flows, turbidity currents, etc.). They represent deposits on the steep proximal slopes of a submarine fan complex. There are local seals within the fan deposits, and the Upper Brae Formation is sealed by the Kimmeridge Clay (see Turner et al., 1987).

Passing on to an example that models a type (ii) feature, Fig. 10.9 shows a section across the Humbly Grove oilfield, Hampshire, onshore southern England. Although not in the same North Sea setting as the previous examples, this field is on the north-west margin of the Weald Basin, has a horst-block structure (the bounding fault in this case being listric), and a Jurassic reservoir, and so will serve as a model to illustrate this next type of structure that could involve combination trap situations.

Late Triassic/early Jurassic crustal extension in a N–S direction resulted in extensive E–W-striking listric faulting in Jurassic and early Cretaceous times (Hancock and Mithen, 1987). The majority of faults hade to the south with, in places, some north-hading

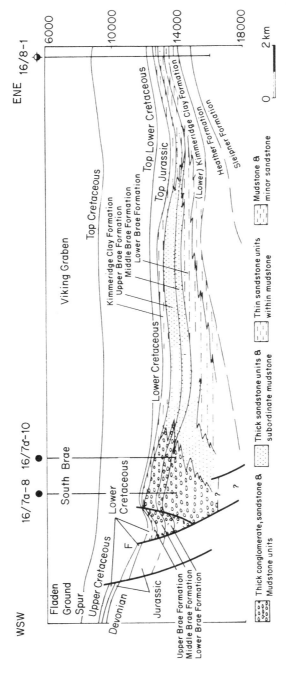

Fig. 10.8. Section through the South Brae Field, again showing a thick pile of sand and conglomerate on the downthrow side of a major shelf-edge fault. These thickened units include the South Brae oilfield reservoir units. The shelf-edge fault is marked F. Reproduced by permission of Graham & Trotman Ltd (from Turner et al., 1987)

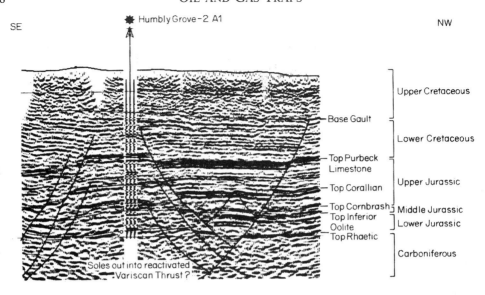

Fig. 10.9. Seismic section across Humbly Grove oilfield on land in the south of England. The structure is a tilted horst block and the main bounding faults of the horst are listric. The reservoir is in oolitic carbonate sands of Bathonian age. Note the strong Purbeck Limestone marker. Reproduced by permission of Graham & Trotman Ltd (from Hancock and Mithen, 1987)

antithetic faults, as in the example shown. The south-hading listric normal faults are thought to sole-out at deep-seated reactivated Variscan thrusts in the underlying Palaeozoic sequence. In Fig. 10.10 a schematic depth-structure map of the field is shown, indicating the GOC, the OWC, and the location of the seismic line (A–A′ approximately) seen in Fig. 10.9. The reservoir sequence is in oolitic carbonate sands of late Bathonian age. There are several strong seismic markers present, the Purbeck Limestone being a notable example. Study of potential source rocks by vitrinite reflectance/TAI together with present depth suggests that a minimum of 2000 feet (610 m) of Cretaceous and Tertiary overburden has been removed by erosion. Again, combination trap situations are possible in this modelled example due to a variety of causes, such as unconformity sealing over the crest of the horst block.

10.2.3 Tilted and Eroded Fault Blocks

In the northern part of the North Sea, UK sector, a very important factor in the formation of Jurassic oil and gas traps is the presence of the major regional unconformity of late Jurassic/early Cretaceous age known as the 'Late Cimmerian' Unconformity. The formation overlying the unconformity surface in the Viking Graben area is a generally argillaceous sequence—sometimes a thin Lower Cretaceous development, overlain by a thicker Upper Cretaceous mudstone series, which is developed in these northern areas instead of the Chalk encountered in the central and southern North Sea zones. These formations present a very efficient seal for underlying oil or gas in the Jurassic.

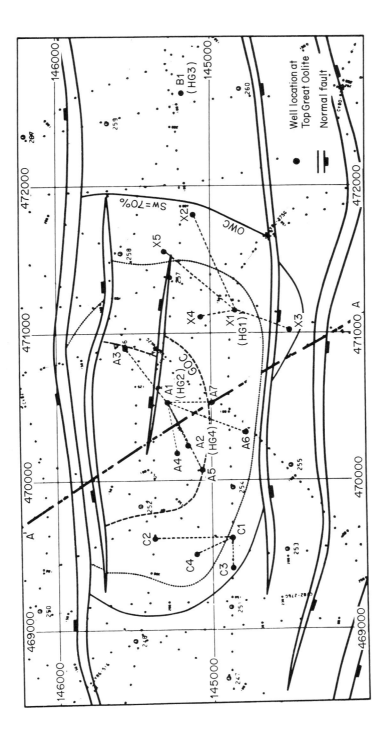

Fig. 10.10. Sketch depth-structure contour map of the Humbly Grove oilfield, showing GOC, OWC, and contours on top of the Great Oolite reservoir. Reproduced by permission of Graham & Trotman Ltd (from Hancock and Mithen, 1987)

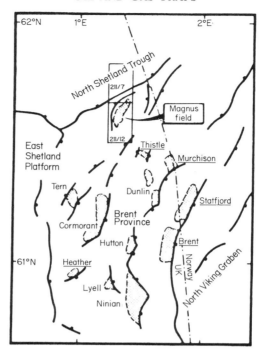

Fig. 10.11. Sketch map of the Brent Province showing locations of many of the tilted and eroded fault-block type fields in this area. Names of fields discussed in the text are underlined. Reproduced by permission of The Institute of Petroleum, London (from De'Ath and Schuyleman, 1981, p. 342)

Many of the hydrocarbon traps in the northern Viking Graben are of the tilted, eroded fault-block type, with the upper surface of the fault block being a buried topographic surface overlain unconformably by the Cretaceous. Some of the fields of this type in the Brent Province—Brent, Ninian, Dunlin, Cormorant, Statfjord, and others—are shown on the sketch map in Fig. 10.11 (from De'Ath and Schuyleman, 1981) and a cross-section through a hypothetical field of this type, endowed with composite characteristics, is shown in Fig. 10.12.

As noted by Walmsley (1975), late Jurassic/early Cretaceous Earth movements resulted in a block-faulted submarine topography only partially eroded before the Cretaceous inundation. Where erosion in post-Jurassic times was not severe, the tilted Jurassic sediments (bituminous marine shales with occasional shallow-water sandstones, which are developed throughout the Jurassic, but primarily in the Middle Jurassic) remain on the flanks and sometimes on the crests of the buried hills formed by the tilted blocks. They are sealed updip by the overlying unconformable Cretaceous.

Depths to the Jurassic vary between about 8000 and 10,500 feet (2600–3200 m). Reservoir thicknesses range up to about 330 feet (100 m) with high porosities and permeabilities. Gas caps are sometimes present—for example, in the Brent field.

Brent, the type structure, is seen in the E–W seismic section (cf. Fig. 10.12) shown in Fig. 10.13(a), with a line-drawn interpretation (b) and a well location linking the two

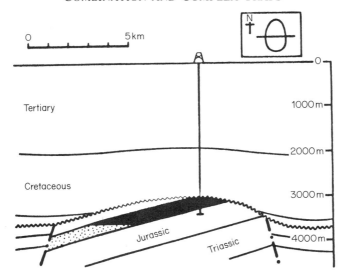

Fig. 10.12. Schematic section and plan view (inset) of a hypothetical field of the Brent-type—an eroded and tilted fault block, with oil in a combination trap in the Jurassic beneath the Jurassic–Cretaceous (Late Cimmerian) unconformity. Reproduced by permission of Elsevier Applied Science Publishers Ltd (from Walmsley, 1975a)

(Blair, 1975). Little detail is shown in this illustration, except that the Jurassic paralic sands that provide the main Brent reservoir are indicated, the main Graben axis is off-section to the east, and the block containing the Hutton Field structure adjoins the Brent block to the west. Blair (1975) mentions the Brent structural style as the 'basic tensional style' common in this part of the North Sea. It is noted parenthetically here that more recently there has been a suggestion (e.g. Frost, 1987) that the style has, ultimately, a compressional origin, but it is not believed that this view has (to date) gained general acceptance.

Another E–W seismic section across the Brent structure, in a slightly different location to the previous example, is shown in Fig. 10.14, and in this case bears rather more stratigraphic detail (Bowen, 1975). Bowen describes the Brent structure as a 'north–south striking, westerly gently dipping, partially eroded fault block with a relief of 3000 feet (910 m) above the structurally deeper areas to east and west. A major eastward-hading fault bounding the Brent block to the east lies some 3.5 km east of the crest of the buried escarpment; this is presumed to have a throw of many thousands of feet. . . . The total displacement of the Brent fault block, relative to the next block to the east probably exceeds 6000 feet (1830 m)'.

Bowen also reports that when the discovery well was drilled, the Upper Cretaceous was found to consist of siltstones, claystones, and marls with occasional thin chalky limestones. This sequence replaces the Chalk that is present in the southern and central North Sea areas. Below this, a thin 250 feet (76 m) Lower Cretaceous (ascribed) section was encountered, consisting mainly of marls, followed by a thin Aptian–Albian marl/limestone sequence overlying the Kimmeridge Clay. The upper main reservoir, the Brent Formation of Middle Jurassic age, has beneath it the Lower Jurassic Dunlin Formation (siltstones,

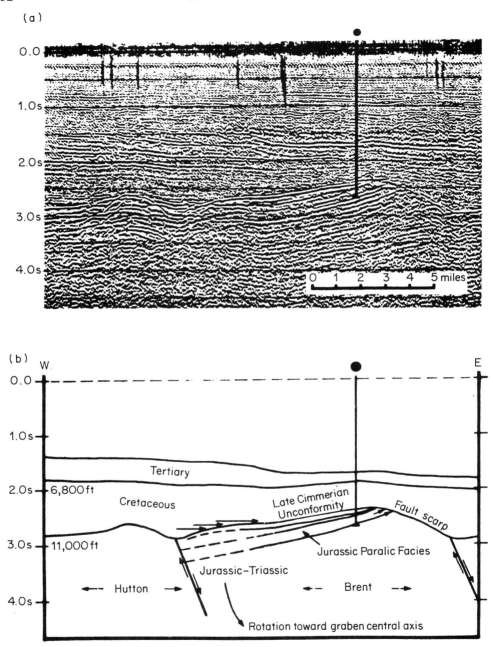

Fig. 10.13. Seismic section (a), with interpretation (b) across the type structure, Brent. The two displays are linked by the drilled well indicated. Note the seismic characteristics of the tilted block, and the events onlapping the eroded unconformity surface. Reproduced by permission of Elsevier Applied Science Publishers (from Blair, 1975)

COMBINATION AND COMPLEX TRAPS 333

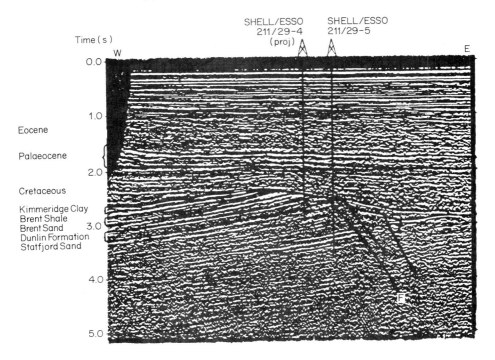

Fig. 10.14. Another seismic section across the Brent structure, but in a different location to that in the Fig. 10.13. The Brent and Statfjord Sands shown are the two principal reservoirs. The major eastward-hading normal fault bounding the structure to the east is marked at F. Reproduced by permission of Elsevier Applied Science Publishers Ltd (from Bowen, 1975)

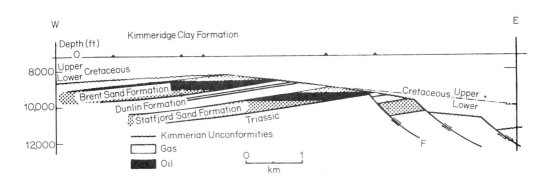

Fig. 10.15. Schematic interpretation of the seismic section in Fig. 10.14, with the major east bounding fault marked F, as in the latter illustration. Reproduced by permission of Elsevier Applied Science Publishers Ltd (from Bowen, 1975)

shales, and thin sands). Beneath the Dunlin Formation is the Lower Jurassic to Rhaetic Statfjord Formation, which is the second main oil reservoir. A schematic cross-section with the locations of some Shell–Esso wells marked, is seen in Fig. 10.15. This corresponds approximately to the seismic example shown in the previous illustration.

The Brent oil is thought to be sourced from Upper Jurassic shales buried in the 'hydrocarbon kitchens' in the troughs to the east and west of the Brent block. On the other hand, the gas found may have originated from the Brent sequence itself, which includes some thin coal seams as well as abundant plant material. Bowen (1975) should be read for more detailed information on the Brent Field.

10.2.4 Other Brent-type Examples

Because of their importance, and frequency of occurrence in the Viking Graben area, three other examples of tilted block fields will be illustrated and discussed briefly. These are the Heather, Thistle, and Murchison Fields, the locations of which are marked on the sketch map of Fig. 10.11.

The Heather Field was discovered in UK block 2/5 just to the east of the East Shetland Platform, in 1973. The structure is basically a north-east striking, westward-tilted block

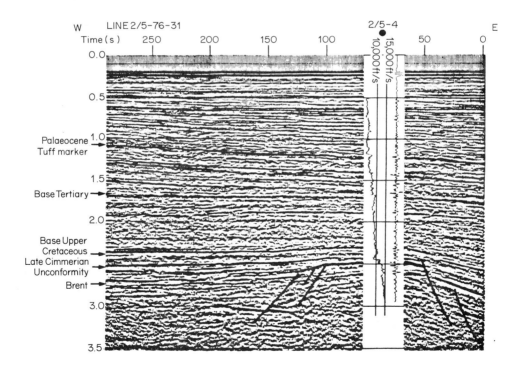

Fig. 10.16. Seismic section across the Heather Field. The axis of the graben is off-section to the east. Note the main east bounding fault, and a secondary fault, at the east end of the section. Reproduced by permission of The Institute of Petroleum, London (from Gray and Barnes, 1981, p. 338)

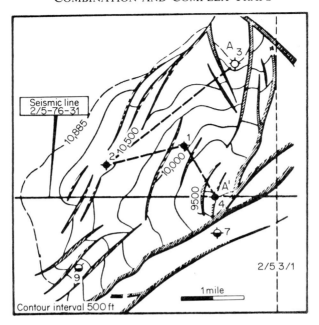

Fig. 10.17. Sketch depth-structure contour map of the Heather Field, with contours on top of the Brent Sandstone. The location of the east–west seismic line in Fig. 10.16 is marked. Reproduced by permission of The Institute of Petroleum, London (from Gray and Barnes, 1981, p. 338)

within the Viking Graben. The Brent Sandstone (Middle Jurassic) reservoir has five sub-units, and there is a 1400-feet oil column covering 6000 acres. However, complex geology—a high incidence of faulting coupled with variable cementation—results in disappointing estimates of in-place and recoverable reserves compared with those of other fields in the area.

A seismic section across the field is seen in Fig. 10.16 and a sketch contour map on top of the Brent Sandstone forms Fig. 10.17 (in which the location of the seismic line in Fig. 10.16 is indicated). The seismic section is notable for the relatively good resolution at Top Brent Sandstone level. The Statfjord Formation appears to be largely absent here—only a questionable 20 feet of calcareous Nansen Member has been logged, although the four sub-units of the Dunlin Formation are present below the Brent Sandstone.

As can be seen, both Lower and Upper Cretaceous rocks rest unconformably on the Jurassic Kimmeridge Clay. The Cretaceous consists of claystone, marl, and thin limestones, similar to lithologies in Brent.

In Fig. 10.18, three schematic W–E sections suggest the structural growth history during the Jurassic. Lower Jurassic shales were deposited on a stable eroded Triassic surface. Little or no erosion took place before deposition of the Brent Group. The major faulting occurred owing to extension of the Viking Graben after deposition of the Brent Sandstone, resulting in the tilted blocks. Positive areas were subjected to varying degrees of erosion. Further faulting took place after deposition of the Lower Kimmeridge Shales. The end of the Jurassic sedimentation was followed by the hiatus marked by the 'Late Cimmerian' Unconformity. For further details, see the paper by Gray and Barnes (1981).

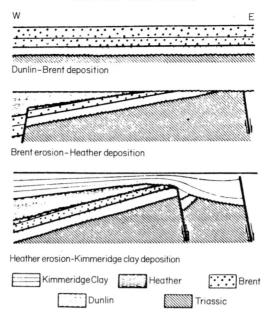

Fig. 10.18. Schematic east–west sections across the Heather Field, suggesting the structural growth history during the Jurassic. Reproduced by permission of The Institute of Petroleum, London (from Gray and Barnes, 1981, p. 338)

Thistle Field is one of the most northerly of the North Sea oilfields, and a schematic section of it is shown in Fig. 10.19. In this case, the structure (also in the northern part of the Viking Graben) is a fault block tilted to the *east*, with the major faulting bounding the feature at the western side of the block. Studies of the Middle Jurassic reservoir sands have revealed that they are part of a deltaic complex, and include such sand bodies as barrier islands, tidal channels, and distributary mouth bars (see also Murchison Field later). The oil column in the field extends over a 716-feet vertical interval (Hallett, 1981). The geology of Thistle has been complicated by synsedimentary (growth) faulting, and in at least one area condensed sequences of the Brent Formation have been produced owing to sea-floor relief, leading to 1:10 reductions in sequence thickness in some cases.

Murchison is a typical Jurassic tilted fault-block structure, with the Middle Jurassic Brent Sandstone as the producing reservoir. In Fig. 10.20, a sketch depth-structure contour map on top of the Brent Sandstone is shown with the reservoir seen dipping to the north-west at 3°. A major normal fault in the east-south-east bounds the feature, throwing the Jurassic down to the south-east. Normal faults along the sides of the block form the boundaries of the structure in these directions. A structural roll-over (dot-shaded) along the east-south-east edge of the fault block is the result of post-depositional erosion of the Brent Formation adjacent to the major east-south-east fault.

The seismic section in Fig. 10.21 shows this roll-over to the east-south-east of Well 211/19-2, where the seismic markers truncate against the major fault (marked F in Figs 10.20 and 10.21). Compare this with the effects seen in the Heather Field structure in

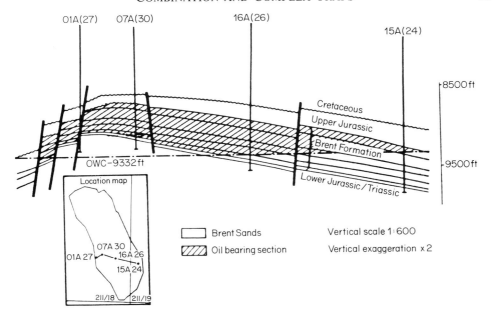

Fig. 10.19. Schematic section approximately east–west (see the inset sketch map) across the Thistle Field. This time, the fault block is tilted to the *east*, and not the west, as in the previous examples. The Middle Jurassic reservoir sands in this field have been identified as delta-related sand bodies. Reproduced by permission of The Institute of Petroleum, London (from Hallett, 1981, p. 321)

Fig. 10.16, and explained by the line drawing of the schematic in Fig. 10.18, which shows how the truncation effects on the Brent Formation cause the roll-over. Simpson and Whitley (1981) describe a study of the Brent Formation reservoir model for Murchison in terms of a series of delta-related sub-units: Lower Brent (pro-delta), Middle Brent (distributary mouth/shore) and Upper Brent (delta top). The latter, named the Ness Member, has been subdivided into environmental zones including crevasse splays, channels, overbank swamps, and lagoon bay units. The study has been carried out by integrating core studies with well-log correlations and petrophysical work.

The Murchison Field structure is fairly typical of the Brent-type fields. Apart from Simpson and Whitley (1981), further information is available in Davies and Watts (1977).

10.3 Complex Traps

10.3.1 Introduction

As mentioned at the beginning of this chapter, the term 'complex traps' in the present context is intended to mean oil and gas traps occurring in highly folded, thrust-faulted belts. The suggestion has been made more than once in this work that it can be a barren exercise to attempt to 'force' oil and gas traps into too rigid a classification, and this certainly holds good for complex traps as much as for others, since they may well fit into other categories, such as anticlinal or fault traps. As always with instances of possible ambiguity,

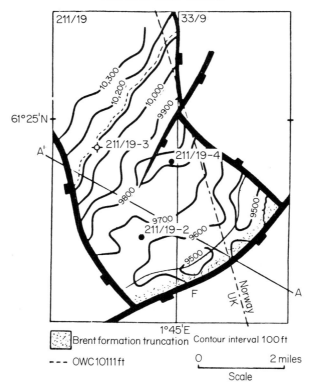

Fig. 10.20. Sketch depth-structure map of the Murchison oilfield; contours are on top of the Brent Sandstone, the producing reservoir. The major normal fault bounding the structure is marked F, and is in the east-south-east. Other normal faults bound the sides of the structure. Dip of the reservoir is gently to the north-west, but there is a small roll-over to south-east dip where the reservoir horizons meet the major fault F (dot shaded). See Fig. 10.21. Reproduced by permission of The Institute of Petroleum, London (from Simpson and Whitley, 1981, p. 311)

the genetic aspects must be considered when deciding what type of trap is in question.

Complex fold/thrust belts are found in many parts of the world, and at many levels in the stratigraphic column, sometimes with associated hydrocarbon traps, and often (although not invariably) closely related to existing or past 'active' continental margins. Modern belts of this type exist down the western side of both North and South America, circum-Caribbean, through Japan and the Philippine Islands, Alaska and round the Arctic, and from the Western Mediterranean through to Indonesia and beyond. Other, older trends exist, some in the same locations as the current belts.

Considering the large expanse of the globe encompassed by the present and past overthrust belts, surprisingly few major producing areas occur within them. Pratsch (1985) lists only 25 such areas, and poses the question—why are major producing areas so scarce in this type of setting?

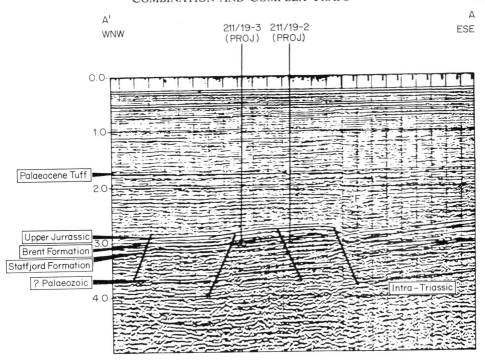

Fig. 10.21. Seismic section across the tilted fault-block structure of the Murchison Field, with the east-south-east bounding fault marked F. Note the roll-over of the Upper Jurassic events where they truncate against F. This is due to the erosional truncation of the Brent Sandstone (see Figs 10.16 and 10.18). Reproduced by permission of The Institute of Petroleum, London (from Simpson and Whitley, 1981, p. 311)

10.3.2 Thrust Belts and Hydrocarbon Traps

Pratsch (1985) also notes that deformation in overthrust belts occurs in three progressive stages—pre-thrust, synthrust, and post-thrust—see Fig. 10.22. He points out that only exceptionally is there any significant preservation of hydrocarbons generated or trapped during the pre- or syn-orogenic (thrust) phases. 'Thrust unit sediments between major thrust faults, or in depocentres that were formed in post-thrust phases, are the most prospective. (Therefore) ... The geological events which occurred after thrusting, and the post-thrusting structures, are valid parameters for the exploration of major and minor oil and gas fields in overthrust belts'. Pratsch (1985, especially the map in Fig. 6) is commended for study, as is also the paper by Butler (1987) discussing the developmental geometry of overthrust belt sequences.

Hydrocarbon traps that form in association with thrust faults, may be above or below the thrust plane. The latter may itself seal a trap, but frequently it is folding related to the thrust that contains the trap. This is seen in the illustration in Fig. 10.23, where the upper sketch is a depth-structure contour map of the Achi-Su Field in the eastern Caucasus

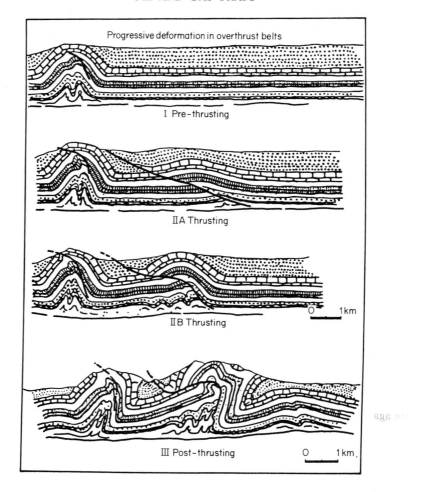

Fig. 10.22. Schematics showing progressive stages of deformation in an overthrust belt. Reproduced by permission of the *Journal of Petroleum Geology* (from Pratsch, 1985)

overthrust belt, USSR. The main reservoirs are found in separate stratigraphic units in an elongate trap formed along the crest of an overthrust fold. In the line drawings of the two sections, the locations of which are marked on the map, section A–A' shows the actual closures in the fold forming the trap, while in section B–B', the thrust (or reverse fault) plane forms the seal of the trap at some levels; occasionally, there are other traps below the fault plane, and sealed by the latter, as seen here. Levorsen (1967) also includes the sketch shown in Fig. 10.24, which generalizes the characteristic traps associated with thrust faulting (cf. Fig. 10.23, and also the following example).

A late 1970s discovery of a giant complex trap field is the Painter Reservoir Field in the Wyoming Thrust Belt, USA as described by Lamb (1980). This field is one of several that produce from the Triassic–Jurassic Nugget Sandstone. The Wyoming Thrust Belt is about 100 miles wide, and forms part of the Western (Cordillera) Thrust Belt of the USA

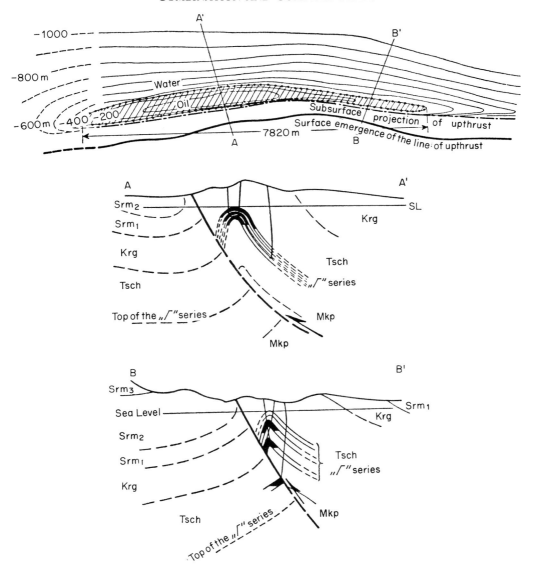

Fig. 10.23. Overthrust fold and fault traps, Achi-Su Field, USSR. From *Geology of Petroleum*, 2nd edition, by A. I. Levorsen. Copyright 1954 ©1967 W. H. Freeman and Company. Reprinted with permission

and Canada. East-to-west shortening occurred through motion in low-angle thrust faults, the thrust sheets trending N–S. The Nugget Sandstone is believed to be aeolian in origin, with porosities varying from 14% to 18%, and permeabilities averaging 22.8 mD (but locally exceeding 1000 mD).

Production is from the crest of an overturned anticline (see Fig. 10.25). The Nugget Sandstone is found again at depth below the overthrust, at about 9000 m below sea level.

342 OIL AND GAS TRAPS

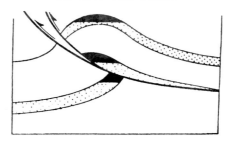

Fig. 10.24. Sketch showing traps characteristic of thrust faulting and folding. From *Geology of Petroleum*, 2nd edition, by A. I. Levorsen. Copyright 1954 © 1967 W. H. Freeman and Company. Reprinted with permission

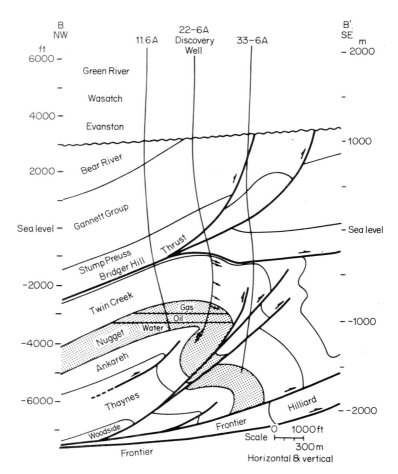

Fig. 10.25. Hydrocarbon trap in an overturned anticline. Section across the Painter Reservoir Field, in the Wyoming Thrust Belt, USA. From Lamb, 1980, reprinted by permission of American Association of Petroleum Geologists

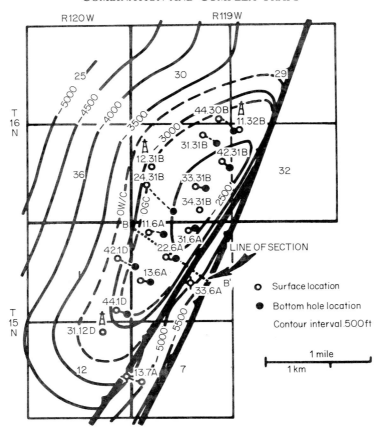

Fig. 10.26. Depth-structure contour map on top of the Nugget Sandstone; B–B' indicates location of section in Fig. 10.25. From Lamb, 1980, reprinted by permission of American Association of Petroleum Geologists

As shown in the figure, Well 33-6A was drilled to investigate the subsynthetic thrust part of the Nugget reservoir, but was a dry hole. Figure 10.26 is a depth-structure contour map on top of the Nugget Sandstone, and indicates the location (B–B') of the line-drawn section in Fig. 10.25.

The trap within the fold, and its relationship to the associated fault plane in Fig. 10.25 should be compared with Figs 10.23 and 10.24 with regard to the position of the trap within the complex structure. It is also instructive to study the schematics of Fig. 10.22 in the light of the generalized sketch of Fig. 10.24, and the other examples shown.

Complex traps normally occur in zones of compressional tectonics. A paper by Gibbs (1984) on the structural evolution of extensional basin margins is of interest in that it helps to explain why extensional tectonics, generally, do not lead to situations in which complex traps (as defined herein) are formed; rather, simple fold and fault traps together with types due to plastic deformation are usually the result of such a structural regime.

Other interesting complex traps worthy of study are: (i) the Point Arguello oilfield, offshore California, located in a large north-north-west-trending anticlinal complex (probably one of the largest offshore fields ever found (refer to Crain *et al.*, 1985); and (ii) the Teak oilfield, located offshore south-east of Trinidad. The Teak structure is a broad, asymmetrical anticline along a compressional fold belt between the Caribbean and South American lithospheric plates. The feature is broken in a highly complex way by many transverse synthetic and antithetic normal faults that compartmentalize the (Pliocene sandstones) multiple reservoirs.

Although without seismic illustrations, Levorsen (1967) is still an excellent work on traps, and it is worthwhile comparing many of the line-drawn sections therein with relevant seismic data when suitable examples are available.

PART V
SOME SPECIAL SITUATIONS

CHAPTER 11

Some Special Situations

11.1 Introduction

There are some interesting potential trap situations and hydrocarbon accumulation phenomena that cannot easily be included in any of the groups considered so far.

This final chapter will be devoted to brief studies of these special situations, which include such effects as pore fluid content variation leading to direct hydrocarbon indicator (DHI) events, the potential for trapping of the phenomena believed to be responsible for the bottom-simulating reflection (BSR), and the intriguing possibility that astroblemes—impact craters on the Earth's surface caused by cosmological debris—may be locations of hydrocarbon accumulation.

The first matter to be discussed will be that of DHI events in seismic data, a subject that has already been referred to from time to time in the foregoing chapters.

11.2 Direct Hydrocarbon Indicator Events

The discussion of DHI events centres on a group of phenomena observed in seismic data as being produced by the presence or absence of pore contents—liquid or gaseous—and the effects on elastic properties of rocks.

Referring to the schematic in Fig. 11.1, when a seismic pulse impinges at normal incidence on an interface, such as X separating the two rock units A and B, some of the seismic energy is transmitted through the interface, and some is reflected back from it towards the surface. The energy partitioning (the ratio of energy reflected to energy transmitted) is known as the *reflection coefficient*, R, and the value of R can be determined from a well-known expression derived from Zoeppritz' equations:

$$R = \frac{\rho(B) \cdot V(B) - \rho(A) \cdot V(A)}{\rho(B) \cdot V(B) + \rho(A) \cdot V(A)}$$

where $\rho(A)$, $\rho(B)$ are the bulk densities of rock units A and B, and $V(A)$, $V(B)$ are the compressional wave (P-wave) seismic velocities in A and B.

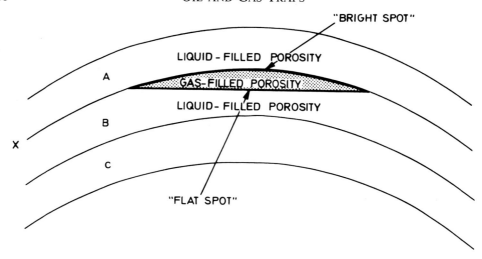

Fig. 11.1. Schematic showing the geometry of bright-spot and flat-spot DHI's

In the schematic of Fig. 11.1, the lens-shaped body of gas-filled porosity indicated within the rock unit B will have an important effect on the term $\rho(B) \cdot V(B)$, which is the *acoustic impedance* of formation B, since the bulk density of the rock will be lowered by the presence of the gas in the pore spaces, relative to the densities due to water-filled porosity in formations A and B above and below the gas accumulation.

This change in acoustic impedance will result in a strong negative reflection coefficient at interface X over the top of the gas-filled porosity, and a strong positive reflection coefficient beneath the gas lens. In the latter location, a reflection event is produced at the gravity controlled bounding interface between the gas- and liquid-filled porosity volumes within formation B (the liquid might be formation brine or oil/condensate), and is referred to as a gas–liquid contact event (GWC), or commonly, a 'flat spot'.

In the terminology current in the industry, where a reflection shows a strong localized lateral amplitude increase (due to lateral change in acoustic impedance contrasts across an interface), the event is called a 'bright spot'. In the ideal case, as in the Fig. 11.1 schematic, both the flat spot and the upper bounding reflection associated with the gas accumulation will be 'bright'.

It is worthwhile remarking that in spite of the term 'flat spot', the event in question may not be horizontal, or even flat (see e.g. Hubbert, 1967). As was noted in the early section on hydrodynamic trapping, a tilt can be imparted to a gas–liquid contact as an effect of hydrodynamic friction related to moving formation fluids. Also, a flat-spot event may be irregular in cross-section owing to minor faulting, or to the variable distribution of the gas-filled intergranular or fracture porosity present (see Jenyon, 1986b).

In Fig. 11.2, a seismic example from the North Sea is shown, this being a migrated section through a known gas accumulation. Within the inset, the flat spot at the base of the gas-filled porosity volume is indicated at A, while the top of the strong white convex-upward event immediately above it marks the strong negative reflection that follows the top of the

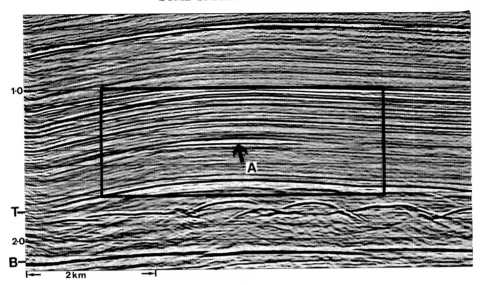

Fig. 11.2. Seismic example of a flat spot (A) related to a proven gas accumulation in the North Sea. Reproduced by permission of Seismograph Service (England) Ltd

gas-filled porosity, and the shape of the seal formation at the top of the interval. There is good correlation between the effects seen here, and those sketched in Fig. 11.1. The lenticular shape of the gas-filled porosity in cross-section is clearly seen, while the basal and top events are bright, showing clear lateral amplitude increases associated with the gas accumulation.

In the example just seen, the flat spot was indeed flat, and approximately horizontal. In Fig. 11.3, another example of a migrated seismic section is seen, exhibiting a very clear, although irregular and slightly tilted, flat spot (lower arrow A within the inset). The upper arrow of the pair within the inset indicates the convex-upward top boundary of the gas accumulation. There is clear evidence of local amplitude increase in a lateral sense by both events. Also here, as in Fig. 11.2, the flat-spot event shows end polarity reversals, which are characteristic of gas–liquid contact events and result from the configuration of the acoustic impedances at the terminations of the flat spot. The event marked X in this example simulates a flat spot, but is actually related to flow-concentration of plastically deformed material (potash salts or red clays) in the crestal zone of a salt pillow. In both Figs 11.2 and 11.3, the interval marked B–T is an interval of mobile salt, while in the Fig. 11.3 example, TY indicates the Base Tertiary event.

The configuration of the gas–liquid contact events—the 'flat spots'—seen in Figs 11.1–11.3 is very much dependent upon the geometry of the situation. This refers particularly to the shape of the trap, and also the thickness of the reservoir interval. If the reservoir unit 'B' in Fig. 11.1 were to be reduced progressively in thickness, at some point the flat-spot event would be tangential to the crestal point of the interface between rock units B and C. If the thickness of B continued to decrease beyond this, then the flat-spot event would perforce divide into two parts, one in each limb of the antiform shape of the reservoir.

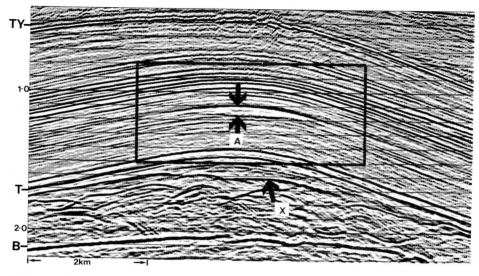

Fig. 11.3. Another seismic example of a North Sea gas accumulation location in cross-section. In this case, the flat spot A is slightly tilted, and irregular. Reproduced by permission of Seismograph Service (England) Ltd

This situation is sketched in Fig. 11.4(a), where a thin interval of gas-filled porosity is indicated with two basal gas–liquid contact flat-spot events, one in each limb of the anticline, shown in A. Clearly, if the volume of gas-filled porosity were to reduce progressively, again at some stage the two flat spots would reunite to become one basal event.

In a similar situation, where such a gas accumulation is present when faulting and erosion takes place, only one limb of the original structure may survive with gas accumulation intact. This is sketched in Fig. 11.4(b), where gas-filled porosity is sealed by fault F and by the overlying stratal unit. In this case, a single basal flat spot A is seen extending across the downdip extremity of the gas-filled porosity.

Three examples of events interpreted as flat spots are seen in Fig. 11.5a indicated by the small arrowheads. The upper section shows a flat-spot event in the crest of a small anticlinal flexure on the upthrow side of a minor listric normal fault. The flat-spot event is itself disrupted slightly by a minor fault within the flexure. The middle and lower seismic section examples show somewhat similar situations, which are comparable with the sketches in Fig. 11.4. Note in both cases how the flat-spot events are extending across an interval, in each case, that is marked by strong upper and lower reflection events that stand out against the background of these rather noisy sections (i.e. they appear to be 'bright', as indicated by the black arrowheads).

All three of these examples of rather subtle gas–water contact events are of considerable interest, since they are located not far from the major Morecambe Bay gasfield in the Manx–Furness Basin. Figure 11.5b is another North Sea example.

It should be noted that while there is a large density difference (and therefore acoustic impedance difference) between gas-filled and liquid-filled porosity, there is only a small difference between porosity filled with oil and formation brine. Therefore, when direct hydrocarbon indicators are referred to, it is really only gaseous hydrocarbons that are

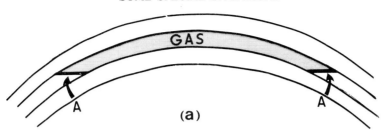

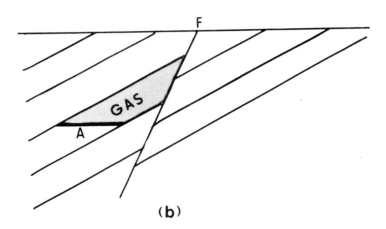

Fig. 11.4. Schematics showing variations in the configuration of flat spots associated with gas-filled porosity

involved. There are isolated cases of an oil–water contact apparently producing a seismic flat-spot type event, but these are rare. Where they occur, it may be that some form of organic metamorphism or diagenetic process has caused alteration to the hydrocarbons in the oil–water contact zone. This has not yet been studied.

There are other phenomena classed as DHI events apart from those already mentioned. One of these is a tendency for high frequencies to be lost from the seismic signal where it passes through gas-filled porosity, leaving a 'frequency shadow' beneath, with only apparent low-frequency events present. Similarly, an 'amplitude shadow' is often seen below the gas zone, resulting from attenuation of a wide band of frequencies and an overall deterioration in the data so that continuous reflections are no longer seen below the gas accumulation.

Because of a lowering of seismic velocity in the gas zone (consequent on the lowering of bulk density in the reservoir rock), there is sometimes seen a 'drooping' of seismic events

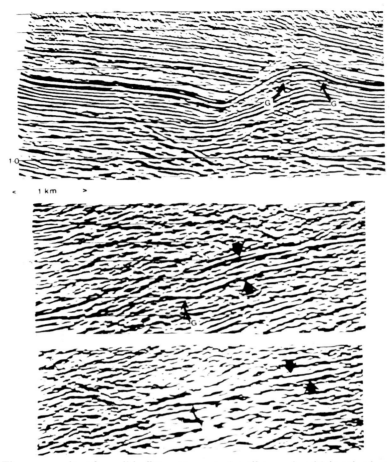

Fig. 11.5a. Three examples of possible flat-spot events (small arrowheads) in seismic sections from the Manx–Furness Basin, UK waters. Reproduced by permission of Seismograph Service (England) Ltd

below the gas zone as they approach its downward extension from the side. This drooping of events beneath the gas is known as 'gas sag'; this can also be seen where there is a so-called 'gas chimney'—with gas leaking upwards through the subsurface, producing a zone on the seismic section showing gas sag on reflections entering the zone. The sag, or negative pull-down anomaly, beneath gas-filled porosity is of considerable significance since it draws attention to the possibility of gas-filled porosity within a reef feature, for instance.

Vertical migration of gas upward from a small Chalk trap, resulting in a 'gas chimney' on the seismic section is seen in Fig. 11.6. Gas has seeped via an extensional fracture system up into the Tertiary as far as a Miocene claystone sequence. In this instance, a Zechstein salt uplift has caused the fracturing of the overburden. The overlying gas sag, pull-down effects, and seismic path distortions are clearly visible in the gas chimney zone vertically above the salt uplift. Insufficient of the Zechstein in the deeper section is shown to enable determination of whether the Carboniferous-sourced gas migrated through the Zechstein section in this location, or elsewhere.

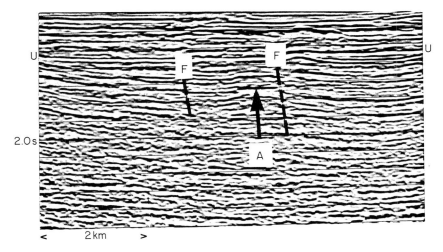

Fig. 11.5b. Seismic example of a flat spot (A) developed in a minor flexure against a fault (right-hand F). There is another minor fault F to the left. This flat spot is 'bright', and shows very well a typical polarity reversal at its left-hand termination (away from the fault). U = unconformity. Reproduced by permission of Seismograph Service (England) Ltd

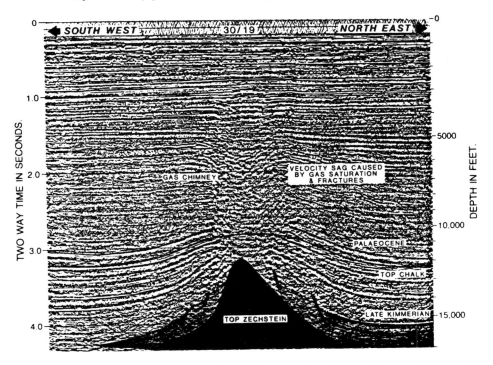

Fig. 11.6. Seismic example showing a 'gas chimney' above a salt uplift. Gas has seeped through from beneath the salt interval, up into an extensional fracture system caused by the uplift in the clastic overburden. Note the 'gas sag' shown by seismic events at the edge of the 'chimney'. Reproduced by permission of Graham & Trotman Ltd (from Cayley, 1987)

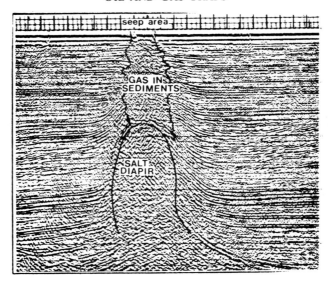

Fig. 11.7. Another example of a gas chimney in seismic data, this time above a salt diapir offshore Norway. Reproduced by permission of the publishers, Butterworth & Co. (Publishers) Ltd © (from M. Hovland and J. H. Sommerville (1985). *Mar. Pet. Geol.*, **2**, 319–325, Fig. 2)

Hovland and Sommerville (1985) have described two gas seepages in the North Sea, one located above a salt diapir in Block 1/9, Norwegian waters (see Fig. 11.7) and the other a shallow gas seepage in Block 15/25, UK waters (see Fig. 11.8). The salt diapir seep was found to consist of about 120 small individual seeps located within a circular area of some 100 m diameter, and was estimated to produce about 24 m^3 of methane gas per day at 75 m water depth. The gas was proved by isotopic methods to be thermogenic and of deep-seated origin. It is producing a marked gas chimney effect in the seismic section.

The UK seep is of the same gas type, and is associated with a very large 'pockmark' (p) depression in the sea floor (about 700 × 450 m, and 17 m deep). As seen in the section, several seismic reflection anomalies are seen at various levels down to a depth of over 1100 m below the sea bed. Many of the anomalies seen are diffractions related to reflector terminations (i.e. due to lateral and sudden changes in porosity content in gas-charged sediments). There has been growing evidence that sea-floor bedform features of this 'pockmark' type are frequently related to gas seepages, although the exact mechanism involved is not yet determined. Plumes of gas bubbles in the water column above these seeps are often in evidence on shallow seismic recordings.

A paper by Carlson *et al.* (1985) suggests that hydrocarbon gases are ubiquitous in the near-surface sediments (down to 250 m (820 feet) depth) of the Navarin continental margin in the northern Bering Sea. There is discussion of the seismic anomalies observable that are related to the presence of shallow gas accumulations, with illustrations.

In order to determine the exact operation of some of the effects seen as DHI events, it is necessary to have a precise knowledge of the acoustic impedances of the rock units involved, and the size and sign of the reflection coefficients along the interfaces. Certain configurations, even where gas is present, can result in no explicit effects being visible.

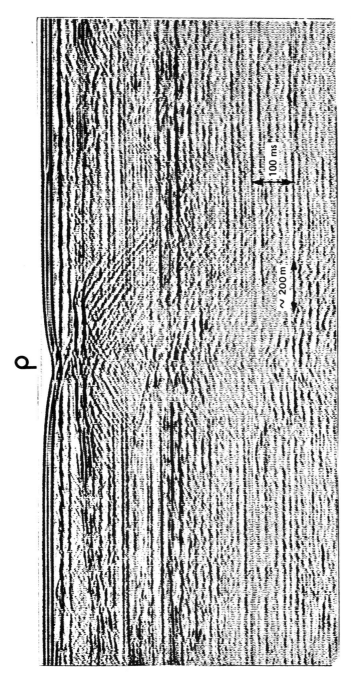

Fig. 11.8. Gas seepage into shallow sediments is shown in this seismic section, which also exhibits a 'pockmark' (p) in the sea bed above the gas anomalies. These depressions in the sea floor frequently indicate the presence of underlying gas. Reproduced by permission of the publishers, Butterworth & Co. (Publishers) Ltd © (from M. Hovland and J. H. Sommerville (1985). *Mar. Pet. Geol.*, **2**, 319–325, Fig. 8)

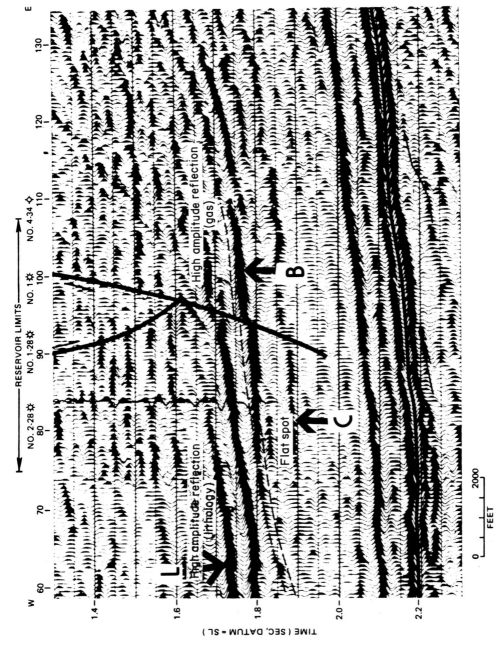

Fig. 11.9. Normal P-wave seismic data, showing DHI events C (a flat spot) and B, an amplitude anomaly ('bright spot') related to a gas accumulation (reservoir limits shown above the section). Event L is a lithology related anomaly (cf. Fig. 11.10) (from Ensley, 1985, *Geophysics*)

Others can produce 'dim spots'—that is, lateral *decreases* in amplitude associated with a gas accumulation.

Most of the phenomena observed are the result of absorption or dispersion affecting seismic signal as it passes through the gas, and also the lowering effect on seismic velocity. Although there is a general understanding of the operation of these effects, a detailed understanding of some of the phenomena has still not been achieved. For a useful discussion of the subject, Anstey (1977) is recommended.

11.2.1 Shear-wave Studies of Direct Hydrocarbon Indicator Events

An interesting recent development is that involving the use of shear-wave seismic acquisition (see Ensley, 1984, 1985). It is well known that shear (S) waves cannot propagate through a liquid. They differ from compressional (P) waves in that their seismic velocity (only approximately half of P-wave velocity in a given medium) is not significantly affected by fluid content changes in rock porosity. Because of this relationship, a 'bright spot' DHI event will not have any comparable presence in shear-wave data. On the other hand, a lithology related P-wave anomaly *will* have a comparable S-wave anomaly (Ensley, 1985).

This can readily be seen in the seismic examples of Figs 11.9 and 11.10. In Fig. 11.9, an expanded part of a P-wave seismic section is seen, with an amplitude anomaly (bright spot) at B, and a flat spot at C, related to a proved gas reservoir (the limits of which are indicated above the section). There is also a synthetic seismogram overlaid, from the Cowell 2–28 Well shown.

Compare this section with Fig. 11.10, an expanded portion of the S-wave (shear) seismic data. Note the queries at the locations of B and C in Fig. 11.9; the DHI characteristics present in the P-wave seismic data are absent in the S-wave data. On the other hand, a strong lithology related anomaly (L) associated with a low-velocity shale overlying a higher velocity siltstone, is present in both P-wave and S-wave data.

Ensley (1985) suggests that from these and other observations, although shear-wave seismic data quality is generally poorer than that of P-wave data, it can be used at least qualitatively to evaluate P-wave DHI events, as here. If S-wave data quality improves (which has been the case since Ensley wrote the papers), the ratio V_P/V_S is sensitive to the type of fluid present in a rock, and is therefore a potential hydrocarbon indicator. In practice, this ratio can be derived from seismic data by computing the S- to P-wave travel time ratio (T_S/T_P) for a given interval. A similar suggestion was made as regards the ratio of the P- and S-wave reflection coefficients along the top of the reservoir, which is said also to be sensitive to the pore-fluid present. Petroleum geologists should note the potential of these methods in the case of gas reservoirs.

11.3 The Bottom Simulating Reflection Event

An anomalous event has been observed in seismic data of many deep-water areas. This event runs below and subparallel to the sea bed event. Katz (1981) shows an example seismic section from the continental slope east of North Island, New Zealand, and notes that: event (i) is strongly reflective, having a high impedance contrast; event (ii) cuts across stratal events from the bedding planes of clastic sediments; event (iii) is subparallel with the sea-

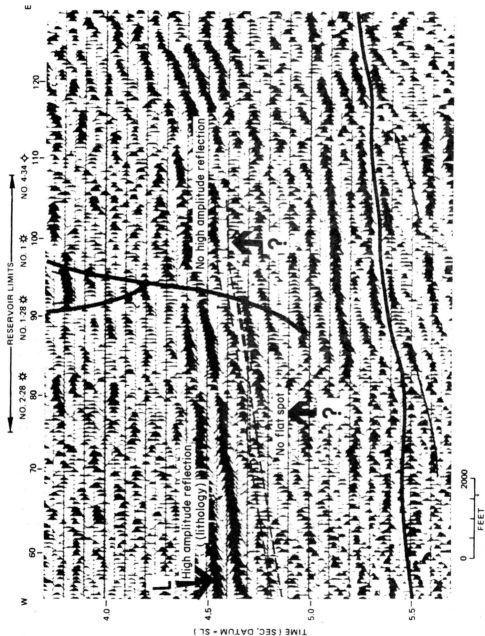

Fig. 11.10. Shear-wave (S) seismic data covering the same zone as the Fig. 11.9 section. Note the absence of events C and B (marked by the queries), although event L is still present. This is good supporting evidence that anomalies C and B are DHI events, related to pore-content variation (from Ensley, 1985, *Geophysics*)

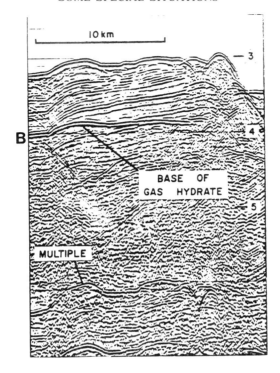

Fig. 11.11. Seismic example from continental slope east of North Island, New Zealand, showing BSR (bottom simulating reflection) at B, marking the base of a zone of gas hydrates in this location. Note the strength of the event, its shape relationship to the sea-bed reflector, and its cross-cutting of the normal stratal events present—all characteristics of a BSR. Reproduced by permission of the *Journal of Petroleum Geology* (from Katz, 1981)

floor reflection; and (iv) increases sub-bottom depth with increasing depth of sea floor (and hence with decreasing temperature of bottom water)—see Fig. 11.11, event B.

Events of this type have been observed in seismic data in many parts of the world. Kvenvolden (1985) shows examples recorded at two sites of the DSDP, one in the Atlantic on a spit-like extension of the continental rise on a passive margin with a water depth of 3191 m, while the other is at 2031 m water depth in upper slope sediments of an active accretionary margin in the Pacific Ocean (Fig. 11.12). The name 'bottom simulating reflection' has been applied to the phenomenon, and several investigators (e.g. Ewing and Hollister, 1972; Shipley *et al.*, 1979) have suggested that BSRs indicate the possible occurrence of gas hydrates, with the event defining the base of these 'inclusion compounds', which are known to occur under certain conditions in ocean-bottom sediments, and in permafrost areas.

Inclusion compounds are substances in which the molecules of one component become trapped within the molecular lattice of another, but without any true chemical bonding between the components. In the case of gas hydrates, gaseous hydrocarbon molecules (typically methane, ethane, and CO_2) become trapped within an expanded lattice of water molecules. In the marine environment, this occurs under conditions of high pressure (great

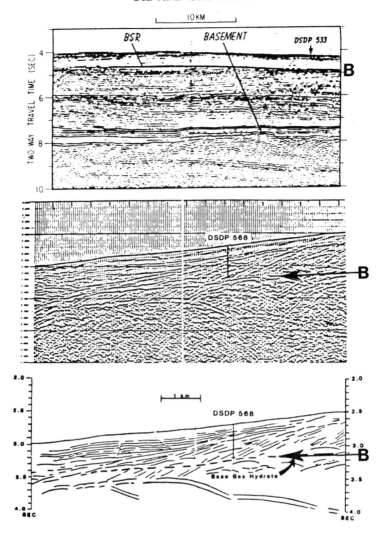

Fig. 11.12. Seismic evidence of BSR events: (a) at B, from a DSDP site in the Atlantic on a spit-like extension of the continental rise, in 3191 m of water; (b) the BSR at B in the seismic data is from another DSDP site in 2031 m of water in a location of upper slope sediments; (c) line-drawn interpretation of the seismic section in the middle. Reproduced by permission of the publishers, Butterworth & Co. (Publishers) Ltd (from K. A. Kvenvolden (1985). *Mar. Pet. Geol.*, **2**, 65, Fig. 2)

depths of water above the sediments) and low temperatures, in related ratio (i.e. the shallower the water, the lower the temperature necessary, and *vice versa*). Therefore they tend to form beneath Arctic waters of 1000 feet or more in depth; beneath subtropical waters, however, water depths greater than about 2000 feet are required. Collins and Watkins (1985), for instance, show an example (Fig. 11.13) of the BSR in slope sediments off south-west Mexico where water depths fall within the range of 2250–4500 m.

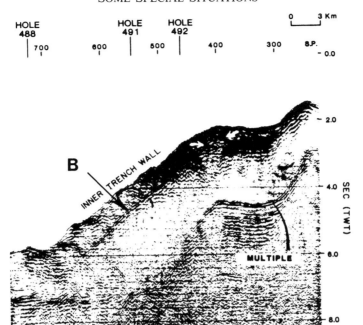

Fig. 11.13. Another seismic example of BSR (B) in water depths of 2250–4500 m in slope sediments off south-west Mexico (from Collins and Watkins, 1985, *Geophysics*)

The physical appearance of gas hydrates is that of ice-like crystals, and these may be disseminated widely through appreciable thicknesses of subsea-bottom sediments. This may cause a 'homogenization' of the acoustic properties in the sediments in some areas, resulting in poor acoustic impedance contrasts and hence weak reflections from original sedimentary bedding planes.

On the other hand, the acoustic velocity in a gas hydrate layer is relatively high in comparison with the undercompacted shallow sediments beneath the hydrate layer—probably because of the increased rigidity imparted to the sediments in which they are present by the cementation effect of the ice-like matrix. This may give a velocity differential across the interface at the base of the gas hydrate layer of 3000 ft/s or more. Such a contrast is capable of generating a strong reflection event below and subparallel to the sea bed, this event showing a strong cross-cutting relationship to the—often weak—reflections produced by the original sedimentary bedding. The reflection time between the sea-bed event and the basal gas hydrate reflector (the BSR) tends to increase fairly rapidly up to a certain value of water depth; with increasing water depth beyond this value, the time differential increases much more slowly. Evrenos *et al.* (1971) demonstrate that the hydrated layer is impermeable to fluids, and that therefore free gas below the layer, being unable to migrate upward, may collect and result in hydrocarbon accumulations, sometimes marked by DHI phenomena ('bright spots', etc.).

The depth to which gas hydrates can remain stable depends on overburden pressure (water-depth dependent in the marine environment) and the geothermal gradient (see Macleod, 1982). Clearly, from the above notes, they must be regarded as important potential seals, and perhaps also sources, for gaseous hydrocarbons.

Gas hydrates may also be important in the transportation of sediments. From what has been said, it is clear that the stability of a hydrate layer could be critically affected by a drop in pressure (e.g. by a lowering of sea level)—or an increase in temperature brought about by changes in the geothermal/hydrothermal gradient (by e.g. igneous activity). The resulting decomposition of the hydrated layer into its constituent components could lead to catastrophic eruptive gassing and burst-out, perhaps on a very large scale, with resultant transport and redistribution of the associated sediments, perhaps by destabilization of a continental slope with the initiation of massive turbidite flows (McIver, 1982). Certainly the presence of a decomposing gas hydrate layer with its over-abundance of gas-cut mud would provide an extremely mobile lubricating layer favouring the inception of large-scale slide phenomena in an area with any relief. See also Jenyon and Fitch (1985) and Kvenvolden and Barnard (1983)—the latter paper has a map of locations of known/inferred gas hydrates and gas hydrate stability field data.

Gas hydrates are not the only possible cause of the BSR. Another phenomenon to which the BSR has been ascribed is the 'opal transition'. This is a diagenetic alteration in diatomaceous sediments, in which with increasing depth of burial, the opal-A form of the diatom frustules undergoes dissolution and reprecipitation as the opal-CT form. The transition brings about a density increase of the order of 0.4–0.5 g/cm^3. This increase at the interface between the two forms of opal would produce an acoustic impedance contrast more than adequate to result in a strong seismic reflection. This is believed to be the explanation of BSR events seen in the shelf areas of the Bering Sea (Hammond and Gaither, 1983), and a seismic section that includes the BSR event is shown in Fig. 11.14 (the BSR event being indicated by the the arrowheads at the extremities of the section). Since the transition is directly pressure/depth related, the higher pressure opal-CT form will exist *beneath* the transition 'front', which will move up through the section with continuing deposition and increasing burial depth. In some locations, the 'front' would be concordant with bedding while in other locations it would be discordant.

Below the transition, the diagenetic effects 'left behind' by the upward-moving front may tend to homogenize the acoustic properties of the sediments, leading to amplitude attenuation and loss of coherence of the original events on the seismic section. A marked increase in seismic velocity *below* the transition is also observed. It is worthwhile noting the difference between this effect, and that of the gas hydrates discussed earlier—with the latter, an analogous process of 'acoustic homogenization' occurs, also causing higher than normal seismic velocity, but in the case of the gas hydrates, this takes place *above* the transition to normal sediments, in the zone impregnated by the hydrates.

It is noted that others (e.g. Isaacs, 1983) have drawn attention to the fact that opal-CT itself transforms to quartz with increasing burial loading, and it is possible that this second transformation is responsible for the appearance of the BSR in the shelf areas of the Bering Sea (as in Fig. 11.14). Alternatively, it could be the cause of some deeper transition not seen on the seismic section in this example. Also, it is not known at the time of writing,

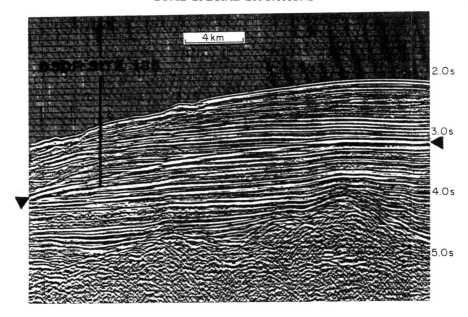

Fig. 11.14. Another possible cause of BSR events—the 'opal transition', indicated here between the two black arrowheads at the ends of the seismic section. This event marks the *top* of the zone of transition to a denser form of opal; cf. the effects of gas hydrates, where the BSR indicates the *base* of the hydrate zone (from Hammond and Gaither, 1983, *Geophysics*)

if, as is the case with gas hydrates, the opal transition zone forms an impermeable barrier; if so, it could also provide an important seal to trap hydrocarbon accumulations below. As the transition is a reprecipitation process, it seems likely that it would form such a barrier.

11.4 Astroblemes

The potential of 'astroblemes'—Dietz's (1961) term for meteoritic impact craters on the Earth's surface—for the accumulation of hydrocarbons, has been recognized for some time. This potential has not, so far, been unequivocally fulfilled for reasons that will become clear, but geologists and interpreters of geophysical data in particular should be aware of the possibilities.

Impact craters are probably the most commonly observed features on planetary bodies in the Solar System. With the exception of the Earth, the inner planets and satellites (Mercury, Venus, Moon, Mars, Phobos and Deimos (the small moons of Mars)), together with a majority of the satellites of the giant gas planets, Jupiter and Saturn, are all seen— or in the case of Venus, known by infrared photography—to have surfaces more or less dominated by impact craters. These craters vary in size from titanic circular basins, many hundreds of kilometres in diameter, to microcraters produced by dust-sized grains.

Based on the most popular of the current ideas (none of which is universally accepted) regarding the origin and development of the planets and satellites, it is believed that the primitive crust of the Earth was as heavily cratered as the other bodies—perhaps looking something like the now-familiar images of the lunar surface in the so-called 'highland' areas.

A study of lunar geological history tells us that the period of large and frequent impacts—the principal period of cratering—was probably coeval with that on the Moon, which took place between about 4600 Ma and 3700 Ma (i.e. during the earlier part of the Archaean Eon on Earth: the oldest Earth rocks that exist today have been isotopically dated to 3600 Ma). The main reasons why no extensive cratering remains to be seen on the Earth's surface include early remelting of the crust, geodynamic effects (such as lithospheric plate movements, igneous, metamorphic, and volcanic processes), and, progressively, the powerful effects of the substantial atmosphere and water circulatory systems that have evolved, allowing active erosion, transportation, and deposition to take place.

Although the main bombardment by cosmic debris ceased so long ago, remnants of the original debris clouds still exist—the meteorite showers such as Perseids and Leonids are believed to represent some of the fragments. Thus, from time to time, meteoritic items large enough to survive the passage through the Earth's atmosphere occur. Relatively small bodies that do not burn up in the atmosphere are nevertheless slowed down to quite low speeds and do little or no damage—often fragments of these are picked up on the surface. Occasionally—perhaps once every few million years—a body large enough to survive the atmospheric passage arrives at the surface at its 'cosmic' velocity, typically 15–17 km/s. Such bodies need to have a mass greater than 1000 tons in order to impact at cosmic velocities, and impacts of such magnitude are fortunately very rare.

Donofrio (1981) in an interesting paper reviews the criteria for the recognition of impact craters on Earth. He notes that by convention, such a feature is classed as 'proved' if meteoritic fragments are found, 'probable' if shock-metamorphic features only are present, and 'possible' if any other criteria (such as morphology) are used to suggest an impact origin. On this basis (Grieve and Robertson, 1979), some 145 features classed as proved, probable, and possible impact craters on Earth had been found. This compares with only about 16 total in 1960, and 100 (50 proved/probable and 50 possible) in 1968 (French, 1968; Guest and Greeley, 1977).

Of the craters on Earth, only about a dozen are 'proved' in the sense that actual meteoritic material has been found *in situ*. (This is not surprising, since the impacting body tends to vaporize). One of these is the best known feature of its type—Meteor Crater, Arizona, which is shown in Fig. 11.15. Meteor Crater is about 1.2 km across and about 200 m deep, with a fairly simple bowl shape and a rim that is slightly raised (about 40 m) above the Arizona desert surface, although the rim rocks have been worn down to some extent by erosion (the impact having occurred during the Pleistocene Period).

The craters, proved and otherwise, mentioned as having been found so far, are *surface* features at present. From the viewpoint of potential hydrocarbon traps (the reasons for this will be discussed later), none of the very few 'proved' impact craters has been found to be an oil or gas trap. The only petroliferous examples are 'probable' or 'possible'.

Impact craters have been classified as 'simple', where they are bowl-shaped features like Meteor Crater, and 'complex' (larger than 4 km diameter in crystalline rocks, or 2–3 km in clastic sediments) where a central uplift or peak develops in the middle of the crater bowl. An example of the latter type is the crater Euler on the Moon, seen in Fig. 11.16. This crater is 28 km in diameter, and is at the lower end of the size range of 'large' craters. Slumping, or terracing, of the crater walls, although poorly developed, can be seen, this

Fig. 11.15. Meteor Crater (also known as Barringer Crater), Arizona, USA. A Pleistocene impact crater, perhaps the best-known and best-preserved of such features. Reproduced from *Geology on the Moon*, Guest and Greeley (1977) by permission of the publisher, Taylor & Francis Ltd, London

being a feature of large craters. The central uplift, which has several separate culminations, is clearly seen.

The reasons that impact craters should form possible hydrocarbon traps relate to both the morphology of the craters, and particularly to the mechanics of impact cratering. The latter subject, although fascinating, is too large a matter to be pursued at any length in this context (those interested are recommended to read, for instance, Guest and Greeley (1977) for an excellent account suited to the non-specialist). However, from the point of view of hydrocarbon exploration, some basic matters are mentioned.

The laboratory experiments of Gault *et al.* (1968) have helped greatly in understanding the phenomena following a cosmic impact. Where an object is large enough (Weatherill (1977) quotes masses greater than 10^8 gm) and travelling fast enough (15–70 km/s) to retain its 'cosmic' kinetic energy, on impact, the latter is converted at the surface to heat and shock waves measured in megabars that move radially outward from the point of impact. Processes occurring include material excavation, cavity formation, slumping, and shock metamorphism. The latter includes high-pressure effects, such as the production of unusual mineral species as high-pressure polymorphs (diamond, stishovite, coesite), high strain-rate effects in quartz, 'shatter cones', and high-temperature effects (melting and fusing of minerals—for example, quartz to lechatelierite). When equilibrium has been reached, the crater floor is covered with a thin layer of superficial fine-grained fall-back material, beneath which is a lens of brecciated material. This passes radially outwards (both horizontally and vertically) into a zone of intense fracturing. The latter is believed to be due to the extensional stress relief that occurs after the intense compressional stress that initially traverses the rocks after impact. As reported by Donofrio (1981) and Pohl *et al.* (1977), the cosmic impact of a 1-km-diameter stony meteorite in Germany (the Ries Basin)

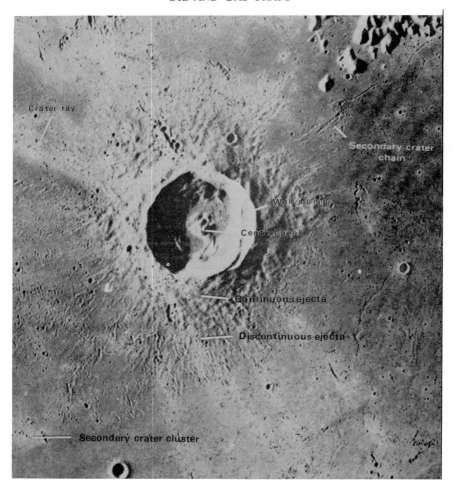

Fig. 11.16. The lunar crater Euler, 28 km in diameter (cf. Figure 11.15—Meteor Crater is 1.2 km in diameter). Euler is a complex crater, having developed a central uplift and some terracing (slumping) of the crater walls. Bowl-shaped features with no central uplift are classed as 'simple' craters. Reproduced from *Geology on the Moon*, Guest and Greeley (1977) by permission of the publisher, Taylor & Francis Ltd, London

caused the body to penetrate 600 m of sedimentary sequence, and another 650 m of crystalline basement rock. The impact and explosion excavated between *124 km³ and 200 km³* of rock, resulting in a complex ringed crater about 22 km in diameter. A seismic survey over the site showed that the crystalline basement rock at the centre of the feature was brecciated and fractured down to a depth of some 6 km.

Observations at various crater sites suggests that the fracture system extends laterally for considerable distances beyond the actual crater. In some instances—for example, at the double crater site of Clearwater Lake, Quebec, Canada—the fracture zone extends outwards to a distance of about 2 crater diameters around the two features.

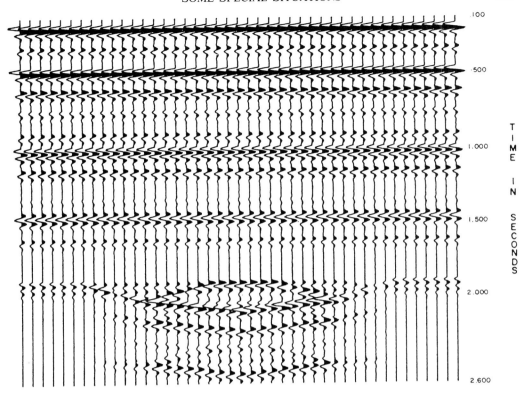

Fig. 11.17. Seismic model of an eroded, simple basement impact crater overlain by a normal marine sequence. Such models, as well as studies of the morphology of actual impact craters, should be made to enable recognition of these features when they occur in seismic and other subsurface data, as it is very probable that some buried impact craters will contain hydrocarbon accumulations.
Reproduced by permission of the *Journal of Petroleum Geology* (from Donofrio, 1981)

Basically, the trapping potential in an impact crater lies in fracture porosity above the spill point, wherever this happens to be in the particular crater. In a simple crater, the traps may be in the breccia lens and fracture system associated with the crater, with particular respect to the upturned crater-rim rocks. In a complex crater, to these can also be added any central uplift features, or wall terraces that may be developed. The rim rocks are somewhat complex stratigraphically, since the force of the impact explosion has usually inverted the sequence of strata there—so that there is a repeat sequence with the upper part being inverted. This has been shown to be the case, for instance, at Meteor Crater.

Earlier, it was noted that the only petroliferous examples so far found were only 'probable' or 'possible' craters. Such, for instance, are the oilfields found in two features in the North American Williston Basin. In Viewfield Oil Pool, for example, a producing Mississippian brecciated limestone forms a rim facies round a deep bowl-shape feature, while at Red Wing Creek, again, brecciated Mississippian rocks produce from a structure that has been interpreted as a central uplift in a faulted, bowl-shaped depression some 16 km in diameter (the Viewfield 'crater' within the rim facies is only 2–3 km in diameter).

A point that must be borne in mind is that all the craters, certain, probable, or possible, recognized on the Earth so far have *all* been found as surface indications. It is therefore unlikely that any would have retained hydrocarbons, even if oil or gas had at some time occupied the features. It seems far more likely that the very large number of impact craters that still remain buried, either in the sedimentary overburden, or in the basement complexes, would include a number of prospective hydrocarbon fields, provided that adequate seals were present. Craters that had been covered by a rapid transgressive depositional phase very shortly after impact would be particularly favourable from the viewpoint of morphology (uneroded rim and central uplift). The possibilities of immense volumes of fracture porosity must be borne in mind—cf. the Ries Basin mentioned above. There is no limitation here to relatively thin layers of a few tens of metres of potential reservoir. Areas 20–40 km in diameter, 6–10 km or more in depth, may be involved. It has been calculated that the reservoir volume for the Brent crater in Canada (3.8 km in diameter, impact age 420–480 Ma), including breccia lens and rim rock, would be of the order of 2 million acre feet.

Clearly, the main problem here is to find such features, and leaving aside serendipitous discoveries when drilling for other targets, remote sensing methods are needed. In particular, seismic data should in many instances be of use in recognizing craters. Donofrio (1981), shows a seismic model (see Fig. 11.17) of an eroded basement crater overlain by a normal marine sequence. The model has been constructed from a credible morphological simulation of a simple crater, with suitable seismic velocities fed into the processing program. Many other crater models built on knowledge of actual examples, both simple and complex, and from lunar or terrestrial sources, could be constructed, and it is important for the seismic interpreter to become acquainted with details of crater morphology, together with rock types involved, and likely velocities in brecciated and fractured rocks (including the matter of anisotropy in fracture systems).

The concept of deep fracturing of the crust resulting from impact cratering has an important bearing on the interesting attempt at the Siljan Ring Complex in Sweden, to detect commercial quantities of abiogenic gas derived from the mantle or crust–mantle boundary, as predicted by Gold and Soter (1980, 1982). The Siljan Ring Complex has been identified as a major impact crater of complex type, and it was conjectured that the fracture zone at depth would extend down to where abiogenic mantle-derived gas in large quantities would migrate into the deep fracture porosity resulting from the impact. Deep drilling at the site has so far not produced any unequivocal evidence for the presence of abiogenic gas in any significant quantity at this location.

There are many excellent books currently available for those wishing to study crater morphology, including Guest and Greeley (1977); Guest *et al.* (1979); Moore and Hunt (1983); Briggs and Taylor (1986). See also Ringwood (1970). For a relatively recent paper on a possible gas accumulation in an impact crater, see Jones (1983). An article by Shirley (1989) gives some most recent statistics of Earth impact craters.

11.5 Igneous Features in Seismic Data

11.5.1 Introduction

Igneous features of a variety of types are frequently present in the seismic data of shelf and depositional basin areas. Usually, they signify little that is not of academic interest

11.5.2 Hypabyssal Intrusions

These features, mainly igneous dykes and sills, are wall-like or sheet-like bodies injected from deep-seated and large bodies of magma into zones of weakness, fault planes, and bedding planes in the overlying crust.

Dykes are vertical or near-vertical walls of typically basalt-composition material (dolerite, tholeiite, etc.) injected into fault planes or other zones of weakness, and are usually indicative that extensional stress was present on a regional scale when the dykes were emplaced.

In seismic data, the effects of a thin dyke (probably 10–20 m thick) are seen in Fig. 11.18 at D. A small bilateral uplift is present at the location of the intrusion, over a substantial time zone on the seismic section. The strong event marked RR is a reflected refraction,

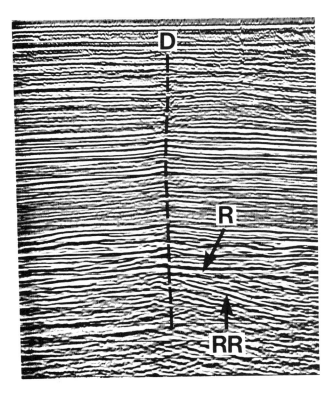

Fig. 11.18. Seismic expression of a small dyke (D) in clastic sediments. Also indicated is one of a series of reflected refractions (RR), this particular example being associated (for the refracted part of its seismic path) with the strong reflection event R. The dyke has probably reflected the seismic energy back along a path identical to its original downward trajectory, to produce events like RR.
Reproduced by permission of Seismograph Service (England) Ltd

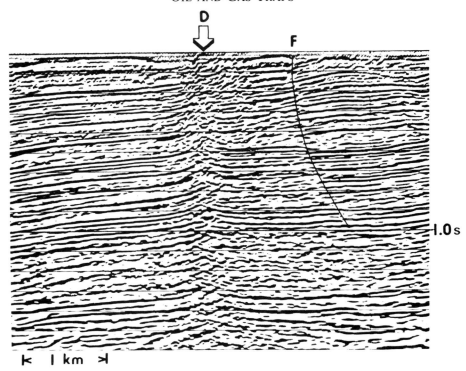

Fig. 11.19. Seismic data showing a larger dyke (D), and a listric normal fault F. Both the dyke and the fault could act as updip trap-forming elements under the right circumstances. Reproduced by permission of Seismograph Service (England) Ltd

the refracted part of the path (of the seismic signal) being related to the strong event marked R—a true stratal reflection (unlike RR which is just coherent noise, and has no physical reality). There are other events of a similar nature (and parallel to) RR above and below the latter, related to other refractors above and below R. Reflected refractions are often seen in the vicinity of faults and other lateral discontinuities, such as salt diapir flanks, as well as igneous intrusions.

Another, larger dyke (D) is shown in Fig. 11.19, adjacent to a slightly listric normal fault F. In this example, the dyke is thicker—perhaps 50–100 m thick—and is having a more disruptive effect on the seismic section than did the previous example. As always, the actual dyke cannot be seen; no vertical surface, or indeed, any surface dipping at greater than 45°, can be shown on a standard seismic section for simple geometrical reasons (energy will be reflected downwards instead of back up to the surface, with the normal seismic detector spread configuration). However, the dyke's location is quite clear from the effects seen. Again, its presence is marked by a bilateral upcurve of the stratal reflections on each side as they approach the location of the intrusion at various levels.

There is some controversy regarding this bilateral upcurve. Some regard the effect as an artifact due to velocity anomaly, perhaps related to contact metamorphism of the 'country rock' in zones on each side of the dyke. However, the author has disagreed with this view, after examining a number of dykes from different locations in seismic data. The opinion has been expressed (Jenyon, 1987b) that the upcurving of the stratal events on both sides of the dyke is probably a real effect, imparted to the sediments by friction from the very high pressure—one might say, explosive—injection of the magmatic material from below.

The suggested explanation involving contact metamorphism is rejected on the grounds that the scale of the effects seen does not match any possible contact metamorphic zones that might be present. Even with quite large dykes, as far as is known, the zone of contact metamorphism on each side of the intrusion is normally very limited in width (from a centimetre or so to a metre, perhaps). It is difficult to reconcile this, and any velocity anomaly deriving from it, to the effects seen, for instance, in Fig. 11.20, where the distance across the whole uplift zone is of the order of 1 km. Also, the uplift effects are *not constant* down the section, as would be expected with an artifact; this suggests a lithology related effect of some kind, in this context.

Another interesting point to consider in this example is the possibility of a dyke (or sill) intrusion forming a seal for hydrocarbon trapping. In Fig. 11.19, the fault F could act as an updip seal in a fault-trap situation—as also, it is suggested, could the dyke D, provided it was injected before migration of any hydrocarbons in the area.

In Fig. 11.20, two (possibly three) examples of fragments of igneous sill intrusions are interpreted as being present. Sills may be concordant (following bedding planes) or discordant (breaking across bedding planes), and in these examples there seems to be some element of discordance over parts of the intrusions. The sills are indicated by three small arrows on the line-drawn interpretation below the seismic section. As with dykes, it is

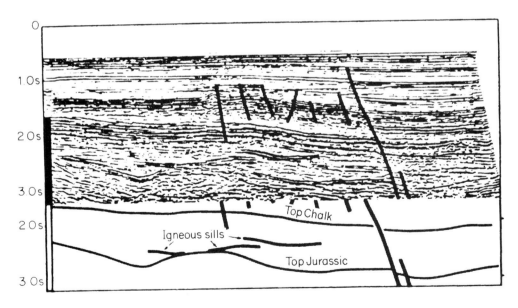

Fig. 11.20. Igneous sill intrusions in a seismic section (above), and indicated by the small arrows on a line-drawn interpretation (below). Reproduced by permission of Graham & Trotman Ltd (from Croker and Shannon, 1987)

possible that in some circumstances, sills could be involved as seal formations in hydrocarbon traps, especially where a discordant sill is involved.

11.5.3 Volcanic Features

A common feature in some areas is the presence of a sequence of flood basalts, sometimes of considerable thickness (hundreds, or even thousands, of metres). This can lead to severe attenuation of seismic data so that little or no information can be obtained from beneath the volcanic sequence. Flood basalts in seismic data often have the appearance of a buried topographic surface, and this is seen in the example of Fig. 11.21. The strong group of events indicated at B–B is the seismic expression of a sequence of Middle Jurassic alkali olivine basalts in the UK North Sea (northern area).

The usually strong acoustic impedance contrast between basalts and clastic sediments (the main reason for the severe attenuation of energy from below, as mentioned in the previous paragraph) means that a strong reflection group can be expected from the sediment-lava interface, as in this example. Note that the top of the lava sequence is irregular, and in places shows where different flow series overlap. At 'a' there is a small graben; here, and at the location of a small fault at the right-hand end of the section, the overburden above the lavas shows no sign of subsidence, differential compaction, etc. However, at 'b', the synform shape in the top of the lavas has subsidence above it in the sedimentary

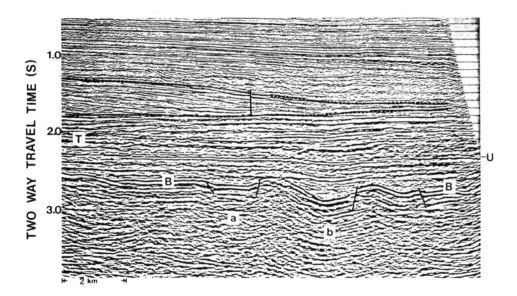

Fig. 11.21. Flood basalts as seen in a seismic section. The strong group of events B–B is at the top of a thick Middle Jurassic basalt sequence. Clear evidence of overlapping of flow series can be seen in some locations. T is the Base Tertiary, and U an unconformity within the Cretaceous. The feature 'a' is a small graben, while 'b' is interpreted as a fissure vent seen in cross-section. See text for further explanation. Reproduced by permission of Seismograph Service (England) Ltd

overburden. This has been interpreted (Jenyon, 1987b) as being due to compaction of gas-charged, vesicular lava as found in the mouth of a vent, and the feature 'b' is therefore interpreted as a fissure vent seen in cross-section.

Again, in this case, a thick basalt sequence (unless very severely weathered) is effectively impermeable, and could act as a seal in a hydrocarbon trap situation where the timing of oil/gas migration was favourable (i.e. later than the time of extrusion and consolidation of the lavas).

References and Bibliography

Amyx, J. W., Bass, D. M., Jr., and Whiting, R. L. (1960). *Petroleum Reservoir Engineering, Physical Properties*. McGraw Hill, New York, 610 pp.

Anderson, R. Y. and Kirkand, D. W. (1980). Dissolution of salt deposits by brine density flow. *Geol. Soc. Am. (Geol.)*, **8**, 66–69.

Anstey, N. A. (1977). *Seismic Interpretation—the Physical Aspects*. IHRDC, Boston, MA.

Armstrong, L. A., ten Have, A., and Johnson, H. D. (1987). The Geology of the Gannet Fields, Central North Sea, U.K. Sector. In Brooks, J. and Glennie, K. W. (Eds), *Petroleum Geology of North West Europe* (Proceedings of Barbican Conference), Vol. 1, 533–548. Graham & Trotman, London.

Arthur, K. R., Cole, D. R., Henderson, G. G. L., and Kushnir, D. W. (1982). Geology of the Hibernia discovery. In Halbouty, M. T. (Ed.), *The Deliberate Search for the Subtle Trap. Mem. Am. Assoc. Pet. Geol.*, **32**, 181–195.

Assefa, G. (1988). Potential hydrocarbon-generating rock units within the Phanerozoic sequence of the Ogaden Basin, Ethiopia: a preliminary assessment using the Lopatin Model. *J. Pet. Geol.*, **11**(4), 461–472.

Atwater, G. I. and Miller, E. E. (1965). The effect of decrease in porosity with depth on future developments of oil and gas reserves in South Louisiana. *Bull. Am. Assoc. Pet. Geol.*, **49**, 334 (abstr.).

Baar, C. A. (1977). *Applied Salt Rock Mechanics 1. The in situ behaviour of salt rocks*. Elsevier, Amsterdam. 294 pp.

Baars, D. L. and Seager, W. R. (1970). Stratigraphic control of petroleum in White Rim Sandstone (Permian) in and near Canyonlands National Park, Utah. *Bull. Am. Assoc. Pet. Geol.*, **54**, 709–718.

Baird, R. A. (1986). Maturation and source rock evaluation of Kimmeridge Clay, Norwegian North Sea. *Bull. Am. Assoc. Pet. Geol.*, **70**, 1–11.

Baker, E. G. (1962). Distribution of hydrocarbons in petroleum. *Bull. Am. Assoc. Pet. Geol.*, **46**, 76–84.

Bane, S. C. and Chanpong, R. R. (1980). Geology and development of the Teak Oil Field, Trinidad, West Indies. In Halbouty, M. T. (Ed.) *Giant Oil and Gas Fields of the Decade 1908–1978, Mem. Am. Assoc. Pet. Geol.*, **30**, 387–398.

Barker, C. (1981). Plate tectonics, organic matter, and basin evaluation for petroleum potential. *Am. Assoc. Pet. Geol., Slide-Tape Ser. in geol.*, Slide Nos. 04, 06, and 09 (Kerogen types).

Barnard, P. C., Cooper, B. S., and Fisher, M. J. (1978). Organic maturation and hydrocarbon generation in the Mesozoic sediments of the Sverdrup Basin, Arctic Canada. In *Proceedings of the 4th International Palynologist Congress, Lucknow, 1976*.

Barton, D. C. (1933). Mechanics of formation of salt domes with special reference to Gulf Coast salt domes of Texas and Louisiana. *Bull. Am. Assoc. Pet. Geol.*, **17**, 1025–1083.
Bates, C. C. (1953). Rational theory of delta formation. *Bull. Am. Assoc. Pet. Geol.*, **37**, 2119–2162.
Beard, D. C. and Weyl, P. K. (1973). Influence of texture on porosity and permeability of unconsolidated sand. *Bull. Am. Assoc. Pet. Geol.*, **57**, 349–369.
Beaudry, D. and Moore, G. F. (1985). Seismic stratigraphy and Cenozoic evolution of West Sumatra Forearc Basin. *Bull. Am. Assoc. Pet. Geol.*, **69**, 742–759.
Berg, O. R. (1982). Seismic detection and evaluation of delta and turbidite sequences: their application to the exploration for the subtle trap. In Halbouty, M. T. (Ed.), *The Deliberate Search for the Subtle Trap. Mem. Am. Assoc. Pet. Geol.*, **32**, 57–75.
Berg, O. R. and Woolverton, D. G. (Eds) (1985). *Seismic Stratigraphy, II. Mem. Am. Assoc. Pet. Geol.*, **39**, 276 pp.
Berg, R. R. (1975). Capillary pressure in stratigraphic traps. *Bull. Am. Assoc. Pet. Geol.*, **59**, 939–956.
Billings, M. P. (1972). *Structural Geology*, 3rd edn. Prentice-Hall, NJ, 606 pp.
Blair, D. G. (1975). Structural styles in North Sea oil and gas fields. In Woodland, A. W. (Ed.), *Petroleum and the Continental Shelf of North West Europe*, Vol. 1: *Geology*, 327–335. Applied Science Publishers, London, for The Institute of Petroleum, London.
Borchert, H. and Muir, R. O. (1964). *Salt Deposits*. Van Nostrand, Wokingham, 338 pp.
Bowen, J. M. (1975). The Brent oil-field. In Woodland, A. W. (Ed.), *Petroleum and the Continental Shelf of North West Europe*, Vol. 1: *Geology*, 353–361. Applied Science Publishers, London, for The Institute of Petroleum, London.
Brady, T. J., Campbell, N. D. J., and Maher, C. E. (1980). Intisar "D" Oil Field, Libya. In Halbouty, M. T. (Ed.) *Giant Oil and Gas Fields of the Decade: 1968–1978. Mem. Am. Assoc. Pet. Geol.*, **30**, 543–564.
Braitsch, O. (1971). *Salt Deposits—Their Origin and Composition*. Springer, New York, 297 pp.
Brennand, T. P. and van Veen, F. R. (1975). The Auk Oil-Field. In Woodland, A. W. (Ed.), *Petroleum and the Continental Shelf of North West Europe*, Vol. 1: *Geology*, 275–281. Applied Science Publishers, London, for The Institute of Petroleum, London.
Brewster, J., Dangerfield, J., and Farrell, H. (1986). The geology and geophysics of the Ekofisk Field waterflood. *Mar. Pet. Geol.*, **3**, 139–169.
Briggs, G. A. and Taylor, F. W. (1986). *Cambridge Photographic Atlas of the Planets*. Cambridge University Press, Cambridge, 256 pp.
Brooks, J. and Glennie, K. W. (Eds), (1987). *Petroleum Geology of North West Europe* (Proceedings of Barbican Conference), Vols. 1 and 2. Graham & Trotman, London, 1219 pp.
Bubb, J. N. and Hatlelid, W. G. (1977). Seismic recognition of carbonate buildups. In Payton, C. E. (Ed.), *Seismic Stratigraphy—applications to hydrocarbon exploration. Mem. Am. Assoc. Pet. Geol.*, **26**, 185–204.
Burley, S. D. and Kantorowicz, J. D. (1986). Thin section and S. E. M. textural criteria for the recognition of cement-dissolution porosity in sandstones. *Sedimentology*, **33**, 587–604.
Butler, J. B. (1975). The West Sole gas-field. In: Woodland, A. W. (Ed.), *Petroleum and the Continental Shelf of North West Europe*, Vol. 1: *Geology*, 213–219. Applied Science Publishers, London, for The Institute of Petroleum, London.
Butler, R. W. H. (1987). Thrust sequences. *J. Geol. Soc. London*, **144**, 619–634.
Cant, D. J. (1986). Diagenetic traps in sandstones. *Bull. Am. Assoc. Pet. Geol.*, **70**, 155–160.
Cardott, B. J. and Lambert, M. W. (1985). Thermal maturation by vitrinite reflectance of Woodford Shale, Anadarko Basin, Oklahoma. *Bull. Am. Assoc. Pet. Geol.*, **69**, 1982–1998.
Carlson, P. R., Golan-Bac, M., Karl, H. A., and Kvenvolden, K. A. (1985). Seismic and geochemical evidence for shallow gas in sediment on Navarin continental margin, Bering Sea. *Bull. Am. Assoc. Pet. Geol.*, **69**, 422–434.
Caselli, F. (1987). Oblique-slip tectonics, Mid-Norway Shelf. In Brooks, J. and Glennie, K. W. (Eds), *Petroleum Geology of North West Europe* (Proceedings of Barbican Conference), Vol. 2, 1049–1063. Graham & Trotman, London.
Cayley, G. T. (1987). Hydrocarbon migration in the central North Sea. In Brooks, J. and Glennie, K. W. (Eds), *Petroleum Geology of North West Europe* (Proceedings of Barbican Conference), Vol. 1, 549–555. Graham & Trotman, London.

Chapman, R. E. (1976). *Petroleum Geology: A Concise Study*. Elsevier, Amsterdam, 302 pp.

Childs, F. B. and Reed, P. E. C. (1975). Geology of the Dan Field and the Danish North Sea. In Woodland, A. W. (Ed.), *Petroleum and the Continental Shelf of North West Europe*, Vol. 1: *Geology*, 429–438. Applied Science Publishers, London, for The Institute of Petroleum, London.

Clifford, H. J., Grund, R., and Musrati, H. (1980). Geology of a stratigraphic giant: Messla Oil Field, Libya. In Halbouty, M. T. (Ed.) *Giant Oil and Gas Fields of the Decade: 1968–1978. Mem. Am. Assoc. Pet. Geol.*, **30**, 507–524.

Cloos, H. (1930). Zur Experimentellen Tektonik. *Die Naturwissenschaften*, **18**(34), 741–747.

Coelewij, P. A. J., Haug, G. M. W., and Van Kuijk, H. (1978). Magnesium-salt exploration in the northeastern Netherlands. *Geol. Mijnbouw*, **57**, 487–502.

Cole, F. W. (1969). *Reservoir Engineering Manual*. Gulf Publishing Company, Houston, TX, 385 pp.

Collins, B. P. and Watkins, J. S. (1985). Analysis of a gas hydrate off southwest Mexico using seismic processing techniques and Deep Sea Drilling Project Leg 66 results. *Geophysics*, **50**, 16–24.

Connan, J. (1974). Time–temperature relation in oil genesis. *Bull. Am. Assoc. Pet. Geol.*, **58**, 2516–2521.

Cooper, B. S. (1977). Estimation of the maximum temperature attained in sedimentary rocks. In Hobson, G. D. (Ed.), *Developments in Petroleum Geology*, Vol. 1, 127–146. Applied Science Publishers, London.

Cornford, C. (1986). Source rocks and hydrocarbons of the North Sea. In K. W. Glennie (Ed.), *Introduction to the Petroleum Geology of the North Sea*, 2nd edn, 197–236. Blackwell Scientific, Oxford.

Coveney, R. M., Jr., Goebel, E. D., Zeller, E. J., Dreschhoff, G. A. M., Angino, E. E. (1987). Serpentination and the origin of hydrogen gas in Kansas. *Bull. Am. Assoc. Pet. Geol.*, **71**, 39–48.

Cowan, P. E. and Harris, P. M. (1986). Porosity distribution in San Andres Formation (Permian), Cochran and Hockley Counties, Texas. *Bull. Am. Assoc. Pet. Geol.*, **70**, 888–897.

Crain, W. E., Mero, W. E., and Patterson, D. (1985). Geology of the Point Arguello Discovery. *Bull. Am. Assoc. Pet. Geol.*, **69**, 537–545.

Crampin, S. (1985). Evaluation of anisotropy by shear-wave splitting. *Geophysics*, **50**, 142–152.

Crampin, S. (1986). Anisotropy and transverse isotropy. *Geophys. Prospect.*, **34**, 34–99.

Crampin, S. (1987). Geological and industrial implications of extensive-dilatancy anisotropy. *Nature*, **328**, 491–496.

Crampin, S., Bush, I., Neville, C., and Taylor, D. B. (1986). Three-component VSP modelling of shear waveforms. *The Leading Edge*, **5**, 35–39.

Crans, W. and Mandl, G. (1980–81). On the theory of growth faulting. *J. Pet. Geol.*, **2**(3) and **3**(2–4).

Croker, P. F. and Shannon, P. M. (1987). The evolution and hydrocarbon prospectivity of the Porcupine Basin, offshore Ireland. In Brooks, J. and Glennie, K. W. (Eds), *Petroleum Geology of North West Europe*, (Proceedings of Barbican Conference) Vol. 2, 633–642. Graham & Trotman, London.

Cumming, A. D. and Wyndham, C. L. (1975). The geology and development of the Hewett Gasfield. In Woodland, A. W. (Ed.), *Petroleum and the Continental Shelf of North West Europe*, Vol. 1: *Geology*, 313–325. Applied Science Publishers, London, for The Institute of Petroleum, London.

Dahlberg, E. C. (1982). *Applied Hydrodynamics in Petroleum Exploration*. Springer, New York, 171 pp.

Davies, E. J. and Watts, T. R. (1977). The Murchison oil-field. In Finstad, K. G. and Selley, R. C. (Eds), *Proceedings of the Mesozoic Northern North Sea Symposium 1977* (MNNSS-77). Norwegian Petroleum Society.

Davis, R. W. (1987). Analysis of hydrodynamic factors in petroleum migration and entrapment. *Bull. Am. Assoc. Pet. Geol.*, **71**, 643–649.

De'Ath, N. G. and Schuyleman, S. F. (1981). The geology of the Magnus oilfield. In Illing, L. V. and Hobson, G. D. (Eds), *Petroleum Geology of the Continental Shelf of North-West Europe*, Vol. 1: *Geology*, 342–351. Heyden, London, for The Institute of Petroleum, London.

Dennison, J. M. (1971). Petroleum related to Middle and Upper Devonian deltaic facies in Central Appalachians. *Bull. Am. Assoc. Pet. Geol.*, **55**, 1179–1193.

De Righi, M. R. and Bloomer, G. (1975). Oil and gas developments in the Upper Amazon Basin—Colombia, Ecuador and Peru. In: *Proceedings of the 9th World Petroleum Congress*, Vol. 3, 181–192. Applied Science Publishers, London.

Dickey, P. A. (1975). Possible primary migration of oil from source rock in oil phase. *Bull. Am. Assoc. Pet. Geol.*, **59**, 337–345.

Dietz, R. S. (1961). Astroblemes. *Sci. Am.*, **205**, 50–58.

Dixon, S. A. and Kirkland, D. W. (1985). Relationship of temperature to reservoir quality for feldspathic sandstone of southern California. *Am. Assoc. Pet. Geol., Southwest Sec. Trans.*, 82–99.

Donofrio, R. R. (1981). Impact craters: implications for basement hydrocarbon production. *J. Pet. Geol.*, **3**, 279–302.

Downey, M. W. (1984). Evaluating seals for hydrocarbon accumulations. *Bull. Am. Assoc. Pet. Geol.*, **68**, 1752–1763.

Dunnington, H. V. (1967). Aspects of diagenesis and shape change in stylolitic limestone reservoirs. In *Proceedings of the 7th World Petroleum Congress*, Vol. 2, 339–352. Elsevier Applied Science Publishers, London.

Edmunds, R. H. (1980). Salt removal and oil entrapment. *Mem. Can. Soc. Pet. Geol.*, **6**, 988.

Elzarka, M. H. and Younes, M. A. A. (1987). Generation, migration and accumulation of oil of El-Ayun Field, Gulf of Suez, Egypt. *Mar. Pet. Geol.*, **4**, 320–333.

England, W. A., *et al.* (1987). The movement and entrapment of petroleum fluids in the subsurface. *J. Geol. Soc. London*, **144**, 327–347.

Ensley, R. A. (1984). Comparison of P- and S-wave seismic data: A new method for detecting gas reservoirs. *Geophysics*, **49**, 1420–1431.

Ensley, R. A. (1985). Evaluation of direct hydrocarbon indicators through comparison of compressional- and shear-wave seismic data: a case study of the Myrnam gas field, Alberta. *Geophysics*, **50**, 37–48.

Epstein, A. G., Epstein, J. B., and Harris, L. (1977). Conodont colour alteration—an index to organic metamorphism. *U.S. Geol. Surv. Prof. Pap.*, **995**, 1–27.

European Continental Shelf Guide (1986–1987). Oilfield Publications Limited, Ledbury, 577 pp.

Evamy, B. D., Haremboure, J., Kamerling, P., Knaap, W. A., Molloy, F. A., and Rowlands, P. H. (1978). Hydrocarbon habitat of Tertiary Niger delta. *Bull. Am. Assoc. Pet. Geol.*, **62**, 1–39.

Evans, C. R. and Staplin, F. L. (1971). Regional facies of organic metamorphism. *Can. Inst. Min. Metall., Spec. Rep.*, **11**, 517–520.

Evrenos, A. I., Heathman, J., and Ralstin, J. (1971). Impermeation of porous media by forming hydrates *in situ*. *J. Pet. Tech.*, **23**, 1059

Ewing, J. L. and Hollister, C. H. (1972). Regional aspects of deep sea drilling in the western North Atlantic. In Hollister, C. H., *et al.* (Eds), *Initial Reports of the Deep Sea Drilling Project*, **11**, 951–973. US Government Printing Office, Washington, DC.

Fagerland, N. (1984). A rift-related stratigraphic trap. *J. Pet. Geol.*, **7**, 227–236.

Ferguson, J. (1988). Oil generation and migration within marine carbonate sequences—a review. *J. Pet. Geol.*, **11**, 389–402.

Fleuty, M. J. (1964). The description of folds. *Proc. Geol. Assoc.*, **75**, 461–489.

Fontaine, J. M., Cussey, R., Lacaze, J., Lanaud, R., and Yapaudjian, L. (1987). Seismic interpretation of carbonate depositional environments. *Bull. Am. Assoc. Pet. Geol.*, **71**, 281–297.

Fowler, C. (1975). The geology of the Montrose Field. In Woodland, A. W. (Ed.), *Petroleum and the Continental Shelf of North West Europe*, Vol. 1: Geology, 467–476. Applied Science Publishers, London, for The Institute of Petroleum, London.

Frankl, E. J. and Cordry, E. A. (1967). The Niger Delta oil province—recent developments onshore and offshore. In *Proceedings of the 7th World Petroleum Congress*, Vol. 2, 195–209. Applied Science Publishers, London.

Fraser, W. W. (1967). Geology of the Zelten Field, Libya, North Africa. In *Proceedings of the 7th World Petroleum Congress*, Vol. 2, 259–264. Applied Science Publishers, London.

French, B. M. (1968). Shock metamorphism as a geological process. In B. M. French and N. M. Short (Eds), *Shock Metamorphism of Natural Materials*, 1–17. Mono Book Corporation, Baltimore, MD.

Frost, R. E. (1987). The evolution of the Viking Graben tilted fault block structures; a compressional origin. In Brooks, J. and Glennie, K. W. (Eds), *Petroleum Geology of North West Europe* (Proceedings of Barbican Conference), Vol. 2, 1009–1024. Graham & Trotman, London.

Fryberger, S. G. (1986). Stratigraphic traps for petroleum in wind-laid rocks. *Bull. Am. Assoc. Pet. Geol.*, **70**, 1765–1776.

Fuller, J. G. C. M. (1982). Oil and gas source rocks: generation and migration. In K. W. Glennie (Ed.), *Introduction to the Petroleum Geology of the North Sea*, Course Notes No. 7, Joint Association for Petroleum Exploration Courses (UK), London.

Galloway, W. E. (1974). Deposition and the genetic alteration of sandstone in northeast Pacific arc-related basins: implications for graywacke genesis. *Geol. Soc. Am. Bull.*, **85**, 379–390.

Galloway, W. E. (1975). Evolution of deltaic systems. In *Deltas, Models for exploration*, 87–89. Houston Geological Society, TX.

Ganz, H., Kalkreuth, W., Oner, F., Pearson, M. J., and Small, J. S. (1987). Enhanced geochemical screening analysis—accurate determination of petroleum source-rock characteristics and comparison with standard techniques. In Brooks, J. and Glennie, K. W. (Eds), *Petroleum Geology of North West Europe*, Vol. 2, 847–851. Graham & Trotman.

Gault, D. E., Quaide, W. L., and Oberbeck, V. R. (1968). Impact cratering mechanics and structures. In French, B. M. and Short, N. M. (Eds), *Shock Metamorphism of Natural Materials*, 87–99. Mono Book Corporation, Baltimore, MD.

Gerlach, T. (1980). Chemical characteristics of the volcanic gases from the Nyirangongo lava lake and the generation of CH-rich fluid inclusions in alkaline rocks. *J. Volcanic Geotherm. Res.*, **9**, 177–189.

Giardini, A. A. and Melton, C. E. (1983). A scientific explanation for the origin and location of petroleum accumulations. *J. Pet. Geol.*, **6**, 117–138.

Gibbs, A. D. (1984). Structural evolution of extensional basin margins. *J. Geol. Soc. London*, **141**, 609–620.

Glennie, K. W., Mudd, G. C., and Nagtegaal, P. J. C. (1978). Depositional environment and diagenesis of Permian Rotliegendes sandstones in Leman Bank and Sole Pit areas of the U.K. Southern North Sea. *J. Geol. Soc. London*, **135**, 25–34.

Goff, J. C. (1983). Hydrocarbon generation and migration from Jurassic source rocks in the East Shetland Basin and Viking Graben of the northern North Sea. *J. Geol. Soc. London*, **140**, 445–474.

Gold, T. and Soter, S. (1980). The deep earth-gas hypothesis. *Sci. Am.*, **242**, 154–161.

Gold, T. and Soter, S. (1982). Abiogenic methane and the origin of petroleum. *Energy Explor. Exploit.*, **1**, 2.

Goodarzi, F. and Higgins, A. C. (1987). Optical properties of scolecodonts and their use as indicators of thermal maturity. *Mar. Pet. Geol.*, **4**, 353–359.

Goudswaard, W. and Jenyon, M. K. (Eds). (1988). *Seismic Atlas of Structural and Stratigraphic Features* (with Supplementary Glossary). European Association of Exploration Geophysicists, The Hague.

Grace, J. D. and Hart, G. F. (1986). Giant gas fields of Northern West Siberia. *Bull. Am. Assoc. Pet. Geol.*, **70**, 830–852.

Gray, I. (1975). Viking gas field. In Woodland, A. W. (Ed.), *Petroleum and the Continental Shelf of North West Europe*, Vol. 1: *Geology*, 241–247. Applied Science Publishers, London, for The Institute of Petroleum, London.

Gray, W. D. T. and Barnes, G. (1981). The Heather oil field. In Illing, L. V. and Hobson, G. D. (Eds), *Petroleum Geology of the Continental Shelf of North-West Europe*, 335–341. Heyden, London, for the Institute of Petroleum, London.

Grieve, R. A. F. and Robertson, P. B. (1979). The terrestrial cratering record. I—current status of observations. *Icarus*, **38**, 212–229.

Grunau, H. R. (1981). Worldwide review of seals for major accumulations of natural gas. *Bull. Am. Assoc. Pet. Geol.*, **65**, 953 (asbstr.).

Grunau, H. R. (1983). Abundance of source rocks for oil and gas worldwide. *J. Pet. Geol.*, **6**, 39–54.

Guest, J. E. and Greeley, R. (1977). *Geology on the Moon*. Taylor & Francis Ltd, London, 233 pp.

Guest, J. E., Butterworth, P., Murray, J., and O'Donnell, W. (1979). *Planetary Geology*. David & Charles, Newton Abbot, 208 pp.

Guthrie, J. M., Hauseknecht, D. W., and Johns, W. D. (1986). Relationships among vitrinite reflectance, illite crystallinity, and organic geochemistry in Carboniferous strata, Ouachita Mountains, Oklahoma and Arkansas. *Bull. Am. Assoc. Pet. Geol.*, **70**, 26–33.

Halbouty, M. T. (1967). *Salt Domes—Gulf Region, United States and Mexico*. Gulf Publishing Corporation, Houston, TX.

Halbouty, M. T. (Ed.) (1980). *Giant Oil and Gas Fields of the Decade: 1968–1978. Mem. Am. Assoc. Pet. Geol.*, **30**, 596 pp.

Halbouty, M. T. (Ed.) (1982). *The Deliberate Search for the Subtle Trap. Mem. Am. Assoc. Pet. Geol.*, **32**, 361 pp.

Hallett, D. (1981). Refinement of the geological model of the Thistle Field. In Illing, L. V. and Hobson, G. D. (Eds), *Petroleum Geology of the Continental Shelf of North West Europe*. Heyden, London, for The Institute of Petroleum, London.

Ham, W. E., et al. (1973). Regional Geology of the Arbuckle Mountains, Oklahoma. *Okla. Geol. Surv. Spec. Publ.*, **73-3**, 61 pp.

Hammond, R. D., and Gaither, J. R. (1983). Anomalous seismic character—Bering Sea shelf. *Geophysics*, **48**, 590–605.

Hancock, F. R. P. and Mithen, D. P. (1987). The geology of the Humbly Grove oilfield, Hampshire, U.K. In Brooks, J. and Glennie, K. W. (eds), *Petroleum Geology of North West Europe* (Proceedings of Barbican Conference), Vol. 1, 161–170. Graham & Trotman, London.

Harding, T. P. (1985). Seismic characteristics and identification of negative flower structures, positive flower structures and positive structural inversion. *Bull. Am. Assoc. Pet. Geol.*, **69**, 582–600.

Hardman, R. F. P. and Eynon, G. (1977). Valhall Field-a structural/stratigraphic trap. *NPF Seminar, 1977*, Paper 14.

Harms, T. C., Tackenberg, P., Pickles, E., and Pollock, R. E. (1981). The Brae oilfield area. In Illing, L. V. and Hobson, G. D. (Eds), *Petroleum Geology of the Continental Shelf of North West Europe*. Heyden, London, for The Institute of Petroleum, London.

Hass, W. H. and Huddle, J. W. (1965). Late Devonian and Early Mississippian age of the Woodford Shale in Oklahoma as determined by conodonts. In *Geological Survey Research. U.S. Prof. Pap.*, **525-D**, 124–132.

Hauseknecht, D. W. and Matthews, S. M. (1985). Thermal maturity of Carboniferous strata, Ouachita Mountains. *Bull. Am. Assoc. Pet. Geol.*, **69**, 335–345.

Hawle, H., Kratochvil, H., Schmied, H., and Wieseneder, H. (1967). Reservoir geology of the carbonate oil and gas reservoirs of the Vienna Basin. In *Proceedings of the 7th World Petroleum Congress*, Vol. 2, 371–395. Applied Science Publishers, London.

Hedberg, H. D. (1967). Geologic controls on petroleum genesis. In *Proceedings of the 7th World Petroleum Congress*, Vol. 2, 3–11. Applied Science Publishers, London.

Héritier, F. E., Lossel, P., and Wathne, E. (1981). The Frigg gas field. In Illing, L. V. and Hobson, G. D. (Eds), *Petroleum Geology of the Continental Shelf of North-West Europe*, 380–391. Heyden, London for The Institute of Petroleum, London.

Heylmun, E. B. (1964). Shallow oil and gas possibilities in east and south-central Utah. *Utah Geol. Mineral. Surv. Spec. Stud.*, **8**(39), 4 pp.

Hobson, G. D. (1954). *Some Fundamentals of Petroleum Geology*. Oxford University Press, London.

Hobson, G. D. (Ed.) (1977). *Developments in Petroleum Geology*, Vol. 1. Applied Science Publishers, London, 335 pp.

Hobson, G. D. (1980). Musing on migration. *J. Pet. Geol.*, **3**, 237–240.

Hoffman, P., Dewey, J. F., and Burke, K. (1974). Aulacogens and their genetic relation to geosynclines, with a Proterozoic example from Great Slave Lake, Canada. In *Modern and Ancient Geosynclinal Sedimentation. Soc. Econ. Paleontol. Mineral. Spec. Publ.*, **19**, 38–55.

Holland, D. S., Nunan, W. E., Lammlein, D. R., and Woodhams, R. L. (1980). Eugene Island Block 330 Field, Offshore Louisiana. In M. T. Halbouty (Ed.), *Giant Oil and Gas Fields of the Decade: 1968–1978. Mem. Amer. Assoc. Pet. Geol.*, **30**.

Hood, A., Gutijahr, C. C. M., and Hancock, R. L. (1975). Organic metamorphism and the generation of petroleum. *Bull. Am. Assoc. Pet. Geol.*, **59**, 986–996.

Hornabrook, J. T. (1975). Seismic interpretation of the West Sole gas field. *Nor. Geol. Unders. (Publ.)*, **Offprint NGU 316**, 121–135.

Houseknecht, D. W. (1987). Assessing the relative importance of compaction processes and cementation to reduction of porosity in sandstones. *Bull. Am. Assoc. Pet. Geol.*, **71**, 633–642.

Hovland, M. and Sommerville, J. H. (1985). Characteristics of two natural gas seepages in the North Sea. *Mar. Pet. Geol.*, **2**, 319–325.

Hubbert, M. K. (1953). Entrapment of petroleum under hydrodynamic conditions. *Bull. Am. Assoc. Pet. Geol.*, **37**, 1954–2026.

Hubbert, M. K. (1967). Application of hydrodynamics to oil exploration. In *Proceedings of the 7th World Petroleum Congress*, Vol. 1, 59–75. Applied Science Publishers, London.

Hunt, J. M. (1967). The origin of petroleum in carbonate rocks. In Chillingar, G. V., Bissel, H. J., and Fairbridge, R. W. (Eds), *Carbonate Rocks*, 225–251. Elsevier, New York.

Ibe, A. C., Ferguson, J., Kinghorn, R. R. F., and Rahman, M. (1983). Organic matter content in carbonate sediments in relation to petroleum occurrence. *J. Pet. Geol.*, **6**, 53–70.

Illing, L. V. and Hobson, G. D. (Eds) (1981). *Petroleum Geology of the Continental Shelf of North-West Europe.* Heyden, London, for The Institute of Petroleum, London, 521 pp.

Illing, L. V., Wood, G. V., and Fuller, J. G. C. M. (1967). Reservoir rocks and stratigraphic traps in non-reef carbonates. In *Proceedings of the 7th World Petroleum Congress*, Vol. 2. Applied Science Publishers, London.

Illing, V. C. (Ed.) (1953). *The Science of Petroleum,* Vol. vi, *The World's Oilfields,* Part I, *The Eastern Hemisphere.* Oxford University Press, London.

Isaacs, C. M. (1983). Porosity reduction during diagenesis of the Monterey formation, Santa Barbara Coastal Area, California. In *The Monterey Formation and Related siliceous Rocks of California,* 257–271. Society of Economic Paleontologists and Mineralogists, Tulsa.

Jenyon, M. K. (1984a). Seismic response to collapse structures in the Southern North Sea. *Mar. Pet. Geol.*, **1**, 27–36.

Jenyon, M. K. (1984b). Upper Carboniferous gas indications and Zechstein features in the Southern North Sea. *Oil Gas J.*, **14 May**, 135–144.

Jenyon, M. K. (1985a). Basin-edge diapirism and updip salt flow in the Zechstein of the Southern North Sea. *Bull. Am. Assoc. Pet. Geol.*, **69**, 53–64.

Jenyon, M. K. (1985b). Fault-associated salt flow and mass movement. *J. Geol. Soc. London*, **142**, 547–553.

Jenyon, M. K. (1985c). Differential movement in salt rock. *Oil Gas J.*, **30 September**, 73–75.

Jenyon, M. K. (1986a). Some consequences of faulting in the presence of a salt rock interval. *J. Pet. Geol*, **9**, 29–52.

Jenyon, M. K. (1986b). DHI events and a possible gas prospect in open acreage in the North Sea. *Oil Gas J.*, **11 August**, 108–112.

Jenyon, M. K. (1986c). Seismic characteristics of shale and evaporite cap rock, with North Sea examples. In *Cap Rock Report*, 8–11. Petroconsultants S.A., Geneva.

Jenyon, M. K. (1986d). *Salt Tectonics.* Elsevier Applied Science Publishers, London, 191 pp.

Jenyon, M. K. (1986e). Salt—with a pinch of water. *Nature*, **324**, 515–516.

Jenyon, M. K. (1987a). Seismic expression of real and apparent buried topography. *J. Pet. Geol.*, **10**, 41–58.

Jenyon, M. K. (1987b). Characteristics of some igneous extrusive and hypabyssal features in seismic data. *Geol. Soc. Am. (Geol.)*, **15**, 237–240.

Jenyon, M. K. (1987c). The development by salt diapirs of superficial overhang features, and effects on associated sediments. In Lerche, I. and O'Brien, J. J. (Eds), *Dynamical Geology of Salt and Related Structures.* Academic Press, Orlando, FL, 832 pp.

Jenyon, M. K. (1987d). Regional salt movement in the British Southern Zechstein Basin. In Peryt, T. M. (Ed.), *The Zechstein Facies in Europe* (Lecture Notes in Earth Sciences). Springer, Berlin, 272 pp.

Jenyon, M. K. (1988a). Overburden deformation related to the pre-piercement development of salt structures in the North Sea. *J. Geol. Soc. London*, **145**, 445–454.

Jenyon, M. K. (1988b). Some deformation effects in a clastic overburden resulting from salt mobility. *J. Pet. Geol.*, **11**, 309–324.

Jenyon, M. K. (1988c). Seismic response to an evaporite/carbonate association. *Carbonates and Evaporites*, Vol. 2, 89–93. Northeastern Science Foundation Inc., CUNY, New York.

Jenyon, M. K. (1988d). Seismic expression of salt dissolution-related features in the North Sea. *Bull. Can. Pet. Geol.*, **36**, 274–283.

Jenyon, M. K. (1988e). Re-entrants of Zechstein Z2 salt in the Mid North Sea High. *Mar. Pet. Geol.*, **5**, 352–358.

Jenyon, M. K. (1988f). Fault–salt wall relationships, Southern North Sea. *Oil Gas J.*, **5 September**, 76–81.

Jenyon, M. K. (1989). Plastic flow and contraflow in salt layers separated by a competent band. *J. Pet. Geol.*, in press.

Jenyon, M. K. and Cresswell, P. M. (1987). The Southern Zechstein salt basin of the British North Sea, as observed in regional seismic traverses. In Brooks, J. and Glennie, K. W. (Eds), *Petroleum Geology of North West Europe* (Proceedings of Barbican Conference), Vol. 1, 277–292. Graham & Trotman, London.

Jenyon, M. K. and Fitch, A. A. (1985). *Seismic Reflection Interpretation*. Gebrueder Borntraeger, Stuttgart, 318 pp.

Jenyon, M. K. and Taylor, J. C. M. (1983). Hydrocarbon indications associated with North Sea Zechstein shelf features. *Oil Gas J.*, **5 December**, 155–160.

Jenyon, M. K. and Taylor, J. C. M. (1987). Dissolution effects and reef-like features in the Zechstein across the Mid North Sea High. In Peryt, T. M. (Ed.), *The Zechstein Facies in Europe*, 51–75. Springer, Berlin.

Jenyon, M. K., Cresswell, P. M., and Taylor, J. C. M. (1984). The nature of the connection between the Northern and Southern Zechstein Basins across the Mid North Sea High. *Mar. Pet. Geol.*, **1**, 355–363.

Johnson, A. and Eyssautier, M. (1987). Alwyn North Field, and its regional geological context. In Brooks, J. and Glennie, K. W. (Eds), *Petroleum Geology of North West Europe* (Proceedings of Barbican Conference), Vol. 2, 963–977. Graham & Trotman, London.

Johnston, D. D. and Johnson, R. J. (1987). Depositional and diagenetic controls on reservoir quality in First Wilcox Sandstone, Livingstone Field, Louisiana. *Bull. Am. Assoc. Pet. Geol.*, **71**, 1152–1161.

Jones, W. B. (1983). A proposed gas pool in the Pleistocene Bosumtwi Impact crater, Ghana. *J. Pet. Geol.*, **5**, 315–318.

Karus, E. V., Ryabinkin, L. A., Galperin, E. I., Teplitskiy, V. A., Demidenko, Yu. B., Mustafayev, K. A., and Rapoport, M. B. (1975). Detailed investigations of geological structures by seismic well surveys. In *Proceedings of the 9th World Petroleum Congress*, Vol. 3, 247–257. Applied Science Publishers, London.

Katz, H. -R. (1981). Probable gas hydrate in continental slope east of the North Island, New Zealand. *J. Pet. Geol.*, **3**, 315–324.

Kennedy, C. L., Miller, J. A., Kelso, J. B., Lago, O. K., and Peterson, D. S. (1971). *The Deep Anadarko Basin*. Petroleum Information, Denver, CO, 359 pp.

Kent, P. E. (1953). Great Britain. In Illing, V. C. (Ed.), *The Science of Petroleum*, Vol. vi, *The World's Oilfields*, Part I, *The Eastern Hemisphere*, 54–57. Oxford University Press, London.

Kent, P. E. (1979). The emergent Hormuz salt plugs of Southern Iran. *J. Pet. Geol.*, **2**, 117–144.

Khan, I., et al. (1986). *Bull. Am. Assoc. Pet. Geol.*, **70**(4), 396–414.

Kidwell, A. L. and Hunt, J. M. (1958). Migration of oil in Recent sediments of Pedernales, Venezuela. In Weeks, L. G. (Ed.), *Habitat of Oil*, 790–817. American Association of Petroleum Geologists, Tulsa, OK.

Kinghorn, R. R. F. and Rahman, M. (1983). Specific gravity as a kerogen type and maturation indicator, with special reference to amorphous kerogens. *J. Pet. Geol.*, **6**, 179–194.

Kirkland, D. W. and Evans, R. (1981). Source rock potential of evaporite environment. *Bull. Am. Assoc. Pet. Geol.*, **65**, 181–190.

Kiser, G. D. (1967). Guanarito erosional pinchout, Venezuela. In *Proceedings of the 7th World Petroleum Congress*, Vol. 2, 501–509. Applied Science Publishers, London.

KOC Staff (1953). Kuwait. In Illing, V. C. (Ed.), *The World's Oilfields*. Oxford University Press, Oxford.

Koestler, A. G. and Ehrmann, W. U. (1987). Fractured chalk overburden of a salt diapir, Laegerdorf, N. W. Germany—exposed example of a possible hydrocarbon reservoir. In Lerche, I. and O'Brien, J. J. (Eds), *Dynamical Geology of Salt and Related Structures*, 457–477. Academic Press, Orlando, FL.

Krumbein, W. C. and Monk, G. D. (1942). Permeability as a function of the size parameters of unconsolidated sand. *Pet. Tech., Am. Inst. Min. Eng. Tech. Publ.*, **1492**, 1–11.

Kubler, B. (1968). Évaluation quantitative du métamorphisme par la cristallinité de l'illite; état des progrès realisés ces dernières années. *Bulletin du Centre de Recherches de Pau*, **2**, 385–397.

Kvenvolden, K. A. (1985). Comparison of marine gas hydrates in sediments of an active and passive continental margin. *Mar. Pet. Geol.*, **2**, 65–71.

Kvenvolden, K. A. and Barnard, L. A. (1983). Hydrates of natural gas in continental margins. In Watkins, J. S. and Rake, C. L. (Eds), *Studies in Continental Margin Geology. Mem. Am. Assoc. Pet. Geol.*, **34**, 631–640.

Lamb, C. F. (1980). Painter Reservoir Field—giant in the Wyoming Thrust Belt. In Halbouty, M. T. (Ed.), *Giant Oil and Gas Fields of the Decade: 1968–1978. Mem. Am. Assoc. Pet. Geol.*, **30**, 281–288.

Lambert-Aikhionbare, D. O. and Ibe, A. C. (1984). Petroleum source-bed evaluation of Tertiary Niger Delta: discussion. *Bull. Am. Assoc. Pet. Geol.*, **68**, 387–389.

Landes, K. K. (1966). Coal reflectance. *Oil Gas J.*, **64, 2 May**, 172–177.

Langstroth, W. T. (1971). Seismic study along a portion of the Devonian salt front in North Dakota. *Geophysics*, **36**, 330–338.

Le Blanc, R. J. (Ed.) (1976a). *Modern Deltas*, Reprint Series No. 18. American Association of Petroleum Geologists, Tulsa, OK.

Le Blanc, R. J. (Ed.) (1976b). *Ancient Deltas*, Reprint Series No. 19. American Association of Petroleum Geologists, Tulsa, OK.

Leder, F. and Park, W. C. (1986). Porosity reduction in sandstone by quartz overgrowths. *Bull. Am. Assoc. Pet. Geol.*, **70**, 1713–1728.

Lees, G. M. (1953). The Middle East. In Illing, V. C. (Ed.), *The Science of Petroleum*, Vol. iv, *The World's Oilfields*, Part I, *The Eastern Hemisphere*. Oxford University Press, London.

Lerche, I. and O'Brien, J. J. (Eds) (1987). *Dynamical Geology of Salt and Related Structures*. Academic Press, Orlando, USA. 832 pp.

Levorsen, A. I. (1967). *Geology of Petroleum*, 2nd edn. W. H. Freeman, San Francisco, CA, 724 pp.

Lijmbach, G. W. M. (1975). On the origin of petroleum. In *Proceedings of the 9th World Petroleum Congress*. Applied Science Publishers, London.

Linsley, P. N., Potter, H. C., McNab, G., and Racher, D. (1980). The Beatrice Field, Inner Moray Firth, U.K. North Sea. In Halbouty, M. T. (Ed.), *Giant Oil and Gas Fields of the Decade: 1968–1978. Mem. Am. Assoc. Pet. Geol.*, **30**, 117–129.

Lohmann, H. H. (1972). Salt dissolution in subsurface of British North Sea as interpreted from seismograms. *Bull. Am. Assoc. Pet. Geol.*, **56**, 472–479.

Lohmann, H. H. (1980). Seismic differentiation between salt diapirs, shale diapirs, and similar bodies. In *Symposium on Interpretational Information from Seismic Data*. Norsk Petroleumsforening, Kristiansand.

Longman, M. W., Fertal, T. G., and Glennie, J. S. (1983). Origin and geometry of Red River dolomite reservoirs, Western Williston Basin. *Bull. Am. Assoc. Pet. Geol.*, **67**, 744–771.

Lopatin, N. V. (1971). Temperature and geologic time as factors in coalification. *Akad. Nauk SSSR Izvestiya, Ser. Geol.*, **3**, 95–106 (English translation by N. H. Bostick, Illinois State Geological Survey, 1972).

Lopatin, N. V. and Bostick, N. H. (1973). Geologicheskiye faktory katageneza ugley (The geological factors in coal catagenesis). In *Symposium Volume, Piroda organicheskogo veschestva sovremennykh i iskpaemykh osadkov* (Nature of Organic Matter in Recent and Fossil sediments), 79–90. Nauka Press, Moscow (Illinois Geological Survey Reprint Series, 1974-Q, 16 pp.

Lucia, F. J. and Murray, R. C. (1967). Origin and distribution of porosity in crinoidal rock. In *Proceedings of the 7th World Petroleum Congress*, Vol. 2, 409–424. Applied Science Publishers, London.

Lynn, H. B. (1986). Seismic detection of oriented fractures. *Oil Gas J.*, 54–55.
MacDonald, G. J. (1983). The many origins of natural gas. *J. Pet. Geol.*, **5**, 341–362.
Mackenzie, A. S., Price, L. C., Leythaeuser, D., Müller, P., Radke, M., and Schaefer, R. G. (1987). The expulsion of petroleum from Kimmeridge Clay source-rocks in the area of the Brae Oilfield, UK continental shelf. In Brooks, J. and Glennie, K. W. (Eds), *Petroleum Geology of North West Europe*, 865–877. Graham & Trotman, London.
Mackenzie, D. B. (1972). Primary stratigraphic traps in sandstones. In King, R. E. (Ed.), *Mem. Am. Assoc. Pet. Geol.*, **16/SEG Spec. Pub.**, **10**, 47–63.
Macleod, M. K. (1982). Gas hydrates in ocean bottom sediments. *Bull. Am. Assoc. Pet. Geol.*, **66**, 2649.
Magara, K. (1977). Petroleum migration and accumulation. In G. D. Hobson (Ed.), *Developments in Petroleum Geology*, Vol. 1, 83–126. Applied Science Publishers, London.
Magara, K. (1980). Comparison of porosity–depth relationships of shale and sandstone. *J. Pet. Geol.*, **3**, 175–185.
Magara, K. (1981). Hydrodynamics—does it trap oil? *J. Pet. Geol.*, **4**, 177–186.
Magara, K. (1987). Fluid flow due to sediment loading—an application to the Arabian Gulf region. In Goff, J. C. and Williams, B. P. J. (Eds), *Fluid Flow in Sedimentary Basins and Aquifers. Geol. Soc. Spec. Publ.*, **34**, 19–28.
Manzur, A. (1985). Delineation of salt masses using borehole seismics. *Oil Gas J.*, **7 October**, 147–149.
Martin, R. G. (1978). Northern and eastern Gulf of Mexico continental margin: stratigraphic and structural framework. In Bouma, A. H., Moore, G. T., and Coleman, J. M. (Eds), *Framework, Facies, and oil trapping Characteristics of the Upper Continental Margin. Am. Assoc. Pet. Geol. Stud. Geol.*, **7**, 21–42.
Martin, R. G. and Bouma, A. H. (1978). Physiography of Gulf of Mexico. In Bouma, A. H., Moore, G. T., and Coleman, J. M. (Eds), *Framework, Facies and Oil Trapping Characteristics of the Upper Continental Margin. Am. Assoc. Pet. Geol. Stud. Geol.*, **7**, 3–19.
Mathews, W. H. and Shepard, F. P. (1962). Sedimentation of Fraser River Delta, British Columbia. *Bull. Am. Assoc. Pet. Geol.*, **46**, 1416–1443.
Maxwell, J. C. (1964). Influence of depth, temperature, and geologic age on porosity of quartzose sandstone. *Bull. Am. Assoc. Pet. Geol.*, **48**, 697–709.
Mayuga, M. N. (1970). Geology and development of California's giant—Wilmington oil field. In Halbouty, M. T. (Ed.), *Geology of Giant Petroleum Fields. Mem. Am. Assoc. Pet. Geol.*, **14**, 158–184.
McCoy, A. W. and Keyte, W. R. (1934). Present interpretations of the structural theory for oil and gas migration and accumulation. In *Problems of Petroleum Geology*, 253–307. American Association of Petroleum Geologists, Tulsa, OK.
McIver, R. D. (1967). Composition of kerogen—clue to its role in the origin of petroleum. In *Proceedings of the 7th World Petroleum Congress*, Vol. 2, 25–36. Applied Science Publishers, London.
McIver, R. D. (1982). Role of naturally occurring gas hydrates in sediment transportation. *Bull. Am. Assoc. Pet. Geol.*, **66**, 789.
McWhae, J. R. (1986). Tectonic history of Northern Alaska, Canadian Arctic, and the Spitsbergen regions since early Cretaceous. *Bull. Am. Assoc. Pet. Geol.*, **70**, 430–450.
Meckel, L. D., Jr. and Nath, A. K. (1977). Geologic considerations for stratigraphic modelling and interpretation. In Peyton, C. E. (Ed.), *Seismic Stratigraphy—Applications to Hydrocarbon Exploration. Mem. Am. Assoc. Pet. Geol.*, **26**, 417–438.
Minor, H. E. and Hanna, M. A. (1941). East Texas oilfield. In *Stratigraphic Type Oil Fields*, 600–640. American Association of Petroleum Geologists, Tulsa, OK.
Mitchum, R. M., Jr. (1985). Seismic stratigraphic expression of submarine fans. In Berg, O. R. and Woolverton, D. G. (Eds), *Seismic Stratigraphy, II. Mem. Am. Assoc. Pet. Geol.*, **39**, 117–138.
Mitchum, R. M., Jr., Vail, P. R., and Sangree, J. B. (1977a). Stratigraphic interpretation of seismic reflection patterns in depositional sequences. In Payton, C. E. (Ed.), *Seismic Stratigraphy—Applications to Hydrocarbon Exploration. Mem. Am. Assoc. Pet. Geol.*, **26**, 117–133.

Mitchum, R. M., Jr., Vail, P. R., and Thompson, S., III (1977b). The depositional sequence as a basic unit for stratigraphic analysis. In Payton, C. E. (Ed.), *Seismic Stratigraphy—Applications to Hydrocarbon Exploration, Mem. Am. Assoc. Pet. Geol.*, **26**, 53–62.

Mitra, S. (1988). Effect of deformation mechanisms on reservoir potential in Central Appalachian Overthrust Belt. *Bull. Am. Assoc. Pet. Geol.*, **72**, 536–554.

Moldowan, J. M., Seifert, W. K., and Gallegos, E. J. (1985). Relationship between petroleum composition and depositional environment of petroleum source rocks. *Bull. Am. Assoc. Pet. Geol.*, **69**, 1255–1268.

Momper, J. A. (1980). Oil expulsion—a consequence of oil generation. *Am. Assoc. Pet. Geol. Slide-Tape Ser. Geol.*, Slide No. 42.

Moore, P. and Hunt, G. (1983). *Atlas of the Solar System*. Book Club Associates, and the Royal Astronomical Society, London, 464 pp.

Moran, W. R. and Gussow, W. C. (1963). The history of the discovery and the geology of Moonie Oil Field, Queensland, Australia. In *Proceedings of the 6th World Petroleum Congress*, Section 1, 595–609. Publ. Verein zur Förderung des 6.Welt-Erdöl-Kongresses, Hamburg.

Morgan, W. A., Copley, J. H., and Schneider, R. E. (1982). Identification of subtle porosity and traps within Frisco Formation, Canadian County, Oklahoma: A geologic, seismic-waveform approach: In Halbouty, M. T. (Ed.), *The Deliberate search for the Subtle Trap. Mem. Am. Assoc. Pet. Geol.*, **32**, 93–114.

Moussa, M. T. (1988). Zooplankton fecal pellets as a source of hydrocarbons in chalk. *J. Pet. Geol.*, **11**, 347–354.

Munns, J. W. (1985). Valhall Field: a geological overview. *Mar. Pet. Geol.*, **2**, 23–43.

Musgrave, A. W. and Hicks, W. G. (1966). Outlining of shale masses by geophysical methods. *Geophysics*, **31**, 711–725.

Nelson, P. H. H. (1980). Role of reflection seismic in development of Nembe Creek Field, Nigeria. In Halbouty, M. T. (Ed.), *Giant Oil and Gas Fields of the Decade: 1968–1978. Mem. Am. Assoc. Pet. Geol.*, **30**, 565–576.

Nettleton, L. L. (1934). Fluid mechanics of salt domes. *Bull. Am. Assoc. Pet. Geol.*, **18**, 1175–1204.

Nettleton, L. L. (1943). Recent experimental and geophysical evidence of mechanics of salt dome formation. *Bull. Am. Assoc. Pet. Geol.*, **27**, 51–63.

Nur, A. (1971). Effects of stress on velocity anisotropy in rocks with cracks. *J. Geophys. Res.*, **76**, 2022–2034.

Nur, A. and Simmons, G. (1969). Stress-induced velocity anisotropy in rock: an experimental study. *J. Geophys. Res.*, **74**, 6667–6674.

Nwachukwu, J. I. and Chukwura, P. I. (1986). Organic matter of Agbada Formation, Niger Delta, Nigeria. *Bull. Am. Assoc. Pet. Geol.*, **70**, 48–55.

Oele, J. A., Hol, A. C. P. J., and Tiemens, J. (1981). Some Rotliegend gas fields of the K and L blocks, Netherlands offshore (1968–1978)—a case history. In Illing, L. V. and Hobson, G. D. (Eds), *Petroleum Geology of the Continental Shelf of North-West Europe*, 289–300. Heyden, London, for The Institute of Petroleum, London.

Oil & Gas Journal (1984). Soviet gas industry hits fast clip, but exports likely to lag goals. **1 October**, 73–76.

Owen, E. W. (1964). Petroleum in carbonate rocks. *Bull. Am. Assoc. Pet. Geol.*, **48**, 1727–1730.

Ozkaya, I. (1988). A simple analysis of primary oil migration through oil-propagated fractures. *Mar. Pet. Geol.*, **5**, 170–174.

Palacas, J. G. (1984). Petroleum geochemistry and source rock potential of carbonate rocks. *Am. Assoc. Pet. Geol. Stud. Geol.*, **18**.

Park, R. G. (1983). *Foundations of Structural Geology*. Blackie, Glasgow, 135 pp.

Parker, J. M. (1967). Salt solution and subsidence structures, Wyoming, North Dakota, and Montana. *Bull. Am. Assoc. Pet. Geol.*, **51**, 1929–1947.

Pennington, J. J. (1975). The Geology of the Argyll Field. In Woodland, A. W. (Ed.), *Petroleum and the Continental Shelf of North West Europe*, Vol. 1: *Geology*, 283–291. Applied Science Publishers, London, for The Institute of Petroleum, London.

Peters, K. E., Moldowan, J. M., Driscole, A. R., and Demaison, G. J. (1989). Origin of Beatrice oil by co-sourcing from Devonian and Middle Jurassic rocks, Inner Moray Firth, United Kingdom. *Bull. Am. Assoc. Pet. Geol.*, **73**, 454–471.

Philippi, G. T. (1965). On the depth, time, and mechanism of petroleum generation. *Geochim. Cosmochim. Acta*, **29**, 1021–1049.
Pohl, J., Stoffler, D., Gall, H., and Ernston, K. (1977). The Ries impact crater. In Roddy, D. J., Pepin, R. O. and Merrill, R. B. (Eds), *Impact and Explosion Cratering*, 343–404. Pergamon, Elmsford, NY.
Porfir'ev, V. (1974). Inorganic origin of petroleum. *Bull. Am. Assoc. Pet. Geol.*, **58**, 3–33.
Posey, H. H., Price, P. E., and Kyle, J. R. (1987). Mixed carbonate sources for calcite caprocks of Gulf Coast salt domes. In Lerche, I. and O'Brien, J. J. (Eds), *Dynamical Geology of Salt and Related Structures*, 593–630. Academic Press, Orlando, Florida.
Pratsch, J. -C. (1983). Gasfields, NW German Basin: secondary gas migration as a major geologic parameter. *J. Pet. Geol.*, **5**, 229–244.
Pratsch, J. -C. (1985). Oil and gas accumulations in overthrust belts—1. *J. Pet. Geol.*, **8**, 129–148.
Price, L. C. (1980). Utilization and documentation of vertical oil migration in deep basins. *J. Pet. Geol.*, **2**, 353–387.
Price, L. C. (1981a). Primary petroleum migration by molecular solution: consideration of new data. *J. Pet. Geol.*, **4**, 89–101.
Price, L. C. (1981b). Aqueous solubility of crude oil to 400°C and 2,000 bars pressure in the presence of gas. *J. Pet. Geol.*, **4**, 195–223.
Price, L. C. (1983). Geologic time as a parameter in organic metamorphism, and vitrinite reflectance as an absolute palaeogeothermometer. *J. Pet. Geol.*, **6**, 5–38.
Purcell, W. R. (1949). Capillary pressures—their measurement using mercury and the calculation of permeability therefrom. *Pet. Trans. Am. Inst. Min. Eng.*, **186**, 39–48.
Pusey, W. C. (1973). Palaeotemperatures in the Gulf Coast using the ESR-kerogen method. Trans. Geol. Socs., Assoc. Gulf Coast, **23**, 195.
Quanmao, C. and Dickinson, W. R. (1986). Contrasting nature of petroliferous Mesozoic–Cenozoic basins in Eastern and Western China. *Bull. Am. Assoc. Pet. Geol.*, **70**, 263–275.
Raffalovich, Rev. F. D. and Daw, T. B. (1984). Use of seismic stratigraphy for Minnelusa exploration, northeastern Wyoming. *Geophysics*, **49**, 715–721.
Rafavich, F., Kendall, C. H. St. C., and Todd, T. P. (1984). The relationship between acoustic properties and the petrographic character of carbonate rocks. *Geophysics*, **49**, 1622–1636.
Rice, D. D. and Claypool, G. E. (1981). Generation, accumulation and resource potential of biogenic gas. *Bull. Am. Assoc. Pet. Geol.*, **65**, 5–25.
Riedel, W. (1929). Zur Mechanik geologischer Brucherscheinungen. *Zentbl. Miner. Geol. Paläont. B.*, 354–368.
Ringwood, A. E. (1970). Origin of the Moon: the precipitation hypothesis. *Earth Planet. Sci. Lett.*, **8**, 131–140.
Russell, W. L. (1960). *Principles of Petroleum Geology*, 2nd edn. McGraw-Hill, New York, 503 pp.
Sangree, J. B. and Widmier, J. M. (1977). Seismic interpretation of clastic facies. In Payton, C. E. (Ed.), *Seismic Stratigraphy—Applications to Hydrocarbon Exploration*. Mem. Am. Assoc. Pet. Geol., **26**, 165–184.
Sanneman, D. (1968). Salt-stock families in north-western Germany. *Am. Assoc. Pet. Geol. Memoir 8, Diapirism and Diapirs*, 261–270. Am. Assoc. Pet. Geol., Tulsa.
Sassen, R. (1987). Organic geochemistry of salt dome cap rocks, Gulf Coast Salt Basin. In Lerche, I. and O'Brien, J. J. (Eds), *Dynamical Geology of Salt and Related Structures*, 631–652. Academic Press, Orlando, Florida.
Scherer, F. C. (1980). Exploration in East Malaysia over the past decade. In Halbouty, M. T. (Ed.) *Giant Oil and Gas Fields of the Decade: 1968–1978*. Mem. Am. Assoc. Pet. Geol., **30**, 423–440.
Scherer, M. (1987). Parameters influencing porosity in sandstones: a model for sandstone porosity prediction. *Bull. Am. Assoc. Pet. Geol.*, **71**, 485–491.
Schmidt, V. and Almon, W. (1983). Development of diagenetic seals in carbonates and sandstones. *Bull. Am. Assoc. Pet. Geol.*, **67**, 545–546(abstr.).
Schmidt, V., McDonald, D. A., and Platt, R. L. (1977). Pore geometry and reservoir aspects of secondary porosity in sandstones. *Bull. Can. Pet. Geol.*, **25**, 271–290.
Schmoker, J. W., Krystinik, K. B., and Halley, R. B. (1985). Selected characteristics of limestone and dolomite reservoirs in the United States. *Bull. Am. Assoc. Pet. Geol.*, **69**, 733–741.

Schowalter, T. T. (1979). Mechanics of secondary hydrocarbon migration and entrapment. *Bull. Am. Assoc. Pet. Geol.*, **63**, 723–760.
Scruton, P. C. (1955). Sediments of the eastern Mississippi delta. In Shepard, F. P. *et al.* (Eds), *Finding Ancient Shorelines. Soc. Econ. Paleontol. Mineral. Spec. Publ.*, **3**, 21–50.
Scruton, P. C. (1960). Delta building and the deltaic sequence. In *Symposium Volume: Recent Sediments, Northwest Gulf of Mexico*, 82–102. American Association of Petroleum Geologists, Tulsa, OK.
Selley, R. C. (1975a). *An Introduction to Sedimentology*. Academic Press, London, 408 pp.
Selley, R. C. (1975b). Genesis, migration, and entrapment of Jurassic oil in the North Sea. *Proceedings of the Society of Petroleum Engineers*, Spring Meeting, London, Paper No. 5269, 7 pp.
Selley, R. C. (1976). The habitat of North Sea oil. *Proc. Geol. Assoc.*, **87**(4), 359–388.
Selley, R. C. (1978). Porosity gradients in North Sea oil-bearing sandstones. *J. Geol. Soc. London*, **135**, 119–131.
Selley, R. C. (1983). *Petroleum Geology for Geophysicists and Engineers*. IHRDC, Boston, MA, 88 pp.
Seni, S. J. (1987). Evolution of Boling Dome cap rock with emphasis on included terrigenous clastics, Fort Bend and Wharton Counties, Texas. In Lerche, I. and O'Brien, J. J. (Eds), *Dynamical Geology of Salt and Related Structures*, 543–591. Academic Press, Orlando, Florida.
Seni, S. J. and Jackson, M. P. A. (1983). Evolution of salt structures, East Texas diapir province, parts 1 and 2. *Bull. Am. Assoc. Pet. Geol.*, **67**, 1219–1274.
Sherbon Hills, E. (1972). *Elements of Structural Geology*, 2nd edn. Chapman & Hall, London, 502 pp.
Sheriff, R. E. (1985). Aspects of seismic resolution. In Berg, O. R. and Wolverton, D. G. (Eds), *Seismic Stratigraphy, II. Mem. Am. Assoc. Pet. Geol.*, **39**, 1–10.
Sheriff, R. E. and Geldart, L. P. (1982). *Exploration Seismology*. Vol. 1: *History, Theory and Data Acquisition*. Cambridge University Press, Cambridge, 253 pp.
Shipley, T. H., *et al.* (1979). Seismic evidence for widespread possible gas hydrate horizons on continental slopes and rises. *Bull. Am. Assoc. Pet. Geol.*, **63**, 2204–2213.
Shirley, K. (1989). Structures have impact on theories. *Am. Assoc. Pet. Geol. Explorer*, **April**, 14–17.
Simpson, R. D. H. and Whitley, P. K. J. (1981). Geological input to reservoir simulation of the Brent Formation (Murchison Field). In Illing, L. V. and Hobson, G. D. (Eds), *Petroleum Geology of the Continental Shelf of North-West Europe*, 310–314. Heyden, London, for The Institute of Petroleum, London.
Smith, D. A. (1966). Theoretical considerations of sealing and non-sealing faults. *Bull. Am. Assoc. Pet. Geol.*, **50**, 363–374.
Smith, D. A. (1980). Sealing and non-sealing faults in Louisiana Gulf Coast Basins. *Bull. Am. Assoc. Pet. Geol.*, **64**, 145–172.
Smith, D. B. (1981). The magnesian limestone (Upper Permian) reef complex of northeastern England. *Soc. Econ. Paleontol. Mineral. Spec. Publ.*, **30**, 161–186.
Smith, D. G. and Pullen, J. B. (1967). Hummingbird structure of south-east Saskatchewan. *Bull. Can. Assoc. Pet. Geol.*, **15**(4), 468–482.
Spiers, C. J., Lister, G. S., and Zwart, H. J. (1982). *The influence of fluid-rock interaction on the rheology of salt rock, and on ionic transport in the salt: first results*, WAS-153-80-7N, 268–280. European Atomic Energy Commission, Vienna.
Spiers, C. J., Urai, J. L., and Lister, G. S. (1986). The effect of brine (inherent or added) on rheology and deformation mechanisms in salt rock. In Hardy and Langer (Eds), *Proceedings of the 2nd Conference on the Mechanical Behaviour of Salt*, Hannover, 1984.
Staplin, F. L. (1969). Sedimentary organic matter, organic metamorphism, and oil and gas occurrence. *Bull. Can. Assoc. Pet. Geol.*, **17**, 47–66.
Stow, D. A. V., Bishop, C. D., and Mills, S. J. (1982). Sedimentology of the Brae oilfield, North Sea: Fan models and controls. *J. Pet. Geol.*, **5**, 129–148.
Tainsh, H. R. (1953). Burma. In Illing V. C. (Ed.), *The Science of Petroleum*, Vol. iv, *The World's Oilfields*, Part I, *The Eastern Hemisphere*. Oxford University Press, London.
Talbot, C. J., Tully, C. P., and Woods, P. J. E. (1982). The structural geology of Boulby (Potash) Mine, Cleveland, United Kingdom. *Tectonophysics*, **85**, 167–204.
Tankard, A. J. and Welsink, H. J. (1987). Extensional tectonics and stratigraphy of Hibernia Oil Field, Grand Banks, Newfoundland. *Bull. Am. Assoc. Pet. Geol.*, **71**, 1210–1232.

Tannenbaum, E., Huizinga, B. J., and Kaplan, I. R. (1986). Role of minerals in thermal alteration of organic matter—II: a material balance. *Bull. Am. Assoc. Pet. Geol.*, **70**, 1156-1165.

Taylor, J. C. M. (1977). Sandstones as reservoir rocks. In: Hobson, G. D. (Ed.), *Developments in Petroleum Geology*, Vol. 1. Applied Science Publishers, London.

Taylor, J. C. M. (1978a). Sandstone diagenesis: state of the art, 1977. *J. Geol. Soc. London*, **135**, 3-5.

Taylor, J. C. M. (1978b). Control of diagenesis by depositional environment within a fluvial sandstone sequence in the northern north Sea basin. *J. Geol. Soc. London*, **135**, 83-92.

Taylor, J. C. M. (1986). Late Permian-Zechstein. In Glennie, K. W. (Ed.), *Introduction to the Petroleum Geology of the North Sea*. Blackwell Scientific, Oxford, 278 pp.

Taylor, S. R. and Lapré, J. F. (1987). North Sea chalk diagenesis: its effect on reservoir location and properties. In Brooks, J. and Glennie, K. W. (Eds), *Petroleum Geology of North West Europe* (Proceedings of Barbican Conference), Vol. 1, 483-495. Graham & Trotman, London.

Thiébaud, C. E. and Robson, D. A. (1981). The geology of the Asl Oilfield, Western Sinai, Egypt. *J. Pet. Geol.*, **4**, 77-87.

Tieh, T. T., Berg, R. R., Popp, R. K., Brasher, J. E., and Pike, J. D. (1986). Deposition and diagenesis of Upper Miocene arkoses, Yowlumne and Rio Viejo Fields, Kern County, California. *Bull. Am. Assoc. Pet. Geol.*, **70**, 953-969.

Tiratsoo, E. N. (1984). *Oilfields of the World*. Scientific Press, Beaconsfield, 392 pp.

Tissot, B. P. (1972). The application of the results of organic geochemical studies in oil and gas exploration. In Hobson, G. D. (Ed.), *Developments in Petroleum Geology*, Vol. 1, 53-82. Applied Science Publishers, London.

Tissot, B. (1977). The application of the results of organic geochemical studies in oil and gas exploration. In *Developments in Petroleum Geology 1* (G. D. Hobson, Ed.), 53-82. Applied Science Publishers, London.

Tissot, B. P. and Welte, D. H. (1978). *Petroleum Formation and Occurrence; A New Approach to Oil and Gas Exploration*. Springer, New York, 521 pp.

Trevena, A. S. and Clark, R. R. (1986). Diagenesis of sandstone reservoirs of Pattani Basin, Gulf of Thailand. *Bull. Am. Assoc. Pet. Geol.*, **70**, 290-308.

Trorey, A. W. (1970). A simple theory for seismic diffractions. *Geophysics*, **35**, 762-784.

Trusheim, F. (1960). Mechanism of salt migration in northern Germany. *Bull. Am. Assoc. Pet. Geol.*, **44**, 1519-1540.

Tschopp, R. H. (1967). Development of Fahud Field. In *Proceedings of the 7th World Petroleum Congress*, Vol. 2, 243-250. Applied Science Publishers, London.

Turner, C. C., Cohen, J. M., Connell, E. R., and Cooper, D. M. (1987). A depositional model for the South Brae oilfield. In Brooks, J. and Glennie, K. W. (Eds), *Petroleum Geology of North West Europe* (Proceedings of Barbican Conference), Vol. 2, 853-864. Graham & Trotman, London.

Urban, J. B. (1960). *Microfossils of the Woodford Shale (Devonian) of Oklahoma*. MS thesis, University of Oklahoma, Norman, 77 pp.

Van den Bark, E. and Thomas, O. D. (1981). Ekofisk: first of the giant oil fields in Western Europe. *Bull. Am. Assoc. Pet. Geol.*, **65**, 2341-2363.

Van der Ploeg, P. (1953). Egypt. In Illing, V. C., (Ed.), *The Science of Petroleum*, Vol. vi, *The World's Oilfields*, Part I, *The Eastern Hemisphere*, 151-157. Oxford University Press, Oxford.

Van Hinte, J. E. (1982). Synthetic seismic sections from biostratigraphy. In Watkins, J. S. and Drake, C. L. (Eds), *Studies in Continental Margin Geology. Mem. Am. Assoc. Pet. Geol.*, **34**, 675-685.

Verdier, A. C., Oki, T., and Suardy, A. (1980). Geology of the Handil Field, (East Kalimantan—Indonesia). In Halbouty, M. T. (Ed.), *Giant Oil and Gas Fields of the Decade: 1968-1978. Mem. Am. Assoc. Pet. Geol.*, **30**, 399-421.

Wallis, W. E. (1967). Offshore petroleum exploration, Gippsland and Bass Basins. In *Proceedings of the 7th World Petroleum Congress*, Vol. 2, 783-791. Applied Science Publishers, London.

Walls, J. D. (1982). Tight gas sands—permeability, pore structure, and clay. *J. Pet. Technol.*, **34**, 2708-2714.

Walls, J. D., Nur, A. M., and Bourbie, T. (1982). Effects of pressure and partial water saturation on gas permeability in tight sands; experimental results. *J. Pet. Technol.*, **34**, 930-936.

Walmsley, P. J. (1975a). The Forties Field. In Woodland, A. W. (Ed.), *Petroleum and the Continental Shelf of North West Europe*, Vol. 1: *Geology*, 131–140. Applied Science Publishers, London, for The Institute of Petroleum, London.

Walmsley, P. J. (1975b). Oil and gas developments in the North Sea. In *Proceedings of the 9th World Petroleum Congress*, Vol. 3, 131–140. Elsevier Applied Science Publishers, London.

Walper, J. L. (1977). Palaeozoic tectonics of the southern margin of North America. *Trans. Assoc. Gulf Coast Geol. Socs.*, **27**, 230–241.

Waters, K. H. (1973). *Reflection Seismology*. Wiley, New York, 377 pp.

Watts, N. L. (1987). Theoretical aspects of cap-rock and fault seals for single- and two-phase hydrocarbon columns. *Mar. Pet. Geol.*, **4**, 274–307.

Waugh, B. (1970). Formation of quartz overgrowths in the Penrith Sandstone (lower Permian) of northwest England as revealed by scanning electron microscopy. *Sedimentology*, **14**, 309–320.

Weatherill, G. W. (1977). The nature of the present interplanetary crater-forming projectiles. In Roddy, D. J., Pepin, R. O., and Merrill, R. B. (Eds), *Impact and Explosion Cratering*, 613–615. Pergamon, Elmsford, NY.

Weaver, C. E. (1960). Possible uses of clay minerals in search for oil. *Bull. Am. Assoc. Pet. Geol.*, **64**, 916–926.

Weeks, L. G. (1958). Habitat of oil and factors that control it. In *Habitat of Oil* (Weeks, L. G. Ed.), 1–60. *Am. Assoc. Pet. Geol. Special Publication*, Am. Assoc. Pet. Geol., Tulsa.

Welham, J. and Craig, H. (1979). Methane and hydrogen in East Pacific Rise hydrothermal fluids. *Geophys. Res. Lett.*, **6**, 829–831.

Welte, D. H., Hagemann, H. W., Hollerbach, A., Leythaeuser, D., and Stahl, W. (1975). Correlation between petroleum and source rock. In *Proceedings of the 9th World Petroleum Congress*, Vol. 2, 179–191. Applied Science Publishers, London.

Westre, S. (1984). The Askeladden gas find—Troms I. In Spencer, A. M. *et al.* (Eds), *Petroleum Geology of the North European Margin*, 33–40. Graham & Trotman, London, for the Norwegian Petroleum Society.

Wheatley, T. J., Biggins, D., Buckingham, J., and Holloway, N. H. (1987). The geology and exploration of the transitional shelf, an area to the west of the Viking Graben. In Brooks, J. and Glennie, D. W. (Eds), *Petroleum Geology of North West Europe* (Proceedings of Barbican Conference), Vol. 2, 979–989. Graham & Trotman, London.

Widess, M. B. (1973). How thin is a thin bed? *Geophysics*, **38**, 1176–1180.

Wilkens, R., Simmons, E., and Carouso, L. (1984). The ratio V_P/V_S as a discriminant of composition for siliceous limestones. *Geophysics*, **49**, 1850–1860.

Williams, J. J., Connor, D. C., and Peterson, K. E. (1975). The Piper Oilfield, U.K. North Sea: a fault-block structure with Upper Jurassic beach-bar reservoir sands. In Woodland, A. W. (Ed.), *Petroleum and the Continental Shelf of North West Europe*, Vol. 1: *Geology*, 363–378. Applied Science Publishers, London, for The Institute of Petroleum, London.

Wilson, G. F. and Metre, W. B. (1953). Assam and Arakan. In Illing, V. C. (Ed.), *The Science of Petroleum*, Vol. iv, *The World's Oilfields*, Part I, *The Eastern Hemisphere*. Oxford University Press, London.

Woodland, A. W. (Ed.) (1975). *Petroleum and the Continental Shelf of North West Europe*, Vol. 1: *Geology*. Applied Science Publishers, London, for The Institute of Petroleum, London, 501 pp.

Wright, N. J. R. (1980). Time, temperature and organic maturation—the evolution of rank within a sedimentary pile. *J. Pet. Geol.*, **2**, 411–425.

Yang Wanli, Li Yongkang, and Gao Ruiqi (1985). Formation and evolution of nonmarine petroleum in Songliao Basin, China. *Bull. Am. Assoc. Pet. Geol.*, **69**, 1112–1122.

Yerkes, R. F., McCulloh, T. H., Schoellhamer, J. E., and Vedder, J. G. (1965). Geology of the Los Angeles basin, California—an introduction. *U.S. Geol. Survey Prof. Pap.*, **420-A**, A1–A57.

Zhang, Yi-gang (1981). Cool shallow origin of petroleum—microbial genesis and subsequent degradation. *J. Pet. Geol.*, **3**, 427–444.

Index

abnormal pore pressure 40
Achi-Su oilfield, USSR 339–341
acoustic impedance 216, 249, 299, 372
adsorption 22, 27
aerated surface waters 21
algae 4, 21
abiogenic gas 4, 22, 23, 60, 368
aeolian rocks, traps in 247
Agbada Formation, Nigeria 20
Akata Formation, Nigeria 20
allochthonous chalks Valhall 91
Alwyn oilfields, East Shetland Basin 219, 220
amplitude-encoded bed thickness 133, 243
Anadarko Basin, USA 17, 19
anaerobic bacteria 22, 23
anhydrite 49, 56
anisotropy
   in rock fabric 52
   source rock 27
anoxic conditions 21
anticlinal nose 97
anticlinal theory 64
anticline 64, 144
   asymmetrical 64
   drape 93 *et seq.*, 103
      passive and active 93, 95
   overturned 64
   recumbent 65
   roll-over 19, 103, 124, 125, 128
   symmetrical 64
   uplift 82 *et seq.*
Appalachian coalfields, USA 68
Appalachian Mountains, USA 122
aquathermal effect 31
aqueous solution 27

Argyll oilfield, North Sea 259, 261, 263, 265, 266
Aruma Shale (seal), Oman 147
'ash marker' (seismic event) 97
Askeladden gasfield, offshore Norway 148–152
Asl oilfield, Western Sinai 152, 153
aspect ratio
   of folds 65, 66
   of pores and cracks 35
Assam 78, 79
astroblemes (cosmic impact craters) 23, 347, 363 *et seq.*
Athabasca Sands 59
Auk oilfield, North Sea 259, 261–264
authigenic minerals 57
authigenic montmorillonite 46
autochthonous chalks, Valhall 91
axial plane (of fold) 64
axial plunge (pitch) 65, 66

back-reef 285, 292
backthrust 122
Bahamas 21
Bajocian–Bathonian reservoir sands 324
Barails Formation, Assam 79
Baronia Field, Eastern Malaysia 159
barrier island/bar deposit 44
basalts in seismic data 372
baselap 221
basic requirements 3
Beatrice Field, UK North Sea 16, 144–146
bedding-plane slip 66
bed thinning in seismic data 240–244
'biochemical fence' concept 253
biogenic origin 4, 22
bioherms 20, 47

biomarkers 16
biostromes 20
biota 21
bittern (Mg-, K-rich) salts 51, 265
bitumen (bituminous) 59, 147
bitumen ratio (bitumen/TOC) 16
Bomu Field, Nigeria 19
Boulby Potash Mine, UK 169
bottom contour currents 125
bottom-simulating reflection (BSR) 347, 357–363
'bow-tie effect' 135, 139
box fold 67
Brae oilfield, North Sea 309
breakaway 123
Brent crater, Canada 368
Brent oilfield, North Sea 331–334
Brent Province, North Sea 330
Brent-type fields, North Sea 144, 324, 330, 331
'bright spot' seismic event 34, 158, 163, 348
brine density flow 281
brittle fracture 69, 125
Broad Fourteens Basin (North Sea) 179, 181
bulk density 249
Bunter Sandstone, offshore UK 106, 107
Bunter Shale, offshore UK 106
buoyancy 25, 30, 31
Burgan Field, Kuwait 16, 93
burial depth 10, 42, 60
buried topography (see also PGM surface) 214
Burma 78, 80–82

C (carbon) content 9
C steranes 16
calcite 22, 86
Caledonian Highlands 144
caliper log 252
cap rock
　salt diapir 201–204
　　'regressive' 204, 205
　seal 49, 55, 57–60
capillary (entry) pressure 25, 26, 29, 30, 55–58, 86, 186, 239
Capitan Reef, USA 281
carbon–carbon bond 30
carbon dioxide CO_2 29
carbon preference index (CPI) 15
carbonate assemblage (source rocks) 16
carbonate environments and seismic facies 260
carbonate reservoirs 37, 47
carbonate-shale sequence 17
carbonate-shale-evaporite sequence 17
carbonates 248 et seq.
　'seismic signature' of 257–259
　continental 21

non-reefal (bedded) 49, 252, 253
reefal 49
shelf-type 21
Carolina Trough, offshore eastern USA 157
Castile Formation, USA 281
cataclasis (cataclasites) 42, 113, 128
catagenesis stage 10, 22
catalytic properties 22
cementation (diagenetic) 35, 37, 40, 42, 58
cement dissolution 45
Central Appalachian Overthrust Belt, USA 42
Chaidamu Basin, PRC 70
chalk 49, 254
　allochthonous 254
　autochthonous 254
　reservoirs 85 et seq., 91
channel sands 234, 235–238
chevron (zig-zag) folds 68
China, eastern and western basins of 233
chlorides 57
clay
　compaction 24
　mineral dehydration 25, 31
　mineral diagenesis 13, 14, 35, 42
　mineralogy 20, 37
clays, authigenic 44, 46
Clearwater Lake, Canada 366
clinoform (seismic event) 226, 301–303
　parallel oblique 228
　shingled 228
closed (tight) fold 66
Clysmic Rift (in Miocene Gulf of Suez) 153
coccolith ooze 87
combination trap 144
compaction 25, 35, 40, 42, 50
　burial depth 37
　'chemical' (pressure solution) 40
　mechanical 40
complex overburden 81
compression structures (faulting) 115, 125
concentric (parallel) folds 67
condensate 10
conjugate folds 67
conodont 14
Conodont Alteration Index (CAI) 14
contact
　gas-liquid 33, 59, 97,
　gas-water (GWC) 34, 148, 175, 349
　liquid-liquid 33, 59, 351
　tilted 33
continuous oil phase 25, 27
'cosmic' velocity 364
cosmology 4, 22
crack density (EDA) 53
cratonic settings 55

crestal collapse (of roll-over) 161
Cretaceous Period 22
crinoidal limestones 51
critical thickness (seismic) 133
cuesta 93, 219, 230, 231
cutin 4
cyclic sediments, seismic effects of 133

Dan Field, Danish North Sea 49, 63, 88–90
Danian Chalk 254, 255
Darcy's Law 36
Dead Sea structure 142
Delaware Basin, USA 51, 281
delta 22, 124, 296
delta lobe 305
depositional drape 287
depositional environments 17, 37, 248
depositional zonation 19
depth conversion (seismic) 178, 179
destructive distillation 6
detachment (decoupling) zone 82, 125
dextral (right-lateral) movement 118
DHI (seismic direct hydrocarbon indicator) 102, 157, 158, 347
diagenetic effects 35, 42, 47
  cessation of 59
diagenesis
  sandstone 42
  stage 10, 22, 23
diapiric shale 207
differential pressure 171
differential salt movement 267
diffusion 25
diffusive mass transfer in halite 170
diffraction pattern (in seismic data) 114, 133–142
Digboi oilfield, Assam 78, 79
dihydrophytol 87
dilatancy 53
dip-slip 122
dislocation/ gliding in halite 170
displacement pressure (*see* capillary pressure) 56
dissolution/collapse 208, 269
distributary channel 305, 306, 308
dolomite 249 *et seq.*, 263
dolomitization 47, 49–51, 257, 288, 293
dome 93
  faulted 63
downhole logs 9
downlap 217, 223, 224, 225
Dowsing Formation, offshore UK 106
drainage (of hydrocarbon source) 9
drape structures 20, 83, 153
drive
  depletion 54
  external gas 54

water 54
DSDP Sites in Atlantic 360
ductile grain deformation 42, 46
ductile shear 69, 70, 125, 128, 153
ductility 55–57
Duke's Wood, UK 75
Dunlin oilfield, North Sea 324
dykes, igneous, in seismic data 369–371
dynamothermal effects 6

Eakring oilfield, UK 72, 73, 75
early introduction of hydrocarbons 86
East African Rift 18, 142
East Pacific Rise 23
East Shetlands Basin, UK waters 222
East Texas oilfield, USA 323
Ecuador (Oriente Province) 103
edge-of-permeability zone 319
'edgewater' limits, Wilmington Field 109
Ekofisk Field, offshore Norway 86–88, 91, 93, 254
Elaterite Basin, USA 247
El-Ayun Field, Egypt 21
electron spin resonance (ESR) 13
Emerald oilfield, North Sea 325
emergent salt diapirs of Hormuz, Iran 171
'erosional anisotropy' 232
'erosional pinchout' 246
Eugene Island Block 330 Field, USA 157
Euler (lunar crater) 364–366
EUR 93
evaporite basins 124, 248
evaporites (as sealing formations) 56
Evergreen Formation, Australia 223
expulsion
  mechanism 20, 21, 27–29, 253
  water 50
extensive-dilatancy anisotropy (EDA) 53

facies-bounded traps 47
facing direction (of fold) 65
Fahud oilfield, Central Oman 146–147
fan, submarine 308 *et seq*
fault
  antithetic 115, 125
  breccia 113, 128
  detachment 122, 123
  drag 103, 105, 106, 128, 135, 136, 145
  gouge 113, 128
  growth 115, 123–128, 155–165
  listric normal 115
  normal 114–115, 137
  plane 113, 130, 133
    non-cylindrical shaped 125
  reverse 115

sealing 57, 78, 113, 115
rotational 121
seal, juxtaposition 57, 78, 113, 115, 152
strike-slip 107, 116–122
synthetic 115, 125
translational movement in 121, 175
trap 112 et seq.
 upthrow 144, 146
faulting
 block 142, 143
  tilted 152
 compartmentalizing 148
 'down-to-basin' type 125, 147, 159
 rotated block 142–144
 step 142, 143
  planar 152, 153
 'towards upthrown block' 147, 159
First Wilcox Sandstone, USA 44
FK migration (of seismic data) 79
'flat spot' seismic event 34, 59, 102, 348
flood basalts 372
Florida Keys, 21
flow
 non-turbulent 36
 turbulent 36
'flow-concentration' 349
flower structure, negative 119–121
fluid inclusion thermometry 13
fluid mechanics (applied to salt) 183
fold
 anticlinal 52
 belts 55
 compressional 63, 82
 flexural (compressional) 52, 83
 rheomorphic 83
 uplift 52, 83
fold-related traps 63
footwall ramp 122, 123
footwall side (of fault) 128
fore-reef 285, 292
Forties Field, offshore UK 95, 97, 99, 100
Forties Field reservoir, North Sea 44
Four Corners area gas, USA 17
fractionation 22
fracture volume 52
fracture systems 52, 112
Fraser River, British Columbia, Canada 299
free water in halite 170
Frigg gasfield (UK/Norway North Sea) 95, 99, 101, 102
Frisco Formation, USA 239

gamma-ray log 9
Gannet oilfield group (North Sea) 199–201

Garzan Field, Turkey 17
gas
 abiogenic 4
 chimney 87, 91, 352–354
 dry 6
 hydrates 22, 55, 359–362
 planets 4
 'pockmark' on sea bed 354, 355
 sag 352,353
 wet 6
 window 7
gas-charged sediment 91
gas-filled porosity 59
gas-in-place 53
gas-prone source rocks 111
geochemical indicators 15
geochemistry 14
geometry (of faults) 112
geomorphic traps (dunes) 47
Gharamul Limestone Formation, Egypt 21
Ghawar Field, Saudi Arabia 99
giant hydrocarbon fields 22,37
Gippsland Basin, Australia 219
glauconite, glauconitic 251
GPOC(gas-prone organic carbon) 6,16
graben 83,135,142,143,145,152, 208
gradient
 geothermal 27, 46, 60, 248
 porosity-depth 42, 46
 pressure 30, 36
 salinity 25
grain orientation 40
grain size 37, 40
graptolites 14
gravity 30
 anomalies due to salt bodies 168
 segregation 55
Great Glen Fault, Scotland 121,129
Green River Shales, USA 9,18
Guanarito, Venezuela 246
Gulf Coast, USA 42,201
gypsum/anhydrite conversion 51

H (hydrogen) 9,23,29
halite
 fluid inclusions in 169
 'inherent brine' 169
 NaCl 55
Handil Field, Indonesia 74,75,78
hangingwall side (of fault) 128
hardgrounds 254
Heart Mountain, Wyoming, USA 122
Heather oilfield, North Sea 334,335,336
Heimdal Formation, offshore Norway 316

heterogeneity (source rock) 27
Hewett gasfield, offshore UK 105–108
Hibernia oilfield, offshore Canada 155
high-temperature oil generation 30
high-wax crude oil 6, 145
H/C ratio 9, 10, 29
H/O ratio 9
Hod Formation (Valhall) 91
horst 116, 145, 147
Humbly Grove oilfield, UK. 75, 326, 328, 329
Hummingbird oilfield, Canada 270, 271, 272
Huntington Beach, Los Angeles Basin, USA 110
Hunton Formation, Oklahoma, USA 218
Hurghada oilfield, Egypt 153, 154
Hutton oilfield, North Sea 324
hydrocarbon
   column 56, 58, 86, 239
   globules 30
   'kitchen' 7
   saturation 59
hydrocarbons, 'heavy' 149
hydrodynamic equilibrium 31
hydrodynamic factors 32, 79
hydrogen-lean amorphous material (HLA) 16
hydrophilic property of clays 37
hydrostatic pressure 26
hydrothermal activity 7
hyperbolas (as seen in seismic data) 133–142
   fault diffraction 128, 134–137
   phase change of 139
   shape-response 128, 135
   truncated bed-end 129
hyperbolic prism concept 141, 142

igneous activity 7
illite 22
   crystallinity index 14, 16
   sharpness ratio 14, 16
Indefatigable gasfield 177
imbricate structures (faulting) 122
impermeable seals 27, 29
increasing salinity 171
inertinite 6, 16
infrared spectroscopy 16
interconnected pore space 36
interfacial tension 56
Intisar 'D' oilfield, Libya 291–293
inversion axis 179, 181, 252
Irati Shales, Brazil 9
irreducible water saturation 45

Jaipur Uplift, Assam 79
joint 112
juvenile origin 23
'juxtaposition' fault seal 58, 78

K Block (Dutch North Sea) 179
Kaltag Fault, Alaska 207
kaolinite 45, 46
Karst topography, karst surface 229, 251
Katalla, Alaska 16
Keg River Formation, Canada 290
Kern County arkoses, USA 45
kerogen 4, 6, 152
   amorphous 4, 14
   density 15, 16
   structured 4, 14
   unstructured 4, 14
   zonation 21
Kimmeridge, UK 75
Kimmeridge Clay Formation 17, 87, 93
Kirkuk Field, Northern Iraq 81–83
Kutei Basin, Indonesia 75

Laegerdorf salt diapir, North-west Germany 91
lag (low-angled exensional fault) 115
Lakes Entrance Formation, Australia 219, 220
lamination planes (in source rock) 27
landslide, or landslip 123, 124
Latrobe Valley Formation, Australia 221
layer-parallel compression 67
leaching 40, 42, 47, 147
levee 102
level of organic metamorphism (LOM) 21
lignin 4
lignite 76
limestones 249 *et seq.*
Lindesnes Ridge, offshore Norway 91
lipids, microbial 23
liptinite 21
lithotectonic position 52
Livingstone Field, USA 44
log
   density 9
   neutron 9
   sonic 9
London–Brabant Ridge 235, 236
Long Beach, Los Angeles Basin, USA 110
Los Angeles Basin, USA 10
Louann Salt (Jurassic), USA 157
Louisiana–Texas Gulf Coast 16
Lower Mannville Shales, Western Canada 9
low seismic velocity zone 86–87

Madison oil, USA 17
Mahakam River delta, Indonesia 75
Manx-Furness Basin, western British waters 143, 240, 241
Marlin Field, Australia 219, 220
material balance equation 55

maturity, thermal 7, 10, 12, 13, 14, 16, 17, 19, 24, 76
Mene Grande field, Venezuela 324
Messla oilfield, Libya 246
metagenesis stage 10, 22
Meteor Crater, USA 364, 365
meteoric water 32
meteorites 4
methane 10, 22, 23
Meulaboh Shelf, Indonesia 300, 301
Mg-rich brines 50, 86
micellar solution 25
micrites 21, 103
microbial genesis 23
microcracks, stress-aligned, fluid-filled 53
microdolomites 49
Middle East 20, 21, 47, 49, 51, 171
Midland Basin, USA 51
Mid North Sea High (MNSH) 250, 276-281
migration 20, 21, 24
  primary 4, 22, 24, 25, 26, 29-32, 60, 76
  route 3
  secondary 4, 22, 25, 30, 31, 60
  short-path 21
milliDarcy (mD) 36
Millstone Grit (Namurian) 72
Minagish Oolite, Southern Kuwait 49, 93
Minnelusa Sands, USA 244, 245
Mississippi Delta, USA 298, 299
Moho 23
monoaromatized steroids 14
montmorillonite/smectite 22
Moonie oilfield, Australia 223-227
Montrose Field, offshore UK 95-99
Montrose Field reservoir, North Sea 43
morphology of reef in seismic data 47
Murchison oilfield, North Sea 336-339
mylonite (mylonitization) 113, 128, 129

Naga Thrust, Assam 78
Nembe Creek Field, Nigeria 162
net slip 122
neutral surface (in fold) 52, 78
Niagara Reef trend, USA 282
Niger Delta, Nigeria 16, 19, 20, 125
Niigata area, Japan 16
non-reefal limestones/dolomites 47
non-turbulent flow 30
Nordland Ridge, offshore Norway 121
North Island, New Zealand (offshore) 359
North Viking gasfield, UK North Sea 175
North-west Highlands, Scotland 122
North-west Shelf, USA 51
Nottinghamshire-Derbyshire coal basin, UK 72
Nugget Sandstone, Utah, USA 40, 340, 341

Nyiragongo lava lake, East Africa 23

oblique-slip 122
O/C ratio 9, 10
oil
  column height 32, 97
  generation mechanism 27
  window 7, 27, 29, 76
oil-in-place 53
oil-prone 21
oil–water contact (OWC) 32, 33
oil-wet laminations 29
Okan Field, Nigeria 19
onlap (overlap) 221, 224
opal transition (as BSR effect) 362, 363
OPOC (oil-prone organic carbon) 6, 16
optical activity 23
Orbitolina Limestone (Burgan) 93
Orcadian Basin, Scotland 18
organic metamorphism 7, 10
orientation (of rock fractures/cracks) 35
Orinoco Delta 24
orogenesis 7
osmosis 25
Ouachita Mountains, USA 7, 13, 16
outgassing (from mantle) 23
Outlook Field, Montana, USA 272, 273
overhang, salt diapir 204
overpressure 11, 27, 28, 40, 87, 93, 208
overstep thrusting 122

Painter Reservoir Field, USA 340-343
palaeogeomorphic features 93, 123
palaeogeothermometry 7, 11, 13
palynomorphs 12
paralic 6, 145, 214
paraffination 79
Parker–McDowell effect 165
passive continental margin 17
Pattani Basin, Gulf of Thailand 46
Pennsylvanian Period 19
pericline 66
peripheral sink 183-185, 192
permeability 35, 37, 89, 97, 147, 219
  absolute 25, 36, 37
  effective 36
  fracture 52
  relative 25, 36, 37
permeameter 36
Permian Basin, USA 51
PGM (palaeogeomorphological) surface 214, 229 et seq.
Phanerozoic 20
Phosphoria Formation, USA 18
'piggy-back thrusts' 122, 123

pinchout traps 240 *et seq.*
pinnate shears 116
Plains-type folds (USA) 69
plant life
  aquatic 4
  terrestrial 4
Playa del Rey, Los Angeles Basin, USA 110
Point Arguello oilfield, offshore USA 344
Pokur 1 Suite, Urengoy gas field, USSR 111
polarity reversals (bright-spot) 349
pop-up structures 122, 123
Porcupine Bank, offshore Eire 147, 149
pore
  fluids 30, 53
  throats 37, 56, 58
poroperm characteristics 21, 42, 60, 78
porosity (horizontal or vertical) 35, 89, 97, 103, 147, 219
  absolute 36
  effective 36
  fracture 51
  intergranular 52
  primary 21, 35, 42, 45, 47, 49
  rate of decline 45
  reduction 45
  sandstone 40
  secondary 21, 40, 42, 47
  traps 20, 239 *et seq.*
porosity–depth relationship, sandstones 40
  shales 40
porphyrins, presence of 23
Port Neches salt dome, Texas 188, 189
post-piercement salt structures 190 *et seq.*
potentiometric surface 32
Potwar Plateau, N.Pakistan 72
Poza Rica field, Mexico 324
Prairie Evaporite salt, USA 272
Precipice Sandstone Formation, Australia 223
preferential salt movement 267
pre-piercement salt structures 185 *et seq*
'pressure cooker' environment 27, 30
pressure solution
  intergranular 40, 42, 46, 57, 91
  tectonically-induced 42
pristane 87
progradation 224
'protopetroleum' 6, 7
proximity surveys 205
pseudotachylite 113, 128
'pull-down' anomaly, seismic 91, 165, 287–289
'pull-up' anomaly, seismic 192, 287–289
'pulsation tectonics' 172
Purcell's equation 56
P-wave (compressional) seismic velocity 249
pyrolysis 20

quartz overgrowths 45, 46, 59, 240

Rangely Field, USA 247
Rann of Kutch, India 171
Recinus Cardium 'A' pool, Canada 58
recovery factor 55
Red River 'C' zone carbonates, USA 50
reducing environment 21
Red Wing Creek structure, USA 367
reef 249, 285
  atoll 285
  barrier 285
core 286
  fringing 285
  patch 286
  pinnacle 285
reflected refraction (seismic event) 128
reflection coefficient, R 347, 348
reflection configuration 221
remnant features (in salt dissolution) 281, 282
reservoir
  drive mechanisms 53
  engineering 53–55
  rock 3, 35
residual carbon 13
reverse faults 70, 71, 79
'reverse flow collapse structure' 189, 190
rheomorphic folds 69, 82
Ricker wavelet (seismic) 243, 244
Riedel shears 118, 119, 214
Ries Basin (astrobleme), Germany 365, 366
rifting 7
rim syncline
  primary 184 *et seq.*
  secondary 184 *et seq.*, 217
  tertiary 184 *et seq.*
River Rhone delta, France 303
Rodessa oilfield, USA 320, 321
Rogaland Group (Valhall seal) 93
rotational movement 121
Rotliegendes play 177, 182
Rotliegendes Sandstone, Southern North Sea 47, 174, 175, 179
'room problem' 83
rounding (of grains) 40
Ruhr coalfields, West Germany 68

sabkha 253
Salado Formation, USA 281
salt
  deformation 169
  depletion/withdrawal 167
  diapir 91, 130, 190, 191, 198, 208
  dissolution/removal 166
  edge dissolution 167, 168, 275, 277, 278

INDEX

geology 166
'glaciers' of Hormuz, Iran 206
  influence of free water on 169, 170
  lateral flow of 167
  lateral injection of 206
  low-stress deformation of 167
  movement, initiation of 171
  pillow 134, 186
  rock 55, 56, 82
    physical properties of 166 et seq.
  swell 184
  tectonics 166 et seq.
  yield strength 169
'Salzspiegel' salt dissolution surface 274
Samarang oilfield, Eastern Malaysia 161
San Andreas Fault, California, USA 121
San Andres Formation, Texas, USA 51
sands
  beach bar 126
  distributary channel 126
  offshore barrier 126
  point bar 126
sandstone
  aeolian dune 38, 45
  alluvial fan 38
  barrier bar
    regressive 39
    transgressive 39
  channel floor deposited 44
  deep sea
    channel, grain flow 39
    fan, turbidite 39
  deltaic
    distributary channel 38
    distributary mouth bar 38
  fluviatile
    braided 38
    meandering 38
  porosity 40
  reservoirs 37
  tidal flat 39
  tidal sand ridge 39
Santa Fe Springs, Los Angeles Basin, USA 110–11
sapropelic material 4, 21
Sarir Sandstone, Libya 246
scarp 219, 231
scolecodonts 14
  absorptive indices of 14
seal
  diagenetic 55, 57
  ductile 173
  hydraulic 56, 57
  membrane 56, 59
  salt rock as 173, 174, 179
  sealing formation 3, 208

secondary recovery operations 35
Seismic Atlas (EAEG publication) 80, 121
seismic data, lack of detail in 77
seismic environments 63
seismic identification of reefs 285
seismic interval velocities 50
seismic resolution 242–244
seismic response to fault geometry 130–133
seismic velocity
  compressional (P-wave) 53
  shear (S-wave) 53, 357
seismic wavelet 133
sequence
  geological 213
  seismic 214
Sgurr Beag Slide, Scotland 70
shale
  maturity 14, 16
  sheath 208, 209
shale–sandstone assemblage (source rocks) 16
shatter cones 365
Shetlands Platform, offshore UK 101
shock metamorphism 364, 365
Siljan Ring Complex, Sweden 4, 368
sills, igneous, in seismic data 371
Simpson–Bromide Sandstone, Oklahoma, USA 40
Singkel, Indonesia 236, 237
Singu (Chauk) oilfield, Burma 81, 82
Sinian reservoirs China 233
sinistral (left-lateral) movement 118, 119
slide 70
smectite/illite conversion 13, 25, 27
SOC (sapropelic organic carbon) 6
sole-out 125
solubility (of oil in water) 25, 29
soluble cements 40
solution (diagenetic) 35
'sombrero structure' 270
Songliao Basin, China 23
sonic logs 245
sorting 37, 40
source rock 3, 9
South Brae oilfield, North Sea 326, 327
sphericity (of grains) 40
Spirit River Formation, Canada 59
spore colour index 12, 13, 14
spore/pollen coloration 12
stable carbon isotope analysis 16
stacked (unmigrated) seismic data 79
steam drilling rig 82
stratigraphic traps 71, 186
stress zones 83, 84, 93, 134
stylolitic limestone reservoirs 51

stylolitization (stylolites) 88, 91, 256, 257
submarine vents 23
subsidence drape 275, 276, 284
sulphates 57
suprafan lobes 309, 312
supratenuous folds 69, 77, 187, 188, 190
Surat Basin, Australia 223
surface geology 146
syndepositional effects 85, 181
system boundary traps (aeolian) 47

TAI (thermal alteration index) 12, 13, 20
Tar Sand Triangle, Utah, USA 247
Teak oilfield, offshore Trinidad 344
'tectonic pulse' in salt movement 172
tectonic stress field 52
tension gash fractures 109, 118, 119
Tertiary 22
tertiary recovery 53
thermal alteration 11
Thistle oilfield, North Sea 324, 336, 337
thrust belts 339
thrusts 70
'time and temperature' 19
Tipam Sandstone stage, Assam 78, 79
TOC (total organic carbon) 6, 9, 13, 20, 21, 29, 76, 152, 253
Top Ranger reservoir, USA 109
Tor Formation, Valhall 91, 255
Torrance Field, Los Angeles Basin, USA 110
Transitional Limestone, Kirkuk Field 82
translational movement 121
trap configuration 3
truncation trap 186
'tuning thickness' 244
turtle structure (salt tectonics) 185, 195–197

Umm as Samim, Oman 146
Umm Shaif Field, Abu Dhabi 97
unconformities 213 *et seq.*
unconformity, angular 231
'uniform' deposits (pre salt movement) 188
Urengoy gasfield, West Siberia, USSR 110, 111
Urengoy megaswell, USSR 111

Valhall Field, offshore Norway 91–93
Van Krevelen diagram 9
velocity hysteresis effects 179, 254
vertical displacement (in faulting) 129
vertical pathways (secondary migration) 31
vesicular lava 373
'vice-jaws' compression 88

Viewfield Oil Pool, USA 367
vitrinite
  in coal 11
  reflectance (%Ro) 7, 8, 11, 14, 16, 17, 19, 20, 76, 87
volcanic features in seismic data 372
volcanic rocks 97
V_P/V_S ratio (seismic velocities) 284, 357
vugular porosity 239

Wareham, UK 75
Wasia Limestone Oman 147
water
  meteoric 32
  saturation of tight sandstones 59
  sediment source 32
water-injection (of reservoir) 82
water-wet systems 30
wave-base 21
West Sole gasfield 178–180
Western Sumatra Forearc Basin 236, 237, 258, 259
Western (Cordillera) Thrust belt, USA 340–341
wet-gas window 7, 10
White Rim Sandstone, USA 247
Wichita Orogeny 19
Wienhausen–Eicklingen salt diapir, Germany 205, 206
Wilcox Sand, USA 16
Williston Basin, Canada/USA 17, 50, 269, 276
Wilmington Field, California, USA 107–111
Woodbine Sand, USA 321, 323
Woodford Shale, USA 17, 19, 239
'working model' 59–60
Wyoming Thrust Belt, USA 340
Wytch Farm Field, UK 75

X-ray diffractograms 14

Yenangyaung oilfield, Burma 80, 81
yield stress, various evaporites 269
Yushashan oilfield, People's Republic of China 70

Zechstein Basin, North-west Europe 20, 34, 51, 68, 169
Zechstein salt 86, 105, 106, 174, 175, 177, 179–182, 267
Zelten Field, Sirte Basin, Libya 103
zeolites 57
zig-zag (chevron) fold 68
Zohar gasfield, Israel 17